Dialogues in Urban and Regional Planning 4

Prize winning papers from the World's Planning School Associations
Dialogues in Urban and Regional Planning offers a selection of the best urban planning scholarship from the world's planning school associations. The award-winning papers presented illustrate some of the concerns and discourse of planning scholars and provide a glimpse of planning theory and practice around the world. All those with an interest in urban and regional planning will find this collection stimulating in opening avenues for research and debate.

Set in context by the editors' introductory chapter, these essays focus on planning concerns within local contexts, but also reflect international issues. The necessity of rising to meet global challenges, the barriers to change, and the characteristics of the new approaches to planning which seem most likely to facilitate change, resonate throughout the papers selected for inclusion in *Dialogues 4*.

First we encounter the problems and opportunities presented by a variety of planning institutions in dealing with *social inclusion and local identity*, ranging from the need to change the planners' culture in Zimbabwe in order to move from government to governance, to theorizing an "intercultural project" that moves beyond contemporary "multiculturalism," in Vancouver, Canada, to the need for flexible strategies to pressure the state to enact collaborative and just planning processes in Johannesburg, South Africa. Next we are exposed to new ways of thinking about the organization and use of *urban spaces*, with an "oil vulnerability" assessment of Australian cities, a rethinking of urban green space in the British transition from modernism's vision to an alternative aimed at active provision of ecosystem services, an examination of urban design strategies aimed at improving public security in Brazilian *favelas*, and a plea for incorporating residents' culture and lifestyles in planning for "*living heritage cities*" like Melaka, Malaysia. At the intersection of social inclusion and local identity with the organization and use of urban spaces is an examination of the plans produced by the Congress for the New Urbanism (CNU) for the rebuilding of eleven American towns in the aftermath of Hurricane Katrina. The realization of social equity goals will require more than physical design. Then we are stimulated to *think strategically* by new analyses of classic issues: Latin American failures to attain economic "maturity" when hindered by divergence between rapid urbanization and industrialization; the mega-project planning problem of pervasive misinformation about the costs, benefits, and risks, and the consequent waste; and the short-comings of cost-benefit analysis in French transportation policy research. Finally, we have an apt summary of the *barriers* to carrying out the new approaches to planning, with an analysis of its demands on planning professionalism in a range of institutional and governance settings in Victoria, Australia.

This book is published in association with the Global Planning Education Association Network (GPEAN), and its nine member planning schools associations, who have selected these papers based on regional competitions. These associations represent over 360 planning schools in nearly 50 countries around the globe.

Editors: **Thomas L. Harper** is Professor, Faculty of Environmental Design, University of Calgary, Canada. **Heloisa Costa** is Associate Professor, Geography Department, Federal University of Minas Gerais, Brazil. **Anthony Gar-On Yeh** is Chair Professor, Head of Department of Urban Planning and Design, Director of Centre of Urban Studies and Urban Planning, Director of GIS Research Centre, University of Hong Kong, Hong Kong SAR. **Michael Hibbard** is Professor, Department of Planning, Public Policy and Management, and Director, Institute for Policy Research and Innovation, University of Oregon

Contributors: **Arturo Almandoz, Zul Azri Bin Abd Aziz, Rose Compans, Heloisa Costa, Jago Dodson, Bent Flyvbjerg, Thomas L. Harper, Michael Hebbert, Michael Hibbard, Syed Z.A. Idid, Christophe Jemelin, Amin Y. Kamete, Vincent Kaufmann, Alan March, Maria Julieta Nunes de Souza, Luca Pattaroni, Géraldine Pflieger, Samira Ramezani, Leonie Sandercock, Neil Sipe, Emily Talen, Tanja Winkler, Anthony Gar-On Yeh.**

Dialogues in Urban and Regional Planning 4

Prize winning papers from World's Planning School Associations
This biennial series is published in association with Global Planning Education
Association Network (GPEAN). The nine members of GPEAN are:

Association of African Planning Schools (AAPS)
Association of Collegiate Schools of Planning (ACSP)
Association of Canadian University Planning Programs (ACUPP)
Association of European Schools of Planning (AESOP)
Association of Latin American Schools of Planning & Urbanism (ALEUP)
National Association of Urban and Regional Postgraduate and Research
Programmes (ANPUR) in Brazil
Australia and New Zealand Association of Planning Schools (ANZAPS)
Association for Development of Planning Education and Research (APERAU)
Asian Planning Schools Association (APSA)

International Editorial Board

Heloisa Costa
Associate Professor, Geography Department, Federal University of Minas Gerais,
Brazil.

Christophe Demazière
Research Centre on Cities, Territories, Environment and Societies, Planning
Department, University of Tours, France.

Thomas L. Harper
Professor, Faculty of Environmental Design, University of Calgary, Alberta,
Canada.

Michael Hibbard
Director, Institute for Policy Research and Innovation and
Professor, Department of Planning, Public Policy & Management
University of Oregon, U.S.A.

Daniel K.B. Inkoom
Director, MPhil/PhD Programmes, Department of Planning, Kwame Nkrumah
University of Science and Technology, Kumasi, Ghana.

Petter Næss
Professor of Urban Planning, Department of Development and Planning, Aalborg
University, Denmark.

Yukio Nishimura
Professor, Department of Urban Engineering University of Tokyo, Japan.

Awais Piracha
Course Advisor, Planning, School of Social Sciences, University of Western Sydney,
Australia.

Roberto Rodriguez-Garza
Professor, Dept. of Architecture and Industrial Design, Tecnologico de Monterrey,
Mexico.

Willem Salet
Professor, Faculty of Behavioural and Social Sciences, University of Amsterdam,
Netherlands.

Dialogues in Urban and Regional Planning 4

Edited by Thomas L. Harper,
Michael Hibbard, Heloisa Costa,
and Anthony Gar-On Yeh

First published 2011
by Routledge
2 Park Square, Milton Park, Abingdon, Oxon, OX14 4RN

Simultaneously published in the USA and Canada
by Routledge
270 Madison Avenue, New York, NY 10016

Routledge is an imprint of the Taylor & Francis Group, an informa business

© 2011 selection and editorial material Thomas L. Harper, Michael Hibbard, Heloisa Costa and Anthony Gar-On Yeh; individual chapters, the contributors

The right of Thomas Harper, Michael Hibbard, Heloisa Costa and Anthony Yeh to be identified as authors of the editorial material, and of the authors for their individual chapters has been asserted by them in accordance with sections 77 and 78 of the Copyright, Designs and Patents Act 1988.

Typeset in Galliard by
Swales & Willis Ltd, Exeter, Devon
Printed and bound in Great Britain by
CPI Antony Rowe, Chippenham, Wiltshire

All rights reserved. No part of this book may be reprinted or reproduced or utilised in any form or by any electronic, mechanical, or other means, now known or hereafter invented, including photocopying and recording, or in any information storage or retrieval system, without permission in writing from the publishers.

British Library Cataloguing in Publication Data
A catalogue record for this book is available from the British Library

Library of Congress Cataloging in Publication Data
A catalog record for this book has been requested

ISBN13: 978–0–415–59334–2 (hbk)
ISBN13: 978–0–203–84202–7 (ebk)

Contents

Contributors	ix
Preface	xix

1 Introduction: rising to the global challenges
*Thomas L. Harper, Michael Hibbard, Heloisa Costa, and
Anthony Gar-On Yeh* — 1

2 Hanging out with "trouble-causers": planning and governance in
urban Zimbabwe
Amin Y. Kamete — 12

3 Towards a cosmopolitan urbanism: from theory to practice — 38
Leonie Sandercock

4 For the equitable city yet to come — 58
Tanja Winkler

5 Shocking the suburbs: urban location, homeownership and oil
vulnerability in the Australian city
Jago Dodson and Neil Sipe — 83

6 The re-enclosure of green space in postmodern urbanism — 112
Michael Hebbert

7 Safe urban spaces: security issues for city design — 136
Maria Julieta Nunes de Souza and Rose Compans

8 Public space and conservation of a historic living city: Melaka,
Malaysia
Samira Ramezani, Zul Azri Bin Abd Aziz, and Syed Z.A. Idid — 155

viii Contents

9 New urbanism, social equity, and the challenge of post-Katrina
rebuilding in Mississippi 171
Emily Talen

10 Immature take-offs: urbanization, industrialization and
development in twentieth-century Latin America 205
Arturo Almandoz

11 Policy and planning for large-infrastructure projects: problems,
causes, and curses 223
Bent Flyvbjerg

12 Socio-political analysis of French transport policies: the state of
practice 249
*Vincent Kaufmann, Christophe Jemelin, Géraldine Pflieger,
and Luca Pattaroni*

13 Institutional impediments to planning professionalism in Victoria,
Australia 270
Alan March

Index 298

Contributors

Arturo Almandoz (Caracas, 1960) is Urbanist *cum laude* (Universidad Simón Bolívar, USB, Caracas, 1982). Urban Technician Diploma (Instituto Nacional de Administración Pública, INAP, Madrid, 1988). MPhil in Philosophy (USB, Caracas, 1992). PhD in Housing and Urbanism (Architectural Association School of Architecture, Open University, London, 1996). Post-doctoral Centro de Investigaciones Posdoctorales (Cipost), Universidad Central de Venezuela (UCV, Caracas, 2003–4). Professor at the Department of Urban Planning, Co-ordinator of the Urbanism programme (1996–98). Professor Adjunct, Pontifical Catholic University of Chile (PUC), since 2009. In addition to more than 45 articles in specialized journals, he has published 10 books in Spanish about the emergence of modern urbanism and metropolitan culture in Venezuela and abroad, which have won local, national and international awards. Having collaborated in 15 other books and 2 encyclopaedias, Professor Almandoz is the editor of *Planning Latin America's Capital Cities, 1850–1950* (London and New York: Routledge, 2002; 2010). He has been lecturer or speaker at more than 90 events worldwide. His current research interests include the relationship between literature and urban cultural history, and Latin America's modernization and urban historiography. He has been a member of the editorial boards of several periodicals in Chile, Mexico, Spain, UK and Venezuela, where he was Director of *Argos* (2006–8).

Zul Azri Bin Abd Aziz has a B.A. in Landscape, Dept. of Landscape Architecture, Faculty of Architecture, Planning & Surveying, Universiti Teknologi MARA, Malaysia (UiTM), and an M.Sc. in Urban Design, Faculty of Built Environment, University of Technology Malaysia (UTM). He is currently teaching at the Department of Landscape Architecture, Faculty of Architecture, Planning and Surveying, Universiti Teknologi MARA. Sri Iskandar, Chenoh, Perak Darul Redzwan, Malaysia.

Rose Compans is a Brazilian architect with a PhD in urban planning. She works in the public administration of the Municipality of Rio de Janeiro, and she teaches theories of urbanism and town planning at the School of Architecture, Bennett

Methodist Institute. She has published several articles about housing and urban policies. Her doctoral thesis won an award from the National Association of Urban and Regional Post-Graduate and Research (ANPUR), resulting in a book entitled *Urban Entrepreneurialism: Between Theory and Practice*. She is currently researching the impacts of recent housing programs, at the federal and local levels, for low-income populations in the city of Rio de Janeiro.

Heloisa Costa is an architect and urbanist, M.Phil. in Urban Planning (Architectural Association, London, 1983). PhD in Demography (UFMG, Brazil, 1995). She was Visiting Scholar at the Department of Geography, UC Berkeley (1997/8). Since 1993 she has been Associate Professor at the Department of Geography of the Institute of GeoSciences of the Federal University of Minas Gerais, Belo Horizonte, Brazil, lecturing in urban geography, planning, urbanization, population and environment at both undergraduate and graduate levels. From 1985 to 1993 she was part-time Lecturer, Architecture and Urbanism, FAMIH in Belo Horizonte. She has coordinated and participated in several research projects leading to publication of journal articles, book chapters and editing, conference papers and lectures on issues related to urban and environmental politics and planning, housing and public policies. She holds an Urban and Regional Planning Productivity Research Grant from CNPq. Recent research projects include: URBANAT – Urbanization, nature, and urban-environmental politics; New peripheries: politics and urban and environmental regulation in the production of metropolitan space; SWITCH project – Sustainable Water Management Improves Tomorrow's Cities Health. She is a member of the Editorial Board of several Brazilian journals, such as *Revista Brasileira de Estudos Urbanos e Regionais, Cadernos de Arquitetura e Urbanismo* (PUC/MG), *Cadernos Metrópole* and *Geografias* (UFMG). She is a former president of ANPUR – National Association of Graduate Programmes and Research in Urban and Regional Planning (2003–5), and is ANPUR's representative at the Coordinating Committee of GPEAN – Global Planning Education Associations Network, and at the National Council of Cities. She worked extensively in planning at the Provincial Planning Secretary of Minas Gerais (1978–89) and with housing at the local government of Belo Horizonte (1993).

Jago Dodson is Senior Research Fellow, Urban Research Program, Griffith University, Brisbane Australia. Dodson received his PhD from the University of Melbourne in 2002, and has since worked in research-intensive positions at RMIT University and at Griffith University. He has worked on an array of urban research problems covering such areas as public housing, urban regeneration, transport planning, metropolitan governance, infrastructure planning and master-planned communities. Much of his recent work has a strong suburban focus, exemplified by his work (with Neil Sipe) on oil vulnerability. His research publications include the books *Government Discourse and Housing* (Ashgate, 2007) and (with Neil Sipe) *Shocking the Suburbs: Oil Vulnerability in the Australian City* (UNSW Press, 2008) and over 35 refereed papers. Jago is also a regular commentator to public debates about Australian cities in both print and broadcast media.

Bent Flyvbjerg is Professor and Chair of Major Programme Management, and Director, BT Centre for Major Programme Management, Oxford University. He was twice a Visiting Fulbright Scholar to the US, where he did research at UCLA, UC Berkeley, and Harvard. He is currently studying rationality and power in megaproject policy, planning, and management. Flyvbjerg's books and articles have been translated into 18 languages and his research is covered by *Science*, *Economist*, *Financial Times*, *New York Times*, the BBC, and many other news media. He is author of *Rationality and Power: Democracy in Practice* (Chicago, 1998) and *Making Social Science Matter* (Cambridge, 2001), principal author of *Megaprojects and Risk* (Cambridge, 2003), and co-editor of *Decision-Making on Mega-Projects* (Elgar, 2008). He has published research articles in the *Journal of the American Planning Association, European Planning Studies, Environment and Planning, British Journal of Sociology, Politics and Society*, and numerous other journals. Flyvbjerg has served as adviser to the EU Commission, the United Nations, and government and business in many countries.

Thomas L. Harper is Professor, and former Director of the Planning Program, Faculty of Environmental Design, University of Calgary, Canada. His inter-disciplinary research, done collaboratively with philosopher Stanley Stein, is in the area of normative planning theory. Their 2006 book *Dialogical Planning in a Fragmented Society: Critically Liberal, Pragmatic and Incremental*, summarizes their approach. Their current work focusses on planning and design approaches to complex and wicked problems, and on the adaptation of socio-biological systems concepts to planning. Harper is a past-president of the Association of Canadian University Planning Programs, and has been their representative to the Global Planning Education Associations Network coordinating committee since its inception. He has been a member of the Editorial Board, *Journal of Planning Education and Research* (1997–2006) and *Canadian Public Policy* (2006–10). He is a Member of the Canadian Institute of Planners (a professional designation), and has worked with a variety of clients. He has also served on boards and committees for a number of community, educational, and church organizations.

Michael Hebbert is Professor of Town Planning, School of Environment and Development, University of Manchester. Born in Glasgow, his first degree was in Modern History at Merton College Oxford, and his doctorate in Geography at the University of Reading. Over the past 35 years he has lectured at Oxford Polytechnic,

xii Contributors

LSE and the University of Manchester, and published widely on the history of town planning, metropolises (especially London and Manchester), regionalism, urban design and the uses of natural and social science in the planning process. He currently teaches urban design and is researching the application of urban climatology through city planning, with funding from the UK Economic and Social Research Council. From 2002 to 2010 he edited the Elsevier journal *Progress in Planning*. Hebbert is a chartered town planner (Member of the Royal Town Planning Institute) and an Academician of the Academy of Social Sciences.

Michael Hibbard is Professor in the Department of Planning, Public Policy and Management, and founding Director, Institute for Policy Research and Innovation, University of Oregon. He has been at UO since 1980. He chaired the Department of Planning, Public Policy and Management (PPPM) from 1988 to 1997 and from 2000 to 2004. He is also a participating faculty member in the University's Environmental Studies, Historic Preservation, and International Studies programs. He received his PhD in regional planning from UCLA. Among his involvements outside the University, Hibbard has been the President of the Association of Collegiate Schools of Planning and chair of the Global Planning Educators interest Group; he served as editor of the *Journal of Planning Education and Research* and *Progress in Planning*; and he was the U.S. representative on the steering committee that organized the first World Planning Schools Congress, held in Shanghai in 2001. Hibbard's research focuses on community and regional development, with a special interest in the social impacts of economic change on small towns and rural regions, both in the U.S. and internationally. He has published and consulted widely on those issues. Before entering academia he worked for more than 10 years in planning and community development with migrant farm workers and Indian reservations in California.

Syed Z.A. Idid has a Dip. Town Planning (ITM) 1977, B.A. (Hons) Town Planning (GSA/CNAA) 1983, M.A. Urban Design (JCUD/CNAA) 1985, R.T.P.I. (U.K.), M.M.I.P (Malaysia), Dr. Eng. Urban Design and Conservation, University of Tokyo (Japan). After a short stint as an assistant planner in a state town and country planning agency (1977–79), Idid started teaching town planning at the Faculty of Built Environment, Universiti Teknologi Malaysia (1983), and was promoted to Associate Professor (1995). He is a leader in the field of urban design and urban conservation, through the Urban Design and Conservation Research Group (UDCRU) of the faculty. His contributions towards raising awareness of cultural heritage and urban conservation are numerous. In 1991–93 he was the project field coordinator for the national Inventory on Malaysian Architectural Heritage, covering 247 towns throughout Malaysia, for the purpose of urban conservation. In 1994, he produced the country's first 'Conservation Action Area

Plan for *Jalan Tun Tan Cheng Lock* (Heeren Street) and *Jalan Hang Jebat* (Jonker's Street)' for the Melaka Municipal Council. In 1995, he undertook a conservation master plan study to revitalize the former mining town of Sungai Lembing. He has worked with various agencies such as UNESCO, Asia West Pacific Network for Urban Conservation, the Department of Museum and Antiquity, various states, government agencies and the Town and Country Planning Department. He has drawn up policies, guidelines and actions to facilitate the conservation of urban cultural heritage. Most significantly, he has demonstrated the role and importance of the Town Planning profession as the guardian of urban cultural heritage. Syed Idid has written papers related to: urban identity, human activities and urban images, conservation guidelines, the role of town planning in urban conservation, adaptive reuses, public participation and awareness in urban conservation, places of cultural significance and meaning of special places. He is the author of *Pemeliharaan Warisan Bandar* (Conservation of Townscape Heritage), published by the Malaysia Heritage Trust.

Christophe Jemelin is a geographer, and has carried out research on urban mobility at the Federal Institute of Technology, Lausanne (EPFL), 1995–2008. He is now planner and data analyst at the Public Transit Authority in Lausanne. His PhD (2004) deals with the quality of urban public transport services in France and Switzerland, comparing the goals of operators and the perceptions of customers. His main activities are the statistical analysis of mobility (microcensus, specific polls) and research on modal splits in various urban forms. He is co-author of the Swiss Atlas of Spatial Changes, having developed its commuter flows analysis methodology. In addition to national research programs – PREDIT (France)/PNR (Switzerland) – he has taken part in the European GUIDE (Group for Urban Interchange Development and Evaluation) programme, and in the BEST (Benchmarking European Sustainable Transport) project.

Amin Y. Kamete is Lecturer in Planning, School of Environment, Natural Resources and Geography, Bangor University, United Kingdom. He was previously Senior Lecturer, Department of Rural and Urban Planning, University of Zimbabwe, where he began his academic life. In 2003 he moved to the Nordic Africa Institute (NAI), in Uppsala, Sweden, where he was Senior Researcher and Research Programme Coordinator for the Research Programme, "Gender and Age in African Cities". His current research interests are planning theory and practice, with special emphasis on cities, space and power in the context of development control, and management practice vis-à-vis informality and illegality in urban Africa. His most recent publications focus on spatial planning, governance, youth, and housing in the contested urban spaces of Zimbabwe. Kamete is a Zimbabwean national.

Vincent Kaufmann is Professor of Urban Sociology and Mobility, and Director of the Urban Sociology laboratory (LaSUR) Ecole Polytechnique Fédérale de Lausanne (EPFL). He has a Masters degree in Sociology (Universtiy of Geneva) and a Ph.D. on rationalities underlying transport modal practices at EPFL He has been invited lecturer at Lancaster University (2000), Ecole Nationale des Ponts et Chaussées (2001), Université Catholique de Louvain (2004–10) and Nijmegen University (2010). He is a member of the editorial board of the *Swiss Journal of Sociology*. Fields of research are: mobility and urban life styles, relationships between social and spatial mobility, public policies of land planning and transportation. He is author, co-author or editor of 12 books.

Alan March's research examines the ability of planning systems, and planners, to establish and achieve collective goals. In addition to his teaching and research, he has practised as a planner since 1991 in a broad range of private and public sector settings. His publications include papers and chapters examining the practical governance mechanisms of planning in various national and international settings. His forthcoming book with Ashgate considers the unavoidable tensions of planning being embedded within governance mechanisms, and offers a new theoretical framework to explain and remedy systemic failures in urban planning. He is the recipient of numerous awards for his research, including a Melbourne University teaching award. March's professional planning career has included many roles in statutory and strategic planning, advocacy, and urban design. He has practised in Western Australia, the UK, New South Wales and Victoria. Alan has practised as a planner and urban designer in Victoria since 1997, and has worked in many multidisciplinary teams on diverse projects ranging from inner-city residential redevelopment, beachside residences, strategic planning, Melbourne Docklands, to Mosques in Broadmeadows. Alan continues a small urban planning practice in Melbourne, mainly providing advice on inner-city redevelopment and acting as advocate in planning appeals. He is currently working on a number of research projects that consider the ways that planning systems can successfully manage change, particularly a transition to higher density. These include comparisons of the potential for planners to undertake their roles as professionals within different planning systems, including comparisons between Turkey, Spain and Australia. His current funded project with the CFA is to produce new design and planning design guidelines to improve settlement patterns in bushfire-prone areas. March has recently completed papers establishing design parameters for communal open space in high-density housing developments, and ways that car parking can be replaced effectively with bicycle parking in inner-city locations. As part of a wider project on professional university education, March has contributed to the writing of the latest draft of new accreditation guidelines. His future work will seek to find ways to improve the "re-spatialisation" of urban planning in Australia as a fundamental element of effective planning.

Maria Julieta Nunes de Souza is an architect and urban planner, as well as a Professor at the Institute of Research in Urban and Regional Planning, Federal University of Rio de Janeiro. She works in the field of research on *precarious settlements (favelas* – slums in English) in Rio de Janeiro State. The research focus is on the settings and roles that poverty plays in Brazilian cities in the geographic and urban political sense, including regulations and planning tools in urban projects. More recently, she has examined the relationship between urban policies and public safety issues, examining the architectural and physical elements that support public safety policies in Brazilian metropolises. She is also interested in contributions to the urban design and general architecture of the city, in the building of a public democratic sphere, considering the highly unequal social characteristics of social reality of Brazilian cities. From 1981 until she began teaching in 1995, Nunes worked in Public Administration within various governmental planning institutions in Rio de Janeiro state, and participated in several consultancy jobs in formulating master plans. In 1990 she earned her Master's Degree in Urban and Regional Planning at IPPUR/UFRJ with her work on "Urban Questions in the 1988 Federal Constitution". In 2002 she received her Ph.D. in Communications and Culture (ECO/UFRJ) with a thesis dealing with technological communications, contemporary differences and public space.

Luca Pattaroni is Senior Lecturer, Laboratory of Urban Sociology (LaSUR), Ecole Polytechnique Fédérale de Lausanne (EPFL) and associate researcher at the Institute Marcel Mauss (Ecole des Hautes Etudes en Sciences Sociales, Paris). He is codirector of EspacesTemps.net and editor in chief of *Métropolitiques.* He has a Ph.D. in Sociology from the School of Advanced Studies in Social Sciences (EHESS, Paris) and the University of Geneva. In an attempt to build bridges between individual agencies, collective action and public policies, his research and publications deal with the dynamic interrelations between urban order, pluralism and justice. His current work focus on subjectivation processes within cities, and on the impact on urban development of conflicts and militant movements (squatters, right to the city).

Géraldine Pflieger is Senior Lecturer in Urban and Environmental Policies, at the Institute for Environmental Sciences, and at the Department of Political Science of the University of Geneva. She holds a PhD in urban planning from the Ecole Nationale des Ponts et Chaussées (Paris) and was visiting scholar at the University of California, Berkeley in 2002. As an urban and political scientist, she has undertaken research into the links between the management of network infrastructures, natural resources and the analysis of urban policies, in various contexts (Chile, California, France and Switzerland). Her current research projects combine analyses of the regulation of network and natural resources, and of the transformation of metropolitan governance. She is currently developing a research program on the political adaptation of cities to climate change. She is also a member of the research team on the government of great metropolises, coordinated by the Chair "ville" of Sciences Po Paris.

Samira Ramezani has a B.A. (2006) in Architecture, Shiraz Azad University, Iran, and an M.Sc. (2009) in Urban Design, Faculty of Built Environment, University of Technology Malaysia (UTM). She has contributed to the field of urban design by conducting research on child-friendly neighbourhoods, walkable environments, and socio-cultural issues in urban design and conservation and has delivered a number of related papers in international conferences in the Asian Pacific regions. She is currently involved as Research Assistant, Urban Design and Conservation Research Unit (UDCRU) of the Faculty of Built Environment, University of Technology, Malaysia.

Leonie Sandercock is Professor in Community and Cross-cultural Planning, School of Community & Regional Planning, University of British Columbia. She has published eleven books, including *Towards Cosmopolis: Planning for Multicultural Cities* (1998); the edited collection *Making the Invisible Visible: a multicultural history of planning* (1998); and *Cosmopolis 2: Mongrel Cities of the 21st Century* (2003), which won the Davidoff Award (from the American Collegiate Schools of Planning) in 2005. With Giovanni Attili, she made the award-winning documentary *Where Strangers become Neighbours* (National Film Board of Canada, 2007) and is now focusing on multimedia, storytelling, and planning. Sandercock and Attili's book plus DVD package, *Where Strangers become Neighbours: The Integration of Immigrants in Vancouver*, was published by Springer in January 2009. In 2005 Sandercock received The Dale Prize for Excellence in Urban & Regional Planning, awarded by the Department of Urban & Regional Planning at California State Polytechnic University. The 2005 Dale Prize theme was "Voices in Planning: Transforming Land Use Practice through Community Engagement". In 2007 Leonie shared the First Prize (with Collingwood Neighbourhood House) for the BMW Group Award for Intercultural Learning for her work with the Collingwood Neighbourhood House in Vancouver (portrayed in the film *Where Strangers become Neighbours*) and for her essay 'Cosmopolitan Urbanism'. Her current projects are a documentary with Giovanni Attili, *Finding our Way: a path to healing Native/non-Native relations in Canada*, a forthcoming book, *Multimedia Explorations in Urban Policy and Planning: Beyond the Flatlands* (Springer, February 2010) and an exploration of spirituality, urban life and the urban professions.

Contributors **xvii**

Neil Sipe is Associate Professor and Head, Urban and Environmental Planning, Griffith University, Brisbane, Queensland. Sipe received his PhD from Florida State University in 1996. He has taught in the Griffith School of Environment since 1998 and has served as head of the planning program 2002–6 and 2008 to the present. Sipe has an extensive teaching record in the field of transport planning and spatial analysis. In recent research, Neil Sipe and Jago Dodson have been the first Australian scholars to propose methods for defining and mapping transport exclusion and oil vulnerability. He has a strong track record in empirical research that links issues of spatial access and socio-economic equity in urban contexts, both in the USA and in Australia. He is co-author (with J. Dodson) of *Shocking the Suburbs: Oil Vulnerability in the Australian City* (UNSW Press, 2008).

Emily Talen is Professor at Arizona State University School of Geographical Sciences and Urban Planning, and the School of Sustainability. She has a Ph.D. in urban geography, University of California, Santa Barbara, and a master's degree in city and regional planning, Ohio State University. Prior to ASU, she was a faculty in the Department of Urban and Regional Planning, University of Illinois, Champaign-Urbana for 8 years. Before moving to academia, she was a planner with the City of Santa Barbara for 6 years, and is a member of the American Institute of Certified Planners (AICP). Her research focuses on topics dealing with new urbanism, urban design, and the social implications of community design. In addition to close to 50 scholarly journal articles, she has authored three books: one on the historical lineage of new urbanism (*New Urbanism and American Planning: The Conflict of Cultures* (Routledge, 2005)); a study of the urban design requirements of socially diverse neighbourhoods, in *Chicago, Design for Diversity* (Architectural Press, 2008), and *Urban Design Reclaimed* (Planners Press, 2009), a set of 10 urban design exercises for planners. She is co-editor of the newly established *Journal of Urbanism*, published by Taylor & Francis.

Tanja Winkler is Honorary Senior Lecturer, School of Architecture and Planning, University of the Witwatersrand, Johannesburg, South Africa. She completed a PhD on resident involvement (and led urban regeneration work) in the inner city of Johannesburg in 2006 from the School of Community and Regional Planning, University of British Columbia, Vancouver, Canada. She has worked as an urban design consultant in South Africa, and as a municipal official in Britain.

Anthony Gar-On Yeh is Chair Professor, Head of Department of Urban Planning and Design, Director of Centre of Urban Studies and Urban Planning, Director of GIS Research Centre, and former Dean of Graduate School, University of Hong Kong, Hong Kong SAR. He is an Academician of the Chinese Academy of Sciences and a Fellow of the Hong Kong Institute of Planners (HKIP), Royal Town

Planning Institute (RTPI), Planning Institute of Australia (FPIA), British Computer Society (BCS) and the Chartered Institute of Logistics and Transport (CILT). He received the Hong Kong Croucher Foundation Senior Research Fellowship Award in 2001 and the UN-HABITAT Lecture Award in 2008. He has done consultancy work related to his expertise on projects for the Hong Kong Government, World Bank, Canadian International Development Agency (CIDA), Urban Management Programme (UMP), and Asian Development Bank (ADB). He has served in various government consultation boards of the Hong Kong Government. At present, he is the Secretary-General of the Asian Planning Schools Association and the Asia GIS Association. He is on the editorial boards of key international and Chinese journals and has published over 30 books and monographs, and over 180 academic journal papers and book chapters. His main areas of specialization are urban development and planning in Hong Kong, China, and S.E. Asia, and the applications of geographic information systems (GIS) as planning support systems. His current research is on city competition and development in China, Greater PRD regional development and cooperation, high-rise living environments, elderly adjustment to new towns, short-interval land use change detection using radar remote sensing and real-time transport GIS.

Preface

The book series, *Dialogues in Urban and Regional Planning*, offers a selection of the best urban planning scholarship from every region of the world, intended to help bridge language and regional gaps among planning researchers. The award-winning papers presented illustrate the concerns and the discourse of planning scholarship communities, and provide a glimpse into planning theory and practice by planning academics around the world. Conceptualized at the first World Planning Schools Congress in Shanghai in 2001, this series is produced by the nine member planning school associations of the Global Planning Education Association Network (GPEAN). International distribution through the nine associations and through other means helps to make this wide-ranging material available in parts of the world where it would not normally be accessible. The hope is that it will help to make intellectual connections which will integrate and enrich the body of planning scholarship.

This fourth *Dialogues* volume includes papers originating from six continents. They draw on local concerns but also reflect international issues: problems and opportunities presented by a variety of planning institutions in dealing with *social inclusion and local identity*, new ways of thinking about the organization, evolution, and use of *urban spaces*, *strategic thinking* about new analyses of classic planning issues, and *barriers* to carrying out new approaches to planning.

The papers were nominated as a result of competitive processes in nine planning school associations, and were then selected by an International Editorial Board including representation from all the associations. Taken together, they suggest the issues seen as important today by planning educators, and also the diversity of approaches to research and practice that distinguish the world's planning scholarship communities. We hope all those with an interest in urban and regional planning will find this collection valuable in expanding perspectives and approaches pertaining to their work, and that it will inspire them to rise to meet the challenges presented here.

The *Dialogues* series and this volume would not be possible without the efforts of a large group drawn from nine planning school associations, numerous publishing houses, and universities. The members of the International Editorial Board

xx Preface

were vital to selection of papers and to editing of the volume, including Heloisa Costa, Christophe Demazière, Thomas L. Harper, Michael Hibbard, Daniel K.B. Inkoom, Petter Næss, Yukio Nishimura, Awais Piracha, and Roberto Rodriguez-Garza. They first served as participants in nine (national and regional) editorial committees appointed by their planning school associations to vet nominations for the project. We also owe gratitude to the members of these association editorial committees: Carlos Antônio Brandão, Joel Chaeruka, Luciana Corrêa do Lago, Ana Fernandes, Lilian Fessler Vaz, Pierre Filion, Jupira Gomes de Mendonça, David Gordon, Daniel K.B. Inkoom, S. Kusangaya. Geraldo Magela Costa, A. Mosha, Caroline Miller, Peter Ngau, Tim Perkins, Daniel Phiri, Johanna Rosier, Mark Seasons, Utpal Sharma, Elisabete A. Silva, Alison Todes, Smart Uchegbu, and Vanessa Watson. The project has greatly benefited from the vision and support of Alex Hollingsworth at Routledge, and we have enjoyed the cooperation of the various journals and publishing houses responsible for first publication of many of the papers presented here.

Thomas L. Harper
Calgary, Canada
Heloisa Costa
Minas Gerais, Brazil.
Antony Gar-On Yeh
Hong Kong, China
Michael Hibbard
Eugene, Oregon, U.S.A.

Chapter 1

Introduction
Rising to the global challenges

Thomas L. Harper, Michael Hibbard,
Heloisa Costa, and Anthony Gar-On Yeh

Purpose

The purpose of the *Dialogues in Urban and Regional Planning* series is to make generally available a selection of the best scholarship in urban and regional planning from each of the world's scholarly planning communities. The papers are elected by the *Dialogues* editors from among those nominated by the nine associations that are members of the Global Planning Education Association Network. In this introductory chapter we describe the background of *Dialogues* and present an overview of this volume. We begin by summarizing the emergence of the global Planning Schools Movement and the associated rise of an international scholarly community in planning. We then develop a framework for the volume, drawing on the UN Habitat report, *Planning Sustainable Cities.* Finally, we position the various chapters within this framework.

Background

Dialogues in Urban and Regional Planning, of which this is the fourth volume, grew out of discussions at the first World Planning Schools Congress (WPSC) in Shanghai, in 2001. WPSC 1 was a pivotal event in the Planning Schools Movement (GPEAN n.d.). The first truly global meeting of planning scholars, it was the culmination of a series of events that began in the mid-1980s – spontaneous, un-coordinated actions by planning scholars around the world to form national, regional, and linguistic associations, or to transform existing associations. Their aim was to provide forums for the presentation and critical discussion of planning research and pedagogy that would not be constrained by the concerns for immediate application that preoccupy practicing planners. Through these forums – primarily conferences and scholarly journals – the foundations have been laid for a *discipline* of urban and regional planning that is theoretically informed, empirically rigorous, and fosters advances in professional practice.[1]

Four of the associations, the U.S.-based Association of Collegiate Schools of Planning (ACSP), the Association of European Schools of Planning (AESOP), the

Asian Planning Schools Association (APSA), and the Australia and New Zealand Association of Planning Schools (ANZAPS) joined together to organize WPSC 1. WPSC 1 brought together 650 participants from more than 250 planning schools in sixty countries. Leaders of ten planning school associations from around the world met at WPSC 1 to discuss possibilities for ongoing cooperation, especially with respect to advancing the visibility and improving the quality of planning research and education around the world. Nine associations ultimately endorsed the Statement of Shanghai, which, among other things, established the Global Planning Education Association Network (GPEAN).

Since its formation, GPEAN has provided a formal institutional structure for integration, cross-fertilization, and critical analysis in the global planning academy. The leadership of the member associations meets at least annually to exchange information. There are regular meetings of people in counterpart positions in the associations (conference managers, journal editors, and so on). WPSC 2 was held in Mexico City in 2006, and WPSC 3 is planned for Perth in 2011.

At least as important as GPEAN's formal structures have been the informal, interpersonal networks that have grown up in concert with them. While there has not yet been a systematic accounting, there is ample evidence of rapid growth in international scholarly exchanges and collaborations over the decade since WPSC 1. They range from individual collaborations and exchanges to an exponential increase in international participation in association conferences[2] to a proliferation of joint studios and projects involving students and faculty from different universities on different continents.

Following the recursive pattern with which planners have become familiar, these informal networks have in turn led to more formal collaborations among planning scholars. The most prominent example is probably the UN-Habitat report, *Planning Sustainable Cities* (UN-Habitat 2009). Dozens of planning scholars from around the world authored background papers and served as advisors and consultants for what was surely the most wide-ranging effort ever attempted to assess the effectiveness and potential of urban and regional planning as a tool for dealing with the challenges facing cities and their hinterlands. It was a truly global undertaking that would have been inconceivable without the international network of planning scholars that has emerged in the last twenty-five years.

21st century challenges and initiatives in urban and regional planning

The key contribution of *Planning Sustainable Cities* – what makes it so valuable for global planning scholarship – is its development of an authoritative structure for thinking about planning at the most general level as well as examining its application in specific contexts. As such, it provides a useful framework for thinking about the background of the contributions to this volume of *Dialogues in Urban and Regional Planning*.

To reiterate, the purpose of the *Dialogues* series is to make generally available a selection the best scholarship in urban and regional planning from each of the world's scholarly planning communities. The *Dialogues* chapters are thus context specific. However, as *Planning Sustainable Cities* argues convincingly, the world faces a set of *common* urban challenges, even though they manifest themselves differently in varied specific contexts. Further, it describes the *common* characteristics of the 'new planning initiatives' that are springing up around the world in response to these challenges; again, these characteristics manifest themselves differently in varied specific contexts. A brief review of the common challenges and the characteristics of new approaches to planning put forward in *Planning Sustainable Cities* can help us to understand the relationship of the studies in this volume of *Dialogues* to one another, and to position them in the global planning discussion.

Planning Sustainable Cities delineates five major challenges for planning: environmental, demographic, economic, socio-spatial, and institutional.[3] These are not mutually exclusive; on the contrary, they tend to reinforce one another.

Environmental challenges

Over the past twenty years, since the Brundtland Commission report, *Our Common Future* (Brundtland 1987), environmental sustainability has become a center of attention for urban and regional planning. This is nowhere more manifest than in the current concern with how to mitigate the effects of climate change, which will affect the basic elements of life for people around the world – access to water, food production, health, and the environment (Stern 2007).

An associated environmental challenge is the connection between climate change and global dependency on fossil fuels. Coal-fueled electric power stations are a major contributor to climate change (IPCC 2007). Easy availability of oil has both promoted and permitted the low-density and sprawling urban form that depends on private cars. Beyond that, the global economy rests on the ability to move people and goods quickly and cheaply over long distances. Vehicle and aircraft emissions contribute significantly to greenhouse gas emissions and, hence, global warming.

Urbanization itself is inherently risky. The concentrated nature of settlements entails the modification of natural systems, often done poorly. For example, deforestation and consequent slope instability can result in landslides and flooding. Because of the interdependence of human and natural systems, urban settlements are increasingly prone to disaster risk. As well, urban development results in negative environmental impacts through the consumption of natural assets and the overexploitation of natural resources, and a decrease in the landscape which produces ecological goods and services. The challenge for planning scholarship is to find ways to move toward more efficient, resilient, and integrated urban forms that place less demand on the natural environment.

Demographic challenges

Societies around the world face two major demographic challenges. The first is the global urban transition. In 2008, for the first time in history, over half of the world's population lived in urban areas. According to current projections, this will have risen to 70% by 2050. Almost all of this growth will take place in developing regions. Between 2007 and 2025, the annual urban population increase in developing regions is expected to be 53 million (or 2.27%). By comparison, in developed regions the annual increase is projected to be only 3 million (or 0.49%) per year. At the same time, some parts of the world are facing the challenge of shrinking cities whose industrial base has eroded and whose infrastructures are decaying. Most of these are to be found in the developed and transitional regions of the world. But more recently, city shrinkage has occurred in some developing countries as well.

Both rapid growth and shrinkage test the ability of governments to provide urban infrastructure, and of taxpayers to fund such services. As well, they increase the likelihood of natural disasters and social crises at the same time that they make mitigation of such problems more difficult.

A second demographic challenge is the aging of the population in many places, because of declines in mortality and fertility.[4] The most striking cases are in developed countries. In Italy, for example, 19% of the population is now 65 or older, and the figure is expected to reach 28% by 2030. In Japan, one fourth of the population will be 65 or older by 2015, and by 2050 the Japanese population is projected to shrink from its current level of 128 million to 95 million. However, this is a widespread phenomenon. Average ages elsewhere in the world – particularly in Asia and Latin America – are slowly creeping up. In China, declining mortality rates will cause its elderly population to rise from 88 million to 349 million by 2050.

The aging population creates an interlocking set of challenges. As people have fewer children and the working age population shrinks, there are fewer taxpayers to pay for a growing demand for public pension/social security systems. As well, there are fewer caregivers, particularly family members. The upshot is that the number of elderly people living alone in their homes and communities will increase, placing additional burdens on medical and social support systems and requiring significant investment to equip or renovate private homes and public spaces to facilitate the independence of older people.

Economic challenges

No part of the world has been immune to the processes of global economic restructuring, processes that have been accelerated by the current economic crisis. There is much diversity in the impact of these processes, with the specific outcomes strongly influenced by pre-existing local conditions and policies. Nevertheless, two issues can be singled out. Labor markets show a growing *polarization* of occupational and income structures, with a shrinking middle class and an increasing *gap* between the best and least well off. As well, there has been a rapid growth in the informal economy in all urban centers, particularly in developing countries.

Increasing inequality, poverty, and high levels of informal economic activity are not inevitable. The challenge – in both developed and developing countries – is to respond to the environmental and demographic challenges outlined above in ways that increase economic opportunity and justice.

Socio-spatial challenges

Urban regions increasingly find themselves with new spatial forms that have emerged outside the control of the usual processes of land planning and urban design by local governments. These changes are largely in the direction of the fragmentation, separation and specialization of functions and uses within cities, with labor market polarization (and hence income inequality) reflected in growing differences between wealthier and poorer areas. Highly visible *contrasts* have emerged between affluent gentrified and suburban areas and slums, ethnic enclaves, and ghettos. These differences parallel contrasts between areas built for the advanced service and production sector and luxury retail and entertainment, and older areas with declining industries, sweatshops, and informal businesses.

Much of this represents the playing out of land market forces and the logic of real estate development, but it is also a consequence of local policies in which cities have attempted to position themselves as attractive sites for investment in the global competition for capital.

Institutional challenges

Urban and regional planning is a public responsibility, generally carried out by local government. Originally, planning played an *advisory role*. Its function was to provide local government decision makers with advice and recommendations. It was an internal and largely *technical* and analytic activity. Over the past thirty years, however, there have been significant changes in local government in many parts of the world that have challenged that traditional understanding. Three of those changes are particularly far reaching.

The first change has been the expansion of the urban political system from "government" to "*governance*," the involvement of a range of non-state actors in the process of governing, from business and advocacy groups of various stripes to "civil society." A second, related institutional change has been the rise in structured *public involvement*. Responding to the growing unwillingness of communities to simply accept the recommendations of planners and the planning decisions of politicians, governments are routinely building opportunities for the public to participate into the planning process in various ways.

A third institutional change has been in response to the shift in *scale* of urban decision making. The environmental, demographic, economic and socio-spatial challenges faced by local governments have come to exceed their geographically bounded administrative authority. The biggest challenges are now at the scale of the metropolitan region; they are not responsive to the actions of a single jurisdiction. Attempts at metropolitan-level planning through collaboration amongst governments have become increasingly common in many parts of the world.

In response to these challenges there are vigorous ongoing attempts to develop *new approaches* to urban and regional planning. There is a tendency for these new approaches to address a range of concerns – the environment, the economy, health and safety, and so on – within cross-cutting policies, programs, and activities. The new approaches are still largely experimental and have been tested in a limited number of places. Most of them have several of the following elements in *common* as they try to respond to shortcomings in more traditional planning approaches. They:

- are *strategic* rather than comprehensive
- are *flexible* rather than fixed and end-state oriented
- are *action* and implementation oriented
- are stakeholder or *community driven* rather than (only) expert driven
- contain *new objectives* reflecting emerging urban concerns – for example, city global positioning, environmental protection, sustainable development, social inclusion, and local identity
- play an *integrative role* in policy formulation and in urban management by encouraging government departments to coordinate their plans in space; and
- focus on the planning *process*, with the outcomes being highly diverse and dependent upon stakeholder influence or local policy directions.

Global themes

The challenges and characteristics identified in *Planning Sustainable Cities* resonate throughout the papers selected for inclusion in *Dialogues 4*. These scholars all contribute to our understanding of new ways to address the challenges facing planning. Amin Kamete, Leonie Sandercock, and Tanja Winkler take up the problems and opportunities presented by a variety of kinds of planning institutions in dealing with social inclusion and local identity.

Kamete investigates the attitudes, behaviors, and thinking of planners in urban Zimbabwe with respect to "participatory governance" and planners' interaction with the public. His case-in-point is an outreach program to youths who are using public spaces illegally. He finds that it is difficult for planners to operationalize participatory governance because the relational nature of participation conflicts with their traditional planning role of expert advisor. He concludes that the institutional challenges of moving from government to governance and promoting public participation are more than procedural; they require deep cultural changes on the part of planners.

Sandercock presents a theoretically rich analysis of a neighborhood organization that has been working for twenty years to integrate immigrants into a predominantly Anglo-European neighborhood in Vancouver. She uses the case to ask the question about how urban societies can come to terms with their increasing multicultural realities. She first discusses the challenge of how to think about multiculturalism, and proposes the necessity of a deepened understanding of "difference," and its significance for urban politics. She then suggests an alternative way of theorizing an "intercultural project" that moves beyond contemporary "multiculturalism," using as an illustration her case of a local institution that is a catalyst for living together and bridging across cultural differences.

Winkler investigates the potential of civil society to contribute to democracy and development. She critically analyzes three Johannesburg-based civic associations to understand their ability to reshape the public domain – to effect local government planning policies and financial practices. She finds that all three contribute to pluralism, democratic values, and leadership development by giving voice to disadvantaged groups. But they have only limited influence on public decision-making processes because the state uses its power to maintain unjust planning processes. Flexible and pluralistic strategies are needed to continue pressure on the state to enact collaborative, transformative and just planning processes.

Turning to another set of challenges and characteristics, Jago Dodson and Neil Sipe, Michael Hebbert, Maria Julietta Nunes de Souza and Rose Compans, and Samira Ramezani, Zul Azri Bin Adb Aziz, and Syed Z.A. Idid all address new ways of thinking about the organization, evolution, and use of urban spaces.

Dodson and Sipe take up the issue of urban sprawl and rising fuel prices. They used an "oil vulnerability" assessment methodology based on Australian census data to estimate how rising fuel costs, mortgage interest rates, and general inflation will be spatially distributed within Australian cities. They found broad-scale mortgage and oil vulnerability in the outer suburbs across all Australian cities.

Hebbert's essay posits a rethinking of urban green space in the transition from modern to postmodern urbanism, and looks for its underlying social and environmental rationale. He argues that modernism had an open-space vision of an "unbounded public realm" modeled on 18th-century landscape design. That vision dominated the second half of the 20th century through housing and highway development, open-space policies, and regulation. But as concerns rose about its environmental and social utility, toward the end of the century the modernist vision was replaced by an alternative vision aimed at active provision of ecosystem services through urban green space. Hebbert describes this rethinking and assesses its place within a wider planning theory of postmodern urbanism.

Nunes de Souza and Compans interrogate the adoption in Brazil of urban design strategies aimed at improving public security. They credit Oscar Newman (*Defensible Space*, 1972) with originating these strategies, which have since been diffused around the world, including to Latin America. Nunes de Souza and Compans describe their implementation in Brazil generally and specifically evaluate public satisfaction with them in thirty *favelas* (poor neighborhoods) in Rio de Janeiro during the run-up to the 2007 Pan-American games. They found that "the different interventions . . . were widely accepted," and that the redesigned spaces were being utilized by the community. However, they are hesitant about drawing conclusions regarding the long-term effects of the strategy, noting that their findings do not include a post-occupancy evaluation.

Ramezani, Aziz, and Idid address a different but no less important issue. They use the case of Melaka, Malaysia to study the problem of meeting the needs of local residents who find themselves living in a *"living heritage city."* Like other similarly situated cities, Melaka seeks to conserve its historic urban heritage and enhance tourism. Ramezani, Aziz, and Idid note that cities such as Melaka implement tourism infrastructure projects in an effort to increase tourism revenue but spend very little effort trying to understand local residents' use of that same outdoor environment. They argue that the day-to-day activities of residents are an important part of the urban heritage and must be taken into account. Their culture and lifestyle need to be understood and planned for if the goal is to conserve all pertinent characteristics of the city.

Emily *Talen*'s work lies at the boundary between social inclusion and local identity on the one hand, and organization, evolution, and use of urban spaces on the other. She examines the plans produced by the Congress for the New Urbanism (CNU) for the rebuilding of eleven towns along the Gulf Coast of Mississippi

(USA) in the aftermath of Hurricane Katrina. She is especially interested in the social equity goals of the CNU plans. She finds that while they are both implicitly and explicitly stated throughout the plans, the realization of social-equity goals will require more than physical design. In addition, she calls for policy, institutional, programmatic, and process efforts towards social equity.

Strategic thinking about classic planning issues underpins the papers by Arturo Almandoz, Bent Flyvbjerg, and Vincent Kaufmann, Christophe Jemelin, Geraldine Pflieger, and Luca Pattaroni.

Almandoz offers a new reading on one of the classic questions in development – how to explain Latin America's "take-off" during the 20th century and its failure to attain economic "maturity" (he uses Rostow's well-known, though problematic, terminology). He draws on historical, political, and social factors to develop a comparative panoramic view across Latin America of the divergence between urbanization, which was generally rapid, and industrialization, which was problematic. He argues that although it occurred at different times in different countries, this divergence was subject to a regular pattern. He develops a long-term "periodization" to reveal that pattern.

Flyvbjerg takes up another classic planning issue, the pervasive misinformation about the costs, benefits, and risks involved in very large infrastructure projects, and the consequent cost overruns, benefit shortfalls, and waste. He argues that the best explanation for these problems is deliberate misrepresentation of the costs, benefits, and risks by planners and promoters of the projects, in order to increase the likelihood that they will be built. He ends by presenting measures for reforming policy and planning so that the best projects rather than the best promoted (or most misrepresented) are actually built. He advocates not only for better planning methods but more especially for changes in governance structures.

Kaufmann and his colleagues took stock of French transportation policy research between 1995 and 2002. They conducted a content analysis of the relevant studies and found, among other things, that cost-benefit analysis presents a double obstacle to decision making. First, it is based on "non-consensual ideologies," disagreements about what constitutes costs and benefits. Second, it prevents public involvement because it is "characterized by a 'black box' that is unquestioned by experts." In addition, Kaufmann et al. report that the research illustrates the importance of factors that are difficult to monetize and that go beyond maximization of utility.

Alan *March*'s analysis of the demands on planning professionalism provides a nice précis of the barriers to carrying out the new approaches to planning. He notes that planning is located across a range of institutional and governance settings that require particular understandings of the possibilities and problems of professionalism in each specific setting. He uses the state of Victoria (Australia) to demonstrate how the institutional roles of planners influence the exercise of their professional judgment and action. He argues that planners need to consciously acknowledge and adopt the role of democratic facilitator of knowledge development, and finds that doing so requires them to begin with an appraisal of the institutional impediments to participation in a given setting.

Concluding comments

Reading through the papers in *Dialogues 4*, it is clear that one of the challenges identified in the UN Habitat report rises above all the others: They all identify *institutional* challenges that need to be met if planning is to successfully address the *environmental, demographic, economic, and socio-spatial* challenges facing the world's cities. To mention just a few examples for illustrative purposes, Flyvbjerg, Kamete, Kaufmann, and March all examine aspects of the conflation of the *role* of planners and planning as a source of value-neutral analysis, vs. planners as advocates, and/or as facilitators of democracy. Winkler and Talen both point to the importance of understanding the *politics* of the institutions within which planning operates. And Sandercock and Ramezani, Aziz, and Idid ask us to explicitly consider the *cultural* dimension of planning. These and other institutional challenges raised by the authors are not necessarily new, but they have taken on heightened salience in the 21st century.

In another vein, comparing the papers selected for *Dialogues 4* with those included in the first three volumes of *Dialogues*, it is clear that there are a few elements of convergence in planning scholarship internationally. First, there is a *grounding* in concrete *practice*. Beneath the broad range of methodological approaches is a consistent empiricism. Planning scholarship is "about" the current pressing challenges facing different social groups, communities, cities, and regions. Second, even though it is concrete and empirical, planning scholars seek to *link* their work to broader *theoretical* debates in the field and beyond – to debates in the social sciences and environmental design. And finally, there is evidence – sometimes explicit but more often tacit – of the growing *mutual influence* of planning scholarship across the world. Findings, insights, and ideas may be locally embedded, but they are being globally appropriated, adapted and transformed.

Notes

1   For a more complete description of the development of planning school associations worldwide, see Stiftel and Watson (2005).
2   For example, at the 2008 joint ACSP-AESOP conference in Chicago, about 20% of the registrants were from Asian, Australian, and New Zealand universities, in addition to those from Europe, Canada, and the U.S.

3 The following discussion is drawn from *Planning Sustainable Cities*, except as noted.
4 This section is drawn from Wish (2009).

References

Brundtland, G. (ed.) (1987). *Our Common Future: The World Commission on Environment and Development* (Oxford: Oxford University Press).

GPEAN (n.d.). The planning schools movement. At www.gpean.org/ (accessed February 26, 2010).

IPCC (Intergovernmental Panel on Climate Change) (2007). *Climate Change 2007 Synthesis Report* (Geneva: IPCC).

Stern, Nicholas (2007). *The Economics of Climate Change: the Stern Review* (Cambridge: Cambridge University Press).

Stiftel, Bruce, and Vanessa Watson (2005). Building Global Integration in Planning Scholarship. Chapter 1 in Stiftel and Watson (eds.), *Dialogues in Urban and Regional Planning 1* (London: Routledge).

UN-Habitat (United Nations Human Settlements Programme) (2009). *Planning Sustainable Cities: Global Report on Human Settlements 2009* (London: Earthscan).

Wish, Valdis (2009). Global Population Aging: Growing Old. Allianz Knowledge Partnersite. At http://knowledge.allianz.com/en/globalissues/demographic_change/aging_societies/population_aging_global.html (accessed March 5, 2010).

Chapter 2
Hanging out with "trouble-causers"
Planning and governance in urban Zimbabwe

Amin Y. Kamete

Taking the relational nature of participatory governance as a point of departure, this chapter interrogates the attitude, behaviour and thinking of planners in urban Zimbabwe. Particular emphasis is placed on the planners' interaction with the public. The discussion analyses one city's planning system as it implements an ambitious "governance outreach programme" which involves interacting with youths who are using public space illegally. The central argument of the chapter is that it is difficult to operationalise participatory governance in planning mainly because the relational nature of governance requires planners to act in ways that conflict with their preferred role as technical experts. The discussion exposes how pointless it can be for bureaucrats to interact with the public, when the participants' attitude, means, behaviour and style express no confidence in the institutional framework. The chapter suggests that the transformation from government to governance is not merely procedural, and requires a deep cultural change on the part of planners.

Introduction

April 23, 2004. The "Governance Outreach Task Force" (GOTF) of a large urban centre in Zimbabwe is being debriefed by the Town Clerk.[1] This is an official "report-back session" during which members of the task force are reporting on the status of their consultations with youths who have illegally taken over a public car park.

Set up as an integral part of the "Governance Outreach Programme" whose mandate, according to the Town Clerk, is to "to bring residents . . . on board the governance train," the GOTF is led by a Zimbabwe-trained senior planner. He is

---

Submitted by the Association of African Planning Schools.

Originally published in *Planning Theory & Practice* 10:1, 85–103. © 2009. By kind permission of Taylor & Francis Ltd. www.informaworld.com and the Royal Town Planning Institute.

saying something to the effect that the process has not been going according to original expectations, when the Town Clerk impatiently interrupts, "So, what are the lessons, guys? What are the key lessons? . . . I told you guys that we should be learning by doing, and doing by learning. Come on . . . let's make it work . . . Tell it as it is, as it really is."

The Senior Planner says reflectively, "Maybe we need to have realistic expectations . . . slow down a bit . . . and . . . ask ourselves what exactly we are looking for . . . You see, sir, according to the UNDP—."[2] His words are drowned by raucous laughter from the senior bureaucrats. The Senior Planner's immediate boss, the Deputy Director of Engineering Services (Planning), cuts in, "You want us to be *realistic* . . . and yet you bring in those lazy theories from the UN?" There is more laughter.

According to the Senior Planner's subsequent assessment, the GOTF concluded that "the private sector is basically okay . . . because they know what they want . . . and are very easy to deal with." In the GOTF's view, "the most difficult [people] to handle are . . . those 'trouble-causers' in designated trouble spots." "Trouble-causers" has become an official shorthand term for self-employed youths who "have illegally colonised [urban space] and disregard land-use controls." According to the GOTF, the "trouble-causers" "have no direction and do not know what they want." Cautioning that "it will take more time," the Senior Planner optimistically concludes that they are "getting there" in confronting the problem. Throughout the presentation, the other members of the GOTF have been silently nodding in agreement.

Closing the session 30 minutes later, the Town Clerk asks, "Should we continue with the process?" There is a chorus of agreement. After reminding the GOTF to submit a "comprehensive and uncensored written . . . report," he dismisses his charges, save for the Deputy Director of Engineering Services (Planning). The GOTF then files out of the Town Clerk's office and proceeds to the committee room where a closed meeting is held. There, the Senior Planner asks his colleagues what they think should be reported. One of the two locally trained planning technicians in the GOTF says, "*Varume* [gentlemen] this is not *mahumbwe* [child's play] . . . We need more time. The Town Clerk should come with us to see for himself."

This event occurred during my research on youths in contested urban spaces in Zimbabwe. This chapter presents part of the results of this study, analyzing the role of Zimbabwe's urban planning systems in meeting the challenge of participatory local governance. Focusing on the relational aspect of governance, the chapter explores how planners in one city have attempted to achieve participatory governance by reaching out to "troublesome" youths in the public production of urban space. The specific focus on the relational aspects of participatory governance was inspired by the link between governance and aspects of participatory planning.

14 Amin Y. Kamete

It was also motivated by the enthusiastic embracing of governance by most urban councils in Zimbabwe, especially since 1995 (DRUP 1997). This was buttressed by instructions from the then Ministry of Local Government Public Works and National Housing that "made it a requirement for all local authorities to undertake strategic planning *in order to enhance good governance*" (KCC 2003: 1; emphasis added). The working definition of governance in Zimbabwe appears to be that articulated by the state president, namely,

> a process of involving people in the making of decisions that affect their livelihood, in a transparent and accountable manner. It entails the devolution of power and responsibilities upon lower levels of society, encouraging participation, recognising the diversity of communities and societies, and the promotion of openness and elimination of corruption in managing resources.
>
> (ACPD 2002: 2)

The central argument of the chapter is that, with respect to planning, the quest for participatory governance faces obstacles because it requires planners to act in ways that conflict with their preferred role as technical experts. It exposes how pointless technical experts' interaction with the public may be when the participants' attitude, means, behaviour and style betray their lack of faith in the governance framework. It also shows that the passage from government to participatory governance is not merely procedural, and requires a deep cultural change. Apart from highlighting a particular planning situation, the chapter contributes to an understanding of planning's role (and particularly the difficulties it faces) in bringing about democratic change under "guided democracy."

The study on which the chapter is based examined planners' experience, attitude, behaviour and viewpoints towards the official emphasis on local governance. There was a two-part methodology behind the research. The first was non-participatory observation, scrutinising planners as actors by watching them at work in the Governance Outreach Programme. I attended six "public hearings" involving the GOTF and youths and was present at three closed GOTF meetings and two "report-back sessions," where senior city bureaucrats debriefed the GOTF. The second part of the methodology—inspired by findings from the GOTF—involved interviewing 23 planners and planning technicians nationwide, trying to make sense of their actions. The semi-structured interviews were carried out on a one-to-one basis (12), by telephone (6) and through e-mail (5). The interview schedule sought to capture planners' perceptions of governance, the level of their interest in local governance, and their perception of the implications of the emphasis on local governance on planning practice. It also sought their opinions and/or attitudes on four issues: whether planners have a role to play in local governance; whether planners are sufficiently equipped to meet the challenge of local governance;

whether local governance demands a change in planning practice and approach; and whether, to contribute effectively to local governance, planners may need to be retrained.

The rest of the chapter is organised in four parts. The next part presents the framework of analysis by looking at governance as a relational practice. This is followed by a presentation and discussion of this specific piece of research in two parts: firstly, a case-study of the City and secondly, findings from interviews with planners nationwide. This leads to the analysis and interpretation of the research findings before a conclusion closes the discussion.

Governance and planning practice

Although "governance" has degenerated into a "confusing term" (Pierre and Peters 2000: 14) in theory whose "meaning . . . is not always clear" (Peters and Pierre 1998: 223), in practice, participatory governance, which is the concern of this chapter, is essentially a relational practice. According to Schmitter (2002: 52), participatory governance is "a . . . mechanism for dealing with . . . problems . . . in which actors regularly arrive at mutually satisfactory and binding decisions by negotiating with each other and co-operating in the implementation of these decisions." In the public sphere, participatory governance involves a whole range of relationships (Devas 2001: 393) in the management and administration of public affairs.

Even commentators with a relatively narrow conception of governance embrace its relational nature. For example, Clark (2000: 3), who regards governance as merely "the control and co-ordination of activities to attain a range of specified outcomes," acknowledges the assortment of interacting actors in governance processes. In addition to the state, the interacting parties include "interest groups or figurehead groups that represent the interests of capital and labour" (Clark 2000: 3). To the extent that it deals with relationships, this narrow conception is not dissimilar to that of commentators who have a broader view of governance. Among these is Stöhr (2001: 1, 4), who views governance as broadly encompassing "*cooperation* between the public sector, the private sector, and civil society [who] *work together* as partners in building . . . a better society" (my emphasis).

Notably, agencies that work on governance also treat the practice as relational. The UNDP, which has a programme dedicated to governance, regards the concept as comprising "the mechanisms, processes and institutions through which citizens and groups articulate their interests, exercise their legal rights, meet their obligations and mediate their differences" (UNDP 1997: 2–3). At first glance, the notable exception to the emphasis on the relational aspect of governance is the

World Bank, which considers the concept as merely "the process by which authority is conferred on rulers, by which they make the rules, and by which those rules are enforced and modified" (World Bank 2007). The Bank's operationalisation of governance focuses on public sector management, the legal framework, and human rights (World Bank 1994). However, this "governor-centric" governance does imply relational issues in the conferring of authority as well as the enforcement and modification of rules, where, significantly, the Bank emphasises transparency, accountability and participation (World Bank 1994).

The complexities generated by the diversity of actors involved and the multiple levels at which they relate are compounded by the fact that governance is a dynamic, volatile and fluid process (Kamete 2002). This is so because participatory governance entails encounters and relationships between and within the interacting parties (see Goodwin and Painter 1996: 636), principally those who do the governing and those being governed. In this way, governance comprises what Levy (2004: 2–3) terms "sets of interaction," which include political and economic interests as well as formal political institutions and the bureaucracy. The relationships assume different forms in different contexts. Thus, in its extreme forms "governance can either be authoritarian and despotic . . . or democratic and participative" (Matlosa 2005: 5).

Though the above perspectives refer to the national level, the situation is similar at the local level, where the unit of analysis is primarily decentralised local government, from whence planning operates. Being a public function that deals with issues at the heart of social and economic production and reproduction, urban planning is among governmental activities that feature prominently in the governor–governed relationships. When local government embraces participatory governance, the ways in which local planners interact with citizens in the production, ordering and control of urban space speaks volumes about how seriously and successfully planning is moving with the times. This is so because, for a range of specialised and hitherto closed governmental functions and agencies, participatory governance demands an opening up, a reaching out, and a transparency that makes relations not only possible, but also meaningful and useful (McCarney and Stren 2003). In a way, linking planning to local governance addresses the need for a "social theory of planning that . . . specifically incorporate(s) a politico-economic awareness" (Dear 2000: 122), to complement professional, technical expertise.

Reaching out to and interacting with citizens poses a challenge to practices that are directed by instrumental rationalism, which is concerned with the most efficient or cost-effective means to achieve a specific end without reflecting on the value of that end (Braaten 1991: 12). In parts of southern Africa, spatial planning is treated as a state bureaucratic function that faces these challenges. Apart from Zimbabwe, Zambia and Botswana still rely on planning laws that still retain features of colonial planning legislation. These laws borrowed heavily from the British planning tradition that espoused the "rational process" view of planning, emphasising the method or means of planning, and largely driven by instrumental rationality (Taylor 1998: 71). In Botswana, changes made in 1995 relaxed some controls but did not substantially transform the methods espoused in the 1977 Town and Country Planning Act (Bourennane 2007). Similarly, the 1995 changes to Zambia's 1962 Town and Country Planning Act did not change much in terms of planning style (Mwimba 2002). Although planning processes in post-apartheid South Africa are probably the most democratic and participatory in the region, some scholars and practitioners still observe the persistence of top-down planning approaches (Ambert and Feldman 2002: 7) and modernist tendencies (Watson 2002: 35; Mabin 2002: 47; Oranje 2003: 181) in these processes.

Since the 1970s, the supremacy and dominance of the rational process model, which encapsulates instrumental rationality, have been challenged, its assumptions and effectiveness questioned, and alternatives offered (see Healey 1997; Sandercock 1998a, 1998b; Hillier 2002; Hoch 2007). What makes the rational process model incompatible with participatory governance is the potentially autocratic nature of means-orientated or instrumental rationality. Technical experts sometimes appeal to the scientific method in order to mask tyranny. Being an integral part of what Scott (2003: 125) scathingly labels "authoritarian high modernism," planners can easily disguise "the dark side of planning" (Yiftachel 1998) by presenting it as a benign and incontrovertible science. But scientism is inherently undemocratic; it necessarily entails the exclusion of "unqualified" laypeople from specialised activities such as planning. When science is elevated, government (as opposed to governance) becomes the preferred form of state administration, including the making and management of urban places.

Unsurprisingly, where scientism is embraced, planners occupy pole position in the public production of space where serious planning work is legally the preserve of trained "technician[s] of means committed to the value of scientifically based and rationally deduced policy choices" (Healey 1997: 25). Their interaction with laypeople is limited to data collection—and this using research tools that seek answers to questions the experts have determined are the only ones that are valid. For decades, a strict division of labour has ensured that "the planner's job is to deliver unbiased, professional advice and analysis to elected officials . . . who in turn make decisions" (Innes 1998: 52). The role—and duty—of the generality of the urban populace is to provide information and abide by the diktats emanating from the planning system.

Towards a governance-sensitive planning approach?

Relations in participatory local governance are fruitful to the extent that stakeholders desist from strategically deploying certain advantages that they possess. For planning, this rules out resorting to autocratic methodologies that reduce the role of other participants to ratification, objection or protest. Participatory governance therefore presents a challenge for "the design of governance systems and practices" (Healey 1997: 5). One solution to this is collaborative planning (Harris 2002: 25), a programme that is concerned with the "democratic management and control of urban and regional environments and the design of less oppressive planning mechanisms" (Harris 2002: 22). Propounded by diverse communicative theorists, collaborative planning appears to be the answer for the transformation from government to governance.

Its attraction is enhanced by its promotion of deliberative democracy, a concept that is not lacking in adherents (Dryzek 1990; Cunningham 2002). Deliberative democracy is concerned with "encouraging forums . . . conducive to the pursuit of agreement" (Cunningham 2002: 177), a pursuit that is shared by collaborative planning in its endeavour to address the conflicts that make up "one of democracy's 'problems'" (Cunningham 2002: 177). Collaborative planning is suitable for participatory governance because "the terms and conditions of association proceed through public argument and reasoning among equal citizens" (Cohen 1997: 72). In this "inclusionary argumentation" (Healey 1997: 276) scientism does not hold sway because "analysis is not an abstract technical process but an active social enterprise" (Healey 1997: 276).

However, to some critics this is an impossible exercise. To them, it is unworkable and thus, in practical terms, ends up being as insidiously undemocratic as instrumental rationalism is manifestly autocratic. The reason for this lies in the pervasiveness of power. Since Foucault (1998: 93) insists that power is everywhere, it is not surprising that those who advocate the implanting of his ideas in a planning context contend that rationality is not impervious to power (Flyvbjerg 1998). Highlighting the enormity of this fact, Flyvbjerg and Richardson (2002: 47) warn of the meaninglessness of operating "with a concept of communication in which power is absent." Power differentials, coupled with other inequalities, imply that parties in any collaborative action are far from equal. These distortions do not augur well for collaborative planning. Tacitly confirming this, Innes and Booher (1999: 11) admit that there is a possibility of "tactical exercises by one set of special interests attempting to gain an upper hand over others, or by powerful players attempting to co-opt the weaker ones."

By failing to adequately capture and account for the disruptive role of power in planning, communicative theorists may not adequately equip planners operating in systems of "guided democracy" like Zimbabwe to meet the challenge of participatory local governance (Healey 2003: 113). In such contexts, collaborative planning may prove to be "weak in serving as a basis for effective action and change" (Flyvbjerg and Richardson 2002: 45), the very things that a planning system needs to deliver if it is to contribute to the transformation from government to governance. Arming planners for the world of participatory local governance may also require an appreciation of the insidiously sinister realities of social interaction. It requires "a turn towards the dark side of planning theory— the domain of power" (Flyvbjerg and Richardson 2002: 45).

Flyvbjerg and Richardson (2002: 45) make one of the strongest cases for questioning blind faith in communicative planning when they dismiss communication as being "more typically characterised by non-rational rhetoric and maintenance of interests, than by freedom from domination and consensus seeking." Rather than expecting neat rational argumentation to operate, veiled control, rationalisation, charisma and the deployment of dependency relations— all of which distort rational debate—should be accepted as the basis for success in communication. In the real world, real politics (*realpolitik*) and real rationality (*realrationalität*) call for a non-idealistic point of departure for planning theory (Flyvbjerg and Richardson 2002: 47).

Since participatory governance is a relational practice, it is instructive to interpret how the City's technocrats worked with and related to the so-called "trouble-causing" youths in resolving the spatial dispute. It is helpful to analyse how much of their technical superiority and instrumental rationalism the planners were willing to compromise in order to "collaborate" with the youths, and how, in their turn, the youths responded. In the same vein, the issues of power and rationality need to be explored, to see if they contaminated the relationships, making the whole collaborative planning enterprise as dubious and questionable as modernist planning.

Participatory local governance in action

The context

The City is a large urban centre in Zimbabwe, with a population of more than 200,000. Like the rest of the country it is beset by the economic and political problems that have plagued Zimbabwe since 2000. At the time of the study, inflation had peaked at over 600%; unemployment was more than 60% and the informal sector accounted for 40% of employment. The poverty headcount ratio was 80% (IMF 2005: 28–29). Politically, urban centres had become opposition strongholds, most of them having elected opposition legislators and councillors, a scenario that resulted in incessant wrangling between central government and opposition-controlled councils (Kamete 2006).

Zimbabwe's political system can be described as an authoritarian or "guided democracy," which is basically "a tyranny of the elite" that "borders on authoritarianism" leaving "the governed with only limited control over the government" (Pinkney 2003: 11–12). At the time of the study, the government was increasingly being characterised as repressive, intolerant and authoritarian, a perception supported by the passing of the controversial Non-Governmental Organisations Bill (NGO Bill) in 2004, and the zealous enforcement of the Public Order and Security Act (POSA). The NGO Bill puts restrictions on organisations that work in the fields of governance, democracy and human rights. It explicitly defines "issues of governance" as including "political governance issues" (Parlzim 2004: 5). POSA restricts public gatherings of a "political" nature, a term that has been subjectively stretched by the authorities to include a variety of activities that involve interacting with the public as well as soliciting and expressing public opinion.

Impressions from the City

The City's five-year strategic plan explicitly calls for "improved local governance," and the Town Clerk, a qualified lawyer, is the driving force in the transition from government to governance. In his view "participatory governance is cooperatively addressing the burning issues of the City *with* the community, and business, and other stakeholders" (his emphasis). Explaining the thrust towards governance, the Town Clerk told me that it was "the only way to move with the times, the only way to solve our common problems . . . because our residents are our shareholders." He openly admitted to drawing "a bit of inspiration" from international debates on governance, and mentioned by name the World Bank-supported Municipal Development Partnership (MDP). He, however, hastened to point out that the City's "initiative is a home-grown initiative . . . resourced and operationalised internally [without] any help, persuasion or prodding from any quarter whatsoever." My investigations confirmed this claim.

Scrutinising the Governance Outreach Programme provided valuable insight into how the City's quest for improved local governance was operationalised. In particular, a case where the authorities were dealing with a group of self-employed youths at a suburban shopping centre is revealing. The youths had taken over a third of the centre's partially disused parking lot. Their spokesperson explained that they took over the place because "it was just lying idle . . . making no production [*sic*]." Since the designated market did not have enough space, "the only thing to do [was to] make good use of wasted space and change it into a productive area." However, according to the Senior Planner, the youths were "abusing it [the parking lot] for all sorts of things" that included selling a variety of goods, appliance repairs, car washing and hairdressing. The Town Clerk indicated that numerous complaints had been received from "a very wide spectrum of society." Retailers, the owner of a petrol-filling station, and female vendors at a nearby (legal) market complained about loss or disruption of business; public transport operators and motorists were angry about the obstruction of traffic; the City's cleaning unit, police, school authorities, parishioners, and some residents complained about hustling, loss of amenity and antisocial behaviour. When I started observing them, planners described their task as "working with the 'trouble-causers' and others in the neighbourhood . . . to solve conflict [*sic*] between the trouble-causers and . . . residents at the shopping centre" (interview with Deputy Director of Engineering Services (Planning), 30 April 2005).

The 21 youths, comprising 8 females and 13 males, were aged between 16 and 26, all of them with at least two years of secondary school education. They all originated from the neighbouring low-income areas, with 57% of them being the breadwinners for their respective households, while 45% lived alone or with friends. A third (33%) were born and raised outside the City and had migrated to the urban centre "to try to become less poor" (interview with 21-year-old Bongi). Three years previously, they had formed a workers' committee, introducing mandatory paid membership for everyone operating in the car park. The six-member committee represented the youths' collective interests and provided support in the event of crises such as arrest, sickness or bereavement. It consisted of a chairperson (male), secretary (female), treasurer (female) and three male deputies. When the GOTF approached them, the youths promptly formed a rotating three-member liaison group comprising a spokesperson and two committee members, with the restriction that members of the workers' committee secretariat could not be in the liaison group. When I enquired why they did not simply let the workers' committee constitute the liaison group, Sarudzai, the committee secretary, retorted, "Do you see the whole council—the Mayor, Town Clerk and all—coming to us? They have their task force, so do we. We are too busy to let people be held up in this project . . . so we share the burden." This turned out to be a strategic decision ("a deceptive ruse," according to the Senior Planner) that served the youths well.

The Governance Outreach Task Force, which had been established a year earlier, in line with the City's declared aim of moving towards participatory governance, was tasked with "managing and resolving the brewing conflict between the youth and other residents and stakeholders." The Senior Planner led the GOTF, whose members had been specially chosen by the Deputy Director of Engineering Services (Planning) and the Town Clerk. Other members included the planner responsible for development control and forward planning, two planning technicians, the municipal police chief, the ward councillor's representative, and the local Member of Parliament's agent. Sporadically, a man from the President's Office—an operative of the dreaded Central Intelligence Organisation (CIO)—joined the delegation. However, his presence was not viewed as an intrusion by any of the participants; it hardly affected the interactions and deliberations. Probably, this is because the issues being discussed concerned the local authority and the residents. They rarely touched on the national state, although at times the youths did openly voice their unflattering opinions about central government. In any case, as the Senior Planner correctly observed, the man was harmless because he was "usually of dubious sobriety."[3]

Since the issue had already been defined as a planning problem by the local authority, the Town Clerk delegated the power to "negotiate a workable, lasting settlement . . . and come to an acceptable resolution" to planners in the GOTF. The planners decided to meet the youths separately before organising an "all-stakeholder meeting." When I began observing the Governance Outreach Programme, the GOTF had been interacting with the "trouble-causers" for some three months. Altogether, they had held six public hearings with the youths. According to the Senior Planner, the programme "was actually getting somewhere . . . though moving somewhat slowly." Like his GOTF colleagues, the Senior Planner regarded the "final destination" of the Governance Outreach Programme as a scenario where the youths agreed to "vamoose from the parking lot." The planners were by far the most active in the team, the other team members following their lead and, in most cases, almost instinctively agreeing with them. The exception was the councillor's representative and the MP's agent, who always sided with the youths. The man from the President's Office was mostly quiet, speaking only when spoken to.

Observing the Governance Outreach Programme, it became clear that the bureaucrats espoused a state-centric view of governance. They firmly believed that it was the local authority that was supposed to do the governing, with other groups being brought in and granted legitimate status by the local state. This applied particularly to the youths because they represented what the planner responsible for development control and forward planning defined as "inconsequential individual private interests." The field of governance was to be set out, demarcated and managed by the planning system, which had the full blessing of not only the municipal authority, but also central government, through legislation, policy and ministerial directives. According to the Senior Planner, the difference between the recent emphasis on governance and planning's previous modus operandi was that the planning department was now executing its mandate, not as "an isolated and insulated unit but as part of an enabling local government." There was an official willingness to involve other players in local governance so that they could "present their cases . . . and discuss issues openly, honestly and transparently." However, there was no question that the epicentre of all activity was to be the city council. The Town Clerk aptly summed it up when he said, "As the local authority we own the process . . . We would like . . . the people to have joint ownership of the product."

The planners in the local authority emphasised that there was "no reason to change anything in the way we do things here." The planner responsible for development control and forward planning stressed, "If something works, you do not just dump it in favour of dubious fashions." Naturally, in terms of doing planning, planners saw no reason to change from what the Senior Planner described as "the accepted planning model of the day." He was referring to the rational process model of planning, which the Deputy Director of Engineering Services (Planning) breathlessly outlined as: "Identifying the problem, formulating goals, identifying and appraising alternatives, choosing the best options . . . then implementation, monitoring and evaluation." According to the planner responsible for development control and forward planning, the only "slight addition to the scientific approach is that the Governance Outreach Programme will help in improving our intelligence . . . and ensuring more voices are heard and . . . promoting openness." The planners were keen to emphasise that democracy does not demand less technical rigour, pointing out that the scientific approach was still "a must" in technical bureaucracies like planning. To them, governance was not designed to do away with reason, bureaucracy and science, "otherwise all these developed countries which preach governance and democracy would not be what they are today" (Senior Planner, Interview, 1 May 2005). Hence, local governance was "not a very new way of doing things . . . only a supplement to enhance current practice."

Predictably, in the City, planning was an elitist practice. Convinced about their authority and abilities, planners in the GOTF stressed that they could not be faulted for ensuring that everything was done according to law and regulations. Planning also had to be consistent with the City's Master Plan and with the local plan for the contested area. The Senior Planner said, "These plans spell out the spatial goals of the City; and it is these goals that this exercise [the Governance Outreach Programme] is supposed to pursue." Significantly, the Senior Planner insisted that the planning system could not "operate on the basis of scanty and unreliable" (read unofficial) information. This all but ruled out most of the suggestions by the youths who were proposing "crazy, ultra vires and untried options [*sic*]" (interview with Deputy Director of Engineering Services (Planning)) that were not "adequately backed by the information available." The planner responsible for development control and forward planning argued that because of technical, budgetary and time constraints, the local authority could not "launch another survey to assess the new options" because the planning system already had "enough workable alternatives . . . consistent with the operative local plan . . . to work with." The role of youths in local governance in general and the Governance Outreach Programme in particular was to discuss "the alternatives that the GOTF put on the table." They were not to come up with "new crazy and wild ideas" that went beyond what was proposed in the official agenda, and had been vetted and approved by the Deputy Director of Engineering Services (Planning) with the Town Clerk's authority.

The planners in the GOTF regarded themselves as means–ends specialists. Planners' faith in instrumental rationality served their perception of their job as being "to explore a range of alternatives and evaluate each in light of agreed upon goals to see which will work best" (Innes and Gruber 2005: 180). Every technical person in the GOTF believed in the strict division of labour between the youths, the bureaucrats and the policy makers. "Good governance or not, we are paid to come up with the most efficient way to achieve . . . set goals," the planner responsible for development control and forward planning snapped, in response to a question on why the GOTF was asking the youths to ratify official decisions without asking them what they wanted and how they thought they could get it. Turning around the council's list of governance indicators, which included trust, accountability, reciprocity and transparency, he explained that accountability meant that he ultimately would be "the one to be held accountable if the system fails to bring results." To him, by retaining control of the design of means, the GOTF was fully in keeping with the principle of accountability, that is, "our own accountability to council, to central government . . . and to society." He went on, "Whose heads do you think everyone will come calling for if things go wrong? Ours . . . because we are answerable." It seems as long as they were the ones who were ultimately accountable, planners would not give up control. It was inconceivable for them to "share control with people who cannot share the blame."

Ultimately, the Governance Outreach Programme became a set of structured encounters and stage-managed interactions. The agendas for the hearings were drawn up in the planning office. The GOTF, with the approval of senior bureaucrats, decided the date, time and venue of the public hearings. And all hearings were extremely formal. A secretary seconded from the council's typing pool meticulously took minutes. At times she even asked people to speak up or repeat what they had said so she could record it correctly. Explaining this formalism, the Town Clerk said that the GOTF needed to show that the meetings were "first of all orderly . . . but also open, transparent, free and utterly democratic." The Senior Planner said, "We control the procedures . . . and everything . . . so that things do not go out of hand [*sic*]. This ensures that we take records for our files . . . just in case . . . Isn't that what good governance is all about?" In structuring the encounters and managing the interactions, what the planners were looking for was "some kind of reaction to our proposals . . . as a way of amicably [re]solving the conflict between these trouble-causers and everybody else." According to the planners, they alone decided on the issues to be discussed because they had insight into what the real issues were. The Senior Planner pointed out:

> If these . . . trouble-causers feel they have another agenda, they are free to approach us and arrange a meeting. Then it will be their agenda and their meeting . . . and we will be happy to follow it. We called these hearings for a

reason. We call the shots here because this one is our agenda . . . and everybody must . . . follow it.

Not surprisingly, during the interactions, planners did not hesitate to resort to law and coercion. The planner responsible for development control and forward planning revealed that one reason behind the GOTF's monopolisation of the drafting of the agenda was to make sure that they could hear the voices of the youths on issues that the planners deemed relevant. Using their officially legitimated technical knowledge, the GOTF had rationally come up with options and, as the Senior Planner admitted, the role of the public hearings was "to choose the best one with the help of the trouble-causers." The issue to be discussed was not whether the youth had other ways of resolving the conflicts at the shopping centre, but how and when they would be moved to another site.

However, the youths were neither invisible nor passive. They were aware that their knowledge and ways of knowing were not officially recognised. They knew the GOTF's technical knowledge was the only legitimate knowledge. Tongai, aged 19, captured the sentiments of his peers when he stated, "We are ignorant and without power in this . . . game; but we are not blind, deaf or stupid. We know what is going on. We are able to think like them." Partly as a result of the Senior Planner's constant reference to the importance of joint ownership of the final product, the youths figured out that the GOTF needed them to make their project work. Lengthy discussions with the youths revealed that they were aware that the legitimacy of the outcome was partly contingent on their participation. Noting how desperately the authorities sought this legitimacy, the youths figured out that the hearings would not proceed without them and that in the end the outcome would need their ratification. They also knew that probably they would not be forcibly evicted until an amicable solution was found. They fully exploited this fact.

They tried to manipulate the process in their favour. First, they repeatedly slowed down the hearings by what turned out to be needless consulting, haggling, and stage-managed bickering amongst themselves. Explaining why seemingly trivial issues always sidetracked them, Tongai explained that they were buying time. He explained, "A quick solution is the end of us. The process is our medicine . . . Its conclusion is our end." Susan (aged 19), the then liaison group spokesperson, revealed that slowing down the process was a deliberate ploy, "our own way to make sure these people [the authorities] don't arrive [achieve their goal] too quickly."

Another youth strategy was to disrupt the hearings by endlessly changing the liaison group's membership. They were in the habit of unexpectedly dissolving the body and electing a new one in the middle of a hearing, not on the basis of dissatisfaction with performance, but as 23-year-old James explained, "just to give them [the GOTF] headaches like they do to us." Another reason, according to workers' committee chairperson, Fatso (aged 22), was "so that they [the authorities] do not know whom to bribe, harass or threaten." These disruptions partly explain why, after six public hearings, the GOTF was still in the process of "getting somewhere." Hence, while it can be said that planners controlled the project, the same cannot be true of the process and outcome.

Planners on local governance

Table 2.1 summarises findings from interviews with 23 planners, all of them locally trained at the sole planning school, mostly in the 1980s and 1990s. The table reveals some common attitudinal and viewpoint characteristics among planners vis-à-vis local governance, with particular emphasis on participatory governance. These are that: 1) planners are vaguely aware, or at best not sufficiently interested in participatory governance; 2) planners do not feel that governance should be a concern for technocrats; 3) planners feel that they are not adequately equipped to meet the challenge of local governance; and 4) planners are not fully aware of the implications for their working methods and practices of the growing emphasis on "good" local governance.

In response to an open-ended question ("Governance is . . .") most of the planners professed to know what governance was. The most common words that appeared in the responses were "democracy" and "participation." All but two of the planners dismissed governance as just a theory. One of the city's planners described it as "one of these damn theories from the UN and World Bank."

A minority was convinced that governance was government by another name. Notably, a senior environmental planner viewed the concept through technical lenses, stressing that it was for "greater efficiency and effectiveness in goal

Table 2.1 Perceptions on urban local governance (LG) among planners (N=23)

Issue	Opinion							
	Yes		No		Partial		Not sure	
	No.	%	No.	%	No.	%	No.	%
Aware of what 'governance' is about	2	8.7	4	17.4	12	52.2	5	21.7
Interested in local governance	3	13.0	16	69.6	4	17.4	0	0.0
Aware of implications on practice	5	21.7	0	0.0	16	69.6	2	8.7
Planners have a role in making LG work	6	26.1	14	60.9	0	0.0	3	13.0
Planners are sufficiently equipped for LG	0	0.0	18	78.3	5	21.7	0	0.0
LG demands a change in practice	4	17.4	4	17.4	12	52.2	3	13.0
Planners need retraining	10	43.5	6	26.1	7	30.4	0	0.0

Source: Research data (February, 2005).

achievement." Those who were contacted by e-mail gave the impression of knowing more about governance than their counterparts, possibly because they had consulted some written material.

The apparent lack of awareness of this concept on the part of the planners was partly a result of their not being sufficiently interested in governance to be concerned about exploring and defining it. When asked about the reasons for this lukewarm reception, the most common response was that talk about governance represented a passing phase. Revealingly, in quite a number of cases there was a conviction that, since Zimbabwe was a pariah state, there was no point in being interested in UN and World Bank theories, especially considering that there would be no reward for success. This ambivalence is also a reflection of a belief amongst planners that making local governance work is simply not their responsibility. Many believed that governance was about politics, and therefore the responsibility of politicians. A planner who was keen to emphasise his "quarter century experience in the field" highlighted the division of labour between politicians and planners. He reasoned:

> There were too many useless changes popping up in and around urban planning. These things . . . are not for serious professionals like us . . . There is need for local government to adhere to a strict division of labour. Things that have to do with mobilising and mingling with residents . . . are not the work of serious planners. We leave that to . . . politicians . . . Planners deal with the serious technical stuff.

An established planning consultant argued that "planners are not political commissars," emphasising that "some of these tricky . . . things that involve dealing with the masses are better left with opportunists who want to experiment with everything that promises political gain." To her, politicians are such people.

Most planners complained that they were not sufficiently equipped to meet the challenge of local governance. A planning officer bluntly stated that he was "not trained for that kind of work" and did not believe that it would "pay for me to go into the community and meet every Tom, Jack and Jerry [sic]." A planner working in a housing department maintained that the "kind of community outreach demanded by this governance thing, can only be done by social and community health workers because they have the appropriate training." This was not an isolated sentiment. As Table 2.1 shows, the majority of planners believed that they needed retraining to be effective in local governance. A planner from the City stressed, "Unless we go back to college to take all the appropriate courses we will not make it in this governance." However, others believed they would be able to "meet the challenge with minimal on-the-job training," in the words of a development control officer, while a further group were adamant that they could rise to the occasion by bringing skills they had picked up in the field to bear, often claiming that the planning school had not equipped them for other day-to-day challenges of their job. A senior physical planner boasted, "I did not need retraining to meet the challenges and threats posed by ZANU-PF [the ruling political party]. Why should this governance thing be different?"

Not one of the planners was totally unaware of the implications for their working methods of the growing emphasis on participatory local governance. However, the majority underplayed the significance of the challenge, with many of them dismissing it as requiring no more than what the City's Senior Planner termed "just some simple people skills." There was no indication among the planners that the emphasis on governance might entail a re-examination of their preferred role as technical experts. A private planning consultant who works with local authorities, and who had participated in the preparation of a number of the strategic plans of urban councils, complained: "Tampering with technical expertise implies the abolition of organised local government and the death of professionalism." He reasoned:

> There is no need for opening up planning practice . . . no need for consensus. The very spirit and purpose of planning is to enforce order in the built environment. This confounding and time-wasting outreach negates the reality that somebody has to make a decision which some people will not agree with, which is the beauty and necessity of planning.

In a written response, a newly recruited central government planning officer protested:

> So, should we dump our hard-earned technical know-how and still be expected to deal with rational human beings? It's impossible because people will never agree on anything . . . They cannot agree on what they do not understand . . . This is a recipe for disaster. For this thing to work then every urban resident should go to planning school.

Emerging issues

The study raises some critical issues about planners, planning and participatory local governance in Zimbabwe. The important issues arise from the relational nature of participatory governance, centring on how the youths and planners interacted with each other. Most of these issues have to do with the question of planners as technical experts, and revolve around questions of motivation, conviction, attitude, behaviour, style, agency and power. There are also matters of institutional and structural issues, particularly the role of central government.

The City's Governance Outreach Programme has not made much progress. The difficulties besetting it are of two types, namely, specific issues about planners and the culture of planning, and deeper structural issues about governance and politics in Zimbabwe. The GOTF regarded participatory governance as no more than an add-on to existing governmental processes, bureaucratic routines and technical practices. There was no conscious effort to institutionalise participatory local governance. The GOTF had to execute its mandate within the confines of established administrative structures and technical expertise. By the Town Clerk's own admission, if the GOTF worked outside these parameters, its recommendations could be jeopardised. The whole point in "hanging out" with the youths was to get feedback to official proposals. The Governance Outreach Programme's *raison d'être* was to secure ratification and legitimisation for official planning decisions, not to chart new paths.

On their part, the youths did not take the Governance Outreach Programme seriously. They did not bother "institutionalising" their emerging relationship with it. As was the case with the local authority, to the youths, the Governance Outreach Programme was no more than a bureaucratic body. Instead of working through the workers' committee, they set up a liaison group with no clear mandate. This explains why they could dissolve and reconstitute it at will: it was nothing more than a strategic ploy to buy time and frustrate the authorities.

At the national level, the fact that most planners are vaguely aware of (and not sufficiently interested in) local governance signals a lack of motivation about participation, and a lack of belief in the efficacy of governance. It also suggests that planners are not willing to change existing practices and attitudes, being instead convinced that the thrust towards governance is a transient phenomenon that will soon fade away. This view is reinforced by the absence of disincentives for failure, and incentives for success: planners in the study believed that there was no real benefit in the transition to participatory local governance partly because there is no professional recognition for doing a good job, and no yardstick for determining what counts as "good practice." It is hardly surprising, then, that meaningful interaction with the youths was not a major concern for the career-minded technical experts in the GOTF.

One reason for the lethargy is that, save for decreeing the inclusion of "good" governance in strategic plans, government has not made the concept an implementable priority. For a country where decisions on governance tend to be centralised in the executive (Makumbe 1998), pronouncements issuing from local government do not carry a lot of weight. A community development worker speculated that participatory governance would not work because "[central] government has no interest in improving local governance." Maybe she was right. In Zimbabwe's guided democracy, any public activity that involves interfacing with the public cannot accomplish much without the explicit blessing of the national state. The presence of the man from the President's Office in the Governance Outreach Programme's public hearings supports this assertion, as does the drafting and the passing of the NGO Bill and the zealous enforcement of POSA. Planning is adversely affected by this constriction of the public sphere.

There is also the question of human agency. It is partly incorrect to claim that technical experts in Zimbabwe have absolutely no latitude. They are not totally constrained by existing structures and procedures. For example, the City did grant the GOTF some room for manoeuvre. There was space for innovation in the initiation, organisation, management and reporting of the public hearings. The fact that the Town Clerk emphasised "learning by doing and doing by learning" implies a degree of open-endedness that planners could exploit. The question one needs to ask is why the GOTF failed to exploit these interstitial openings in the intricate web of legal, regulatory and bureaucratic controls.

Part of the answer lies in the fact that while not completely subduing human agency, the local government system subtly imposes limitations on intuition, creativity and innovation. One planner aptly compared this to self-censorship in the media. As the Deputy Director of Engineering Services (Planning) observed, "change can be dangerous if in the end you are still answerable." In a situation where they can still be held accountable, technical experts feel safe in the familiar and are averse to dabbling with the unknown, especially where collective responsibility is not clearly spelt out. So tried and tested instrumental rationality is never tampered with, and the "technical/bureaucratic style" (Innes and Gruber 2005: 180) firmly rooted in the rational process view of planning still dominates planning practice in Zimbabwe.

Another salient point relates to power relations between the youths and planners. Power imbued all relations in the Governance Outreach Programme. But this power, defined here as "the general matrix of force relations, at a given time, in a given society" (Hillier 2002: 49), flowed in different directions. The Governance Outreach Programme was not one unified practice that could be tamed and controlled by any one group. It consisted of a multiplicity of micro-practices, each with different configurations of power. Apart from the constitution of the GOTF, the convening, conduct and reporting of the public hearings and meetings were distinct micro-practices with different matrices of force relations, affording varying advantages to technical experts and "trouble-causers." Mishandled, this could work against meaningful communication and interaction. Whereas the authorities dominated the constitution of the GOTF, the setting of its agenda, and the reporting of its meetings and hearings, they only partially controlled the convening and conduct of the hearings. In these processes, the youths had some advantages that afforded them certain types of power. However, what made the situation problematic was that these configurations of power were neither acknowledged nor accounted for in the encounters and interaction.

Perhaps inevitably, the power relations worked disruptively. The planners consciously sought to dominate the whole process: by their words, attitude and behaviour, they were quite clear that they were in a "more equal" position than the youths, and that the state gave them legitimacy through legislative and regulatory instruments. The Governance Outreach Programme did not alter any of this, and consequently the GOTF had no compelling reason to change established planning practices. During their interactions with the youths, planners did not hesitate to resort to law and coercion, thereby disregarding the defining feature of consensus building, namely, that it is "a process that is truly facilitated, as opposed to merely chaired" (Innes and Booher 1999: 11).

What further disrupted the interactional space was the youths' decision not to be passive recipients of the planners' dictates. Their strategic behaviour confirms Foucault's assertion on the coexistence of power and resistance (Foucault 1998: 95). The legitimacy of the process partly depended on the process being "relatively public, inclusive and procedurally regular" (Young 2000: 3), but the authorities' efforts to meet these basic requirements, albeit in a flawed way, contributed to the success of the youths' tactical resistance. The planners' eagerness to reach out, to ensure joint ownership of the product, while simultaneously brandishing their technical superiority, proved to be the Governance Outreach Programme's Achilles heel. Resentful of the GOTF's domineering attitude, the youths easily sensed their desperate need for legitimacy and used it to their advantage. By repeatedly stalling the process, they signalled to planners that the authorities would not get their decision without effort, would not attain legitimacy without sacrifice, and would not secure ratification without compromise. The planners failed to read the signs, and in the end, the Governance Outreach Programme failed to make significant forward strides as a result of this.

An ingrained conviction that the rational planning process is effective, combined with an unbending adherence to instrumental rationality, and coupled with a lack of belief in other planning frameworks, are the main obstacles to participatory governance in the making and management of urban places in Zimbabwe. The continued privileging of technical knowledge over all other knowledge reinforces the technical experts' unyielding stubbornness. The result is a set of asymmetric power relationships that create a very complex power dynamic that does not meet the conditions which Innes and Booher (2004) suggest are appropriate for participatory governance to flourish. Armed with their privileged technical knowledge, planners possess distinct advantages that they have unhesitatingly deployed to get their way. However, as noted above, power is not the sole preserve of planners. For their part, the youths used their situation as participants to manipulate parts of the process in their favour, thereby frustrating planners. The distortions arising out of asymmetric power relations are thus compounded by the systematic deployment of instrumental rationality on the part of planners, with the youths resorting to strategic rationality to disrupt the process (Alexander 2001). This deployment of advantages has distorted the whole field of play in local governance, and the consequence, a flawed interactional space, hampers participatory local governance in urban Zimbabwe. This contributes to a stalemate in Zimbabwean planning.

Sovereign power, based on the authority of central government (see Hindess 1996: 12), legitimates, upholds and sustains the instrumental rationalism that obstructs participatory local governance. The existing administrative, regulatory and legal instruments confer validity and legitimacy exclusively on planners' technical knowledge. In the Governance Outreach Programme, it was the planners' officially recognised and legitimated technical knowledge that counted as real knowledge and served as the fulcrum for all deliberations. Zimbabwe's urban planning system confirms Foucault's assertion that "power produces knowledge" (Foucault 1977: 27): the styles, attitude and beliefs of planners demonstrate that "power and knowledge directly imply one another," suggesting that even in planning "there is no . . . knowledge that does not presuppose and constitute at the same time power relations" (Foucault 1977: 27).

The preceding case-study confirms findings from literature on participatory and collaborative governance, in particular the tenet that, in the short-term, frameworks that seek to harness the best from all stakeholders may not work in highly distorted contexts. In order to make governance work, perhaps it is prudent to heed Foucault's contention that "we should abandon a whole tradition that allows us to imagine that knowledge can exist only where power relations are suspended and that knowledge can develop only outside its injunctions, its demands and its interest" (Foucault 1977: 27). Hence, while collaborative planning offers a way out of the trap of instrumental rationality, the same can hardly be said of its treatment of power, especially "power outside the dialogue" (Innes and Booher 2004: 12), to which the Governance Outreach Programme participants could easily resort to get their way. To work effectively, communicative action requires the "levelling of power relations in contrast to the typically unequal power distributions of most governance contexts today" (Healey 2003: 113). This study suggests that, in some contexts, this may be possible only as a long-term programme. It hardly works in the logistics of day-to-day practice where officially sanctioned technical knowledge, which is brought to what passes for a negotiating table, has traditionally and inevitably held sway. In the short-term, it is inconceivable to imagine other knowledge (and ways of knowing) achieving the same status and validity as planning's technical knowledge in post-2000 urban Zimbabwe.

Conclusion

Much more than awareness and ambition is required for a successful transition to participatory local governance in contexts like urban Zimbabwe, where the culture of planning is inimical to such a process. Obstacles exist in both the technical and socio-political contexts where participatory local governance is supposed to take place. There are structural constraints presented by the local and national state, and a lack of commitment (or even indications of hostility) from central government. However, the attitude, behaviour and perspectives of planners, in particular their lukewarm reception and desultory attempt at operationalising the framework, also constitutes a significant part of the problem.

In this context, the contribution of planners to participatory governance is problematic because it requires planners to act in ways that conflict with their preferred role as technical experts. The study suggests that it is pointless for technical experts to interact with the public when the attitude, means, behaviour and style of the participants all betray their lack of belief in a participatory framework. For planning to make meaningful contribution in governing the cities, it is necessary for planners not only to be aware of and embrace participatory local governance, but also to recognise and accept the possibility of changing the way they do business. In the quest for participatory local governance, planning hardly accomplishes anything if it sticks to a "rigidly structured sequence" (Bryson and Crosby 1996: 463), following procedures required by the rational process model of planning.

For governance not to degenerate into a dysfunctional charade, a deep cultural change on the part of technical experts is necessary. This requirement is not peculiar to Zimbabwe: it also applies to contexts which are much more conducive to participatory governance. In Zimbabwe, planners operate in a highly charged political environment. Understandably, for the risk-averse, a technical culture provides a safe haven from the perceived risks that might arise from engaging head-on with the social and political issues that imbue their work. Be that as it may, when participatory governance is embraced, a cultural change is a precondition for meaningful and fruitful relationships with the public in the production and management of space.

The acceptance of change entails more than revisiting planners' religious devotion to instrumental rationalism. In the context of Zimbabwe, transformation cannot begin and end in the realm of planning. It must interrogate the power that produces, legitimates and sustains planning's knowledge. In this regard, it is helpful to re-examine that which gives planners the conviction that their role is limited only to technical expertise; to question that which makes them reluctant to change current practices; and to scrutinise that which makes them view new frameworks with trepidation, and/or scepticism, and/or hostility.

Undoubtedly, in countries like Zimbabwe, the transformation to participatory governance is, at best, a long-term programme. In the meantime, it may be helpful to acknowledge that "cities are rarely the sites of disinterested practices . . . that the city is as much a means of shutting down possibility . . . as it is a means of opening it up" (Amin and Thrift 2002: 105). For participatory local governance to work, it may be prudent to accept that, while it is good to hope for the best in people, sometimes the unpalatable realities on the ground betray such faith. The interactional space of governance is not that innocent, and perhaps we need to resign ourselves to the fact that sometimes it can never be made innocent; that change has to come at a cost; that it is not bad faith to expect others to try to get their way; but that it is practical, even prudent, to make them work for it.

Notes

1 In keeping with the respondents' explicit request not to be identified, I have removed all material that might reveal the identity of the people discussed and the city they work in. I will use the term "the City" to refer to the urban centre in question.
2 UNDP stands for United Nations Development Programme.
3 Ironically, when in May 2005 government launched Operation Murambatsvina/ Restore Order, a campaign of evictions and demolition, it was the man from the President's Office who was instrumental in saving the youths from eviction.

References

ACPD (2002) *Local Governance and Participation*, Harare: ACPD.
Alexander, E.R. (2001) "The planner-prince: interdependence, rationalities and post-communicative practice," *Planning Theory and Practice*, 2:3, 311–324.
Ambert, C. and Feldman, M. (2002) "Are IDPs barking up the wrong tree? A story of planning, budgeting, alignment and decentralisation." Paper presented at the Planning Africa 2002 Conference, Durban, 18–20 September. Available at www.saplanners.org.za/SAPC/papers/final%20paper%20planafrica.pdf (Accessed 21 November 2005).
Amin, A. and Thrift, N. (2002) *Cities: Reimagining the Urban*, Cambridge, MA: Polity.
Bourennane, M. (2007) "A legal framework for enabling low-income housing: a study of women's access to home-based enterprises in Botswana." PhD Thesis, Stockholm: Royal Institute of Technology.
Braaten, J. (1991) *Habermas's Critical Theory of Society*, Albany, NY: State University of New York Press.
Bryson, J.M. and Crosby, B.C. (1996) "Planning and the design and use of forums, arenas, and courts," in: S.J. Mandelbaum, L. Mazza and R.W. Burchell (Eds) *Explorations in Planning Theory*, pp. 462–482, New Brunswick, NJ: Center for Urban Policy Research.

Clark, I. (2000) *Governance, the State, Regulation and Industrial Relations*, London: Routledge.

Cohen, J. (1997) "Deliberation and democratic legitimacy," in: J. Bohman and W. Rehg (Eds) *Essays on Reason and Politics*, pp. 67–91, Cambridge, MA: MIT Press.

Cunningham, F. (2002) *Theories of Democracy: A Critical Introduction*, London: Routledge.

Dear, M.J. (2000) *The Postmodern Urban Condition*, Malden, MA: Blackwell.

Devas, N. (2001) "Does city governance matter for the urban poor?" *International Planning Studies*, 6:4, 393–408.

DRUP (Department of Rural and Urban Planning) (1997) Urban Governance Outreach Research Programme: Proceedings of the Stakeholder Workshop, Harare: DRUP.

Dryzek, J.S. (1990) *Discursive Democracy: Politics, Policy and Political Science*, Cambridge: Cambridge University Press.

Flyvbjerg, B. (1998) *Rationality and Power: Democracy in Practice*, Chicago: University of Chicago Press.

Flyvbjerg, B. and Richardson, T. (2002) "Planning and Foucault: in search of the dark side of planning theory," in: P. Allmendinger and M. Tewdwr-Jones (Eds) *Planning Futures: New Directions for Planning Theory*, pp. 44–62, London: Routledge.

Foucault, M. (1977) *Discipline and Punish: The Birth of the Prison* (Trans. A. Sheridan), New York, NY: Vintage.

Foucault, M. (1998) *The Will to Knowledge: The History of Sexuality 1* (Trans. R. Hurley), London: Penguin.

Goodwin, M. and Painter, J. (1996) "Local governance, the crises of Fordism and the changing geographies of regulation," *Transactions of the Institute of British Geographers*, 21:4, 635–648.

Harris, N. (2002) "Collaborative planning: from theoretical foundations to practical forms," in: P. Allmendinger and M. Tewdwr-Jones (Eds) *Planning Futures: New Directions for Planning Theory*, pp. 22–43, London: Routledge.

Healey, P. (1997) *Collaborative Planning: Shaping Places in Fragmented Societies*, London: Macmillan.

Healey, P. (2003) "Collaborative planning in perspective," *Planning Theory*, 2:2, 101–123.

Hillier, J. (2002) *Shadows of Power: An Allegory of Prudence in Land-Use Planning*, London: Routledge.

Hindess, B. (1996) *Discourses of Power: From Hobson to Foucault*, Oxford: Blackwell.

Hoch, C. (2007) "Pragmatic communicative action theory," *Journal of Planning Education and Research*, 26:2, 272–283.

IMF (International Monetary Fund) (2005) Staff Report for the 2005 Article IV Consultation, Washington, DC, IMF.

Innes, J.E. (1998) "Information in communicative planning," *Journal of the American Planning Association*, 64:1, 52–63.

Innes, J.E. and Booher, D.E. (1999) "Consensus building as role-playing and bricolage: towards a theory of collaborative planning," *Journal of the American Planning Association*, 65:1, 9–26.

Innes, J.E. and Booher, D. (2004) "Reframing public participation: strategies for the 21st century," *Planning Theory & Practice*, 5:4, 419–436.

Innes, J.E. and Gruber, J. (2005) "Planning styles in conflict," *Journal of the American Planning Association*, 71:2, 177–188.

Kamete, A.Y. (2002) *Shifting Perceptions and Changing Responses: Governing the Poor in Harare, Zimbabwe*, Uppsala: Nordiska Afrikainstitutet.

Kamete, A.Y. (2006) "The return of the jettisoned: ZANU-PF's crack at 're-urbanising' in Harare," *Journal of Southern African Studies*, 32:2, 255–271.

KCC (Kadoma City Council) (2003) Strategic Plan, 2001–2005 (Revised February 2003), Kadoma, Zimbabwe: KCC.

Levy, B. (2004) "Governance and economic development in Africa: meeting the challenge of capacity building," in: B. Levy and S. Kpundeh (Eds) *Building State Capacity in Africa*, pp. 1–42, Herndon, VA: World Bank.

Mabin, A. (2002) "Local government in the emerging national planning context," in: S. Parnell, E. Pieterse, M. Swilling and D. Wooldridge (Eds) *Democratising Local Government: The South African Experiment*, pp. 40–54, Cape Town: University of Cape Town Press.

Makumbe, J.M. (1998) *Development and Democracy in Zimbabwe: Constraints of Decentralisation*, Harare: SAPES.

Matlosa, K. (2005) The state, democracy and development in southern Africa. Paper prepared for the CODESRIA 11th General Assembly, Maputo, 6–10 December.

McCarney, P.L. and Stren, R.E. (2003) *Governance on the Ground: Innovations and Discontinuities in Cities of the Developing World*, Washington, DC: Woodrow Wilson Center Press.

Mwimba, C. (2002) The colonial legacy of town planning in Zambia. Paper presented at the Planning Africa 2002 Conference, Durban, 18–20 September. Available at www.saplanners.org.za/SAPC/papers/Mwimba-39.pdf (Accessed 31 March 2007).

Oranje, M. (2003) "A time and a space for African identities in planning in South Africa?" in: P. Harrison, M. Huchzemeyer and M. Mayekiso (Eds) *Confronting Fragmentation: Housing and Urban Development in a Democratising Society*, pp. 175–189, Cape Town: University of Cape Town Press.

Parlzim (Parliament of Zimbabwe) (2004) Non-Governmental Organisations Bill, H.B. 13, 2004, Harare, Parlzim.

Peters, B.G. and Pierre, J. (1998) "Governance without government? Rethinking public administration," *Journal of Public Administration Research and Theory*, 8:2, 223–243.

Pierre, J. and Peters, B.G. (2000) *Governance, Politics and the State*, New York: St. Martins.

Pinkney, R. (2003) *Democracy in the Third World*, Boulder, CO: Lynne Rienner.

Sandercock, L. (1998a) "The death of modernist planning: radical praxis for a post-modern age," in: M. Douglass and J. Friedmann (Eds) *Cities for Citizens: Planning and the Rise of Civil Society in a Global Age*, pp. 163–184, Chichester: John Wiley.

Sandercock, L. (1998b) *Towards Cosmopolis: Planning for Multicultural Cities*, Chichester: John Wiley.

Schmitter, P. (2002) "Participation in governance arrangements: is there any reason to expect it will achieve 'sustainable and innovative policies in a multi-level context'?" in: J.R. Grote and B. Gbikpi (Eds) *Participatory Governance: Political and Societal Implications*, 51–69, Opladen: Leske & Budrich.

Scott, J.C. (2003) "Authoritarian high modernism," in: S. Campbell and S.S. Fainstein (Eds) *Readings in Planning Theory* (second edition), 125–141, Oxford: Blackwell.

Stöhr, W.B. (2001) "Introduction," in: W.B. Stöhr, J.S. Edralin and D. Mani (Eds) *New Regional Development Paradigms Volume 3: Decentralization, Governance and the New Planning for Local-Level Development*, pp. 1–19, Westport, CT: Greenwood.

Taylor, N. (1998) *Urban Planning Theory since 1945*, London: Sage.

UNDP (1997) *Governance for Sustainable Human Development*, New York: UNDP.

Watson, V. (2002) *Change and Continuity in Spatial Planning: Metropolitan Planning in Cape Town under Political Transition*, London: Routledge.

World Bank (1994) *Governance—The World Bank's Experience*, Washington, DC: World Bank.

World Bank (2007) *What is Governance?* Available at http://go.worldbank.org/G2CHLXX0Q0 (Accessed 18 September 2007).

Yiftachel, O. (1998) "Planning and social control: exploring the dark side," *Journal of Planning Literature*, 12:4, 395–406.

Young, I.M. (2000) *Inclusion and Democracy*, Oxford: Oxford University Press.

Chapter 3
Towards a cosmopolitan urbanism
From theory to practice

Leonie Sandercock

Most cities today are demographically multicultural, and more are likely to become so in the foreseeable future. The central question of this chapter is how to come to terms – theoretically, philosophically, and practically – with this empirical urban reality. What can the practice of the Collingwood Neighbourhood House contribute to our theoretical understanding of the possibilities of peaceful co-existence in the mongrel cities of the 21st century? My argument proceeds in four stages. First, I discuss the challenge to our urban sociological imaginations in thinking about how we might live together in all of our differences. Second, I propose the importance of a deeper political and psychological understanding of difference, and its significance in urban politics. Third, I suggest a way of theorizing an intercultural political project for 21st century cities, addressing the shortcomings of 20th century multicultural philosophy. And finally, I link all of these with the actual achievement of the Collingwood Neighbourhood House in the integration of immigrants in Vancouver.

Introduction

Arriving and departing travelers at Vancouver International Airport are greeted by a huge bronze sculpture of a boatload of strange, mythical creatures. This 7 metre long, almost 4 metre wide and 4 metre high masterpiece, *The Spirit of Haida Gwaii*, is by the late Bill Reid, a member of the Haida Gwaii First Nations band from the Pacific Northwest. The canoe has thirteen passengers, spirits or myth creatures from Haida mythology.[1] The bear mother, who is part human, and the bear father sit facing each other at the bow with their two cubs between them. The beaver is paddling menacingly amidships, and behind him is the mysterious intercultural dogfish woman. Shy mouse woman is tucked in the stern. A ferociously playful

---

Submitted by the Association of Canadian University Planning Programs.

Originally published in L. Sandercock and G. Attili. *Where Strangers Become Neighbours: The Integration of Immigrants in Vancouver*, 193–232. © 2009 By kind permission of Springer Publishing.

wolf sinks his fangs into the eagle's wing, and the eagle is attacking the bear's paw. A frog (who symbolizes the ability to cross boundaries between worlds) is partially in, partially out of the canoe. An ancient reluctant conscript paddles stoically. In the centre, holding a speaker's staff in his right hand, stands the chief, whose identity (according to the sculptor) is deliberately uncertain. The legendary raven (master of tricks, transformations, and multiple identities), steers the motley crew. *The Spirit of Haida Gwaii* is a symbol of the 'strange multiplicity' of cultural diversity that existed millennia ago and wants to be again (Tully 1995: 18).

Amongst other things, this extraordinary work of art speaks of a spirit of mutual recognition and accommodation; a sense of being at home in the multiplicity yet at the same time playfully estranged by it; and the notion of an unending dialogue that is not always harmonious. For the political philosopher James Tully, the wonderfulness of the piece lies in "the ability to see one's own ways as strange and unfamiliar, to stray from and take up a critical attitude toward them and so open cultures to question, reinterpretation, negotiation, transformation, and non-identity" (Tully 1995: 206).

The near extermination of the Haida by European imperial expansion is typical of how Aboriginal peoples have fared wherever Europeans settled (Sandercock and Attili 2009: chapter 1). The positioning of the sculpture at Vancouver International Airport, and an identical piece at the Canadian Embassy in Washington, D.C., gives a poignant presence on both coasts of North America to indigenous people who are still struggling today for recognition and restitution. *The Spirit of Haida Gwaii* stands as a symbol of their survival, resistance, and resurgence, and also perhaps as a more ecumenical symbol for the mutual recognition and affirmation of all cultures that respect other cultures and the earth.

But this sculpture can also be read as a powerful metaphor of contemporary humanity and of the contemporary urban condition, in which people hitherto unused to living side by side are thrust together in what I have called the 'mongrel cities' of the 21st century (Sandercock 2003). Most societies today are demo-graphically multicultural, and more are likely to become so in the foreseeable future. The central question of this chapter of the book, then, is how to come to terms with this historical predicament: how can we manage our co-existence in the shared spaces of the multicultural cities of the 21st century? What kind of theoretical challenge is this? In the four-stage argument that follows, I suggest that there is first the challenge to our urban sociological imagination of how we might live together in all of our differences.

In order to act within mongrel cities, we must have a theoretical understanding of 'difference' and how it becomes significant in urban politics, in spatial conflicts, in claims over rights to the city. Thus, in the second section, I seek to deepen our psychological and political understanding of the concept of difference and, through this, to explain why a politics of difference is related to basic questions of identity and belonging and therefore cannot be wished away. In the third section, I argue that we need to theorize an intercultural political project for 21st century cities, one that acknowledges and addresses the shortcomings of 20th century multiculturalism and establishes political community rather than ethno-cultural identity as the basis for a sense of belonging in multicultural societies.

And finally, I link all of these with the actual achievement of the Collingwood Neighbourhood House in the integration of immigrants in Vancouver.

How might we live together? Three imaginings

Richard Sennett: togetherness in difference

In Flesh and Stone (1994: 358) Sennett laments that the apparent diversity of Greenwich Village in New York is actually only the diversity of the gaze, rather than a scene of discourse and interaction. He worries that the multiple cultures that inhabit the city are not fused into common purposes, and wonders whether 'difference inevitably provokes mutual withdrawal'. He assumes (and fears) that if the latter is true, then 'a multicultural city cannot have a common civic culture' (Sennett 1994: 358). For Sennett, Greenwich Village poses a particular question of how a diverse civic culture might become something people feel in their bones. He deplores the ethnic separatism of old multi-ethnic New York and longs for evidence of citizens' understanding that they share a common destiny. This becomes a hauntingly reiterated question: nothing less than a moral challenge, the challenge of living together not simply in tolerant indifference to each other, but in active engagement.

For Sennett then, there is a normative imperative in the multicultural city to engage in meaningful intercultural interaction. Why does Sennett assume that sharing a common destiny in the city necessitates more than a willingness to live with difference in the manner of respectful distance? Why should it demand active engagement? He doesn't address these questions, nor does he ask what it would take, sociologically and institutionally, to make such intercultural dialogue and exchange possible, or more likely to happen. But other authors, more recently, have begun to ask, and give tentative answers to, these very questions (Parekh 2000; Amin 2002). In terms of political philosophy, one might answer that in multicultural societies, composed of many different cultures each of which has different values and practices, and not all of which are entirely comprehensible or acceptable to each other, conflicts are inevitable. In the absence of a practice of intercultural dialogue, conflicts are insoluble except by the imposition of one culture's views on another. A society of cultural enclaves and de facto separatism is one in which different cultures do not know how to talk to each other, are not interested in each other's well-being, and assume that they have nothing to learn and nothing to gain from interaction.

This becomes a problem for urban governance and for city planning in cities where contact between different cultures is increasingly part of everyday urban life, in spite of the efforts of some groups to avoid 'cultural contamination' or ethnic mixture by fleeing to gated communities or so-called ethnic enclaves.

A pragmatic argument, then, is that intercultural contact and interaction is a necessary condition for being able to address the inevitable conflicts that will arise in multicultural societies.

Another way of looking at the question of why intercultural encounters might be a good thing would start with the acknowledgement that different cultures represent different systems of meaning and versions of the good life. But each culture realizes only a limited range of human capacities and emotions and grasps only a part of the totality of human existence: it therefore 'needs others to understand itself better, expand its intellectual and moral horizon, stretch its imagination and guard it against the obvious temptation to absolutize itself' (Parekh 2000: 336–7). I'd like to think that this latter argument is what Sennett might have had in mind.

James Donald: an ethical indifference

In *Imagining the Modern City* (1999), James Donald gives more detailed thought to the question of how we might live together. He is critical of the two most popular contemporary urban imaginings: the traditionalism of the New Urbanism (with its ideal of community firmly rooted in the past), and the cosmopolitanism of Richard Rogers, advisor to former Prime Minister Tony Blair and author of a policy document advocating an urban renaissance, a revitalized and re-enchanted city (Urban Task Force 1999).

What's missing from Rogers' vision, according to Donald, is 'any real sense of the city not only as a space of community or pleasurable encounters or self-creation, but also as the site of aggression, violence, and paranoia' (Donald 1999: 135). Is it possible, he asks, to imagine change that acknowledges difference without falling into phobic utopianism, communitarian nostalgia, or the disavowal of urban paranoia?

Donald sets up a normative ideal of city life that acknowledges not only the necessary desire for the security of home, but also the inevitability of migration, change and conflict, and thus an 'ethical need for an openness to unassimilated otherness' (Donald 1999: 145). He argues that it is not possible to domesticate all traces of alterity and difference. 'The problem with community is that usually its advocates are referring to some phantom of the past, projected onto some future utopia at the cost of disavowing the unhomely reality of living in the present' (Donald 1999: 145). If we start from the reality of living in the present with strangers, then we might ask, what kind of commonality might exist or be brought into being? Donald's answer is 'broad social participation in the never completed

42 Leonie Sandercock

process of making meanings and creating values . . . an always emerging, negotiated common culture' (Donald 1999: 151). This process requires time and forbearance, not instant fixes. This is community redefined neither as identity nor as place but as a productive process of social interaction, apparently resolving the long-standing problem of the dark side of community, the drawing of boundaries between those who belong and those who don't.

Donald argues that we don't need to share cultural traditions with our neighbours in order to live alongside them, but we do need to be able to talk to them, while also accepting that they are and may remain strangers (as will we).

This is the pragmatic urbanity that can make the violence of living together manageable. Then, urban politics would mean strangers working out how to live together. This is an appropriately political answer to Sennett's question of how multicultural societies might arrive at some workable notion of a common destiny. But when it comes to a thicker description of this 'openness to unassimilable difference', the mundane, pragmatic skills of living in the city and sharing urban turf, neither Donald nor Sennett has much to say. Donald suggests 'reading the signs in the street; adapting to different ways of life right on your doorstep; learning tolerance and responsibility – or at least, as Simmel taught us, indifference – towards others and otherness; showing respect, or self-preservation, in not intruding on other people's space; picking up new rules when you migrate to a foreign city' (Donald 1999: 167). Donald seems to be contradicting himself here in retreating to a position of co-presence and indifference, having earlier advocated something more like an agonistic politics of broad social participation in the never completed process of making meanings and an always emerging (never congealed), negotiated common culture. Surely this participation and negotiation in the interests of peaceful co-existence requires something like daily habits of perhaps quite banal intercultural interaction in order to establish a basis for dialogue, which is difficult, if not impossible, without some pre-existing trust. I will turn to Ash Amin for a discussion of how and where this daily interaction and negotiation of ethnic (and other) differences might be encouraged.

Ash Amin: a politics of local liveability

Ash Amin's report, *Ethnicity and the Multicultural City. Living with Diversity* (2002), was commissioned by the British government's Department of Transport, Local Government and the Regions in the wake of the (so-called) 'race riots' in three northern British cities in the summer of 2001. This report is a self-described 'think piece' that uses the 2001 riots as a springboard 'to discuss what it takes to combat racism, live with difference and encourage mixture in a multicultural and multiethnic society' (Amin 2002: 2). Amin's paper is in part a critique of a document produced by the British Home Office (Building Cohesive Communities, Home Office 2001). It goes deeper and draws on different sources than the Home Office document. The political economy approach of the Home Office analysis of the riots never once mentions globalization, or the colonial past. That is Amin's starting point. The dominant ethnic groups present in Bradford, Burnley and Oldham are Pakistani and Bangladeshi, of both recent and longer-term migrations. What this reflects is the twin and interdependent forces of post-colonialism and globalization.

As several scholars have pointed out (Sassen 1996; Rocco 2000), the contemporary phenomena of immigration and ethnicity are constitutive of globalization and are reconfiguring the spaces of and social relations in cities in new ways. Cultures from all over the world are being de- and re-territorialized in global cities, whose neighbourhoods accordingly become 'globalized localities' (Albrow 1997: 51). The spaces created by the complex and multidimensional processes of globalization have become strategic sites for the formation of transnational identities and communities, as well as for new hybrid identities and complicated experiences and redefinitions of notions of 'home'. As Sassen has argued:

> What we still narrate in the language of immigration and ethnicity . . . is actually a series of processes having to do with the globalization of economic activity, of cultural activity, of identity formation. Too often immigration and ethnicity are constituted as otherness. Understanding them as a set of processes whereby global elements are localized, international labor markets are constituted, and cultures from all over the world are de- and re-territorialized, puts them right there at the center along with the internationalization of capital, as a fundamental aspect of globalization.
>
> (Sassen 1996: 218)

This is the context for Amin's interpretative essay on the civil disturbances, which he sees as having both material and symbolic dimensions. He draws on ethnographic research to deepen understanding of both dimensions, as well as to assist in his argument for a focus on the everyday urban, 'the daily negotiation of ethnic difference'. Ethnographic research in the UK on areas of significant racial antagonism has identified two types of neighbourhood. The first are old white working class areas in which successive waves of non-white immigration have been accompanied by continuing socio-economic deprivation and cultural and/or physical isolation 'between white residents lamenting the loss of a golden ethnically undisturbed past, and non-whites claiming a right of place'. The second are 'white flight' suburbs and estates that have become the refuge of an upwardly mobile working class and a fearful middle class disturbed by what they see as the replacement of a 'homely white nation' by foreign cultural contamination. Here, white supremacist values are activated to terrorize the few immigrants who try to settle there. The riots of 2001 displayed the processes at work in the first type of neighbourhood, but also the white fear and antagonism typical of the second type (Amin 2002: 2).

What is important to understand is that the cultural dynamics in these two types of neighbourhood are very different from those in other ethnically mixed cities and neighbourhoods where greater social and physical mobility, a local history of compromises, and a supportive local institutional infrastructure have come to support co-habitation. For example, in the Tooting neighbourhood of South London, Martin Albrow's research inquired about the strength of 'locality' and 'community' among a wide range of local inhabitants, from those born there to recent arrivals, and among all the most prominent ethnic groups. His analysis reveals that locality has much less salience for individuals and for social relations than older research paradigms invested in 'community' allow. His study reveals a very liquid sense of identity and belonging. His interviewees' stories suggest the possibility that:

> Individuals with very different lifestyles and social networks can live in close proximity without untoward interference with each other. There is an old community for some, for others there is a new site for community which draws its culture from India. For some, Tooting is a setting for peer group leisure activity, for others it provides a place to sleep and access to London. It can be a spectacle for some, for others the anticipation of a better, more multicultural community.
>
> (Albrow 1997: 51)

In this middle income locality there is nothing like the traditional concept of community based on a shared local culture. Albrow describes a situation of 'minimum levels of tolerable co-existence' and civil inattention and avoidance strategies that prevent friction between people living different lifestyles. The locality is criss-crossed by networks of social relations whose scope and extent range from neighbouring houses and a few weeks' acquaintance to religious and kin relations spanning generations and continents.

This study gives us an important insight into the changing social relations within globalized localities. Where is community here? It may be nowhere, says Albrow, and this new situation therefore needs a new vocabulary.

How meaningful is the newly promoted (by the Home Office) notion of community cohesion, when people's affective ties are not necessarily related to the local place where they live? Where is the deconstruction, and reconstruction, of what 'community' might mean in the globalized localities of mongrel cities? 'Globalization makes co-present enclaves of diverse origins one possible social configuration characterizing a new Europe' (Albrow 1997: 54).

While Albrow's research seems to support the urban imaginings of James Donald, discussed earlier, in terms of the feasibility of an attitude of tolerant indifference and co-presence, the difference between Tooting and the northern mill towns that are the subject of Amin's reflection is significant. In those one-industry towns, when the mills declined, white and non-white workers alike were unemployed. The largest employers soon became the public services, but discrimination kept most of these jobs for whites. Non-whites pooled resources and opened shops, takeaways, minicab businesses. There was intense competition for low-paid and precarious work. Economic uncertainty and related social deprivation has been a constant for over twenty years and 'a pathology of social rejection . . . reinforces family and communalist bonds' (Amin 2002: 4). Ethnic resentment has bred on this socio-economic deprivation and sense of desperation. It is in such areas that social cohesion and cultural interchange have failed.

What conclusions does Amin draw from this? How can fear and intolerance be challenged, how might residents begin to negotiate and come to terms with difference in the city? Amin's answer is interesting. The contact spaces of housing estates and public places fall short of nurturing inter-ethnic understanding, he argues, 'because they are not spaces of interdependence and habitual engagement' (Amin 2002: 12). He goes on to suggest that the sites for coming to terms with ethnic (and surely other) differences are the 'micro-publics' where dialogue and prosaic negotiations are compulsory, in sites such as the workplace, schools, colleges, youth centers, sports clubs, community centers, neighbourhood houses, and the micro-publics of 'banal transgression' (such as colleges of further education) in which people from different cultural backgrounds are thrown together in new settings which disrupt familiar patterns and create the possibility of initiating new attachments. Other sites of banal transgression include community gardens, child-care facilities, community centres, neighbourhood watch schemes, youth projects, and regeneration of derelict spaces. I have provided just such an example (Sandercock 2003: chapter 7), the Community Fire Station in the Handsworth neighbourhood of Birmingham, where white Britons are working alongside Asian and Afro-Caribbean Britons in a variety of projects for neighbourhood regeneration and improvement. The Collingwood Neighbourhood House in Vancouver is an even better example of a successful site of intercultural interaction, as I will argue in the final section of this chapter. Part of what happens in such everyday contacts is the overcoming of feelings of strangeness in the simple process of sharing everyday tasks and comparing ways of doing things. But such initiatives will not automatically become sites of social inclusion. They also need organizational and discursive strategies that are designed to build voice, to foster a sense of common benefit, to develop confidence among disempowered groups, and to arbitrate when disputes arise. The essential point is that 'changes in attitude and behavior spring from lived experiences' (Amin 2002: 15).

The practical implication of Amin's work, then, is that the project of living with diversity needs to be worked at 'in the city's micro-publics of banal multicultures'

(Amin 2002: 13). It is clear from Albrow's work, as well as that of Amin, that in today's globalized localities one cannot assume a shared sense of place and that this is not the best 'glue' for understanding and co-existence within multicultural neighbourhoods. Ethnographic research on urban youth cultures referred to by Amin confirms the existence of a strong sense of place among white and non-white ethnic groups, but it is a sense of place based on turf claims and defended in exclusionary ways. The distinctive feature of mixed neighbourhoods is that they are 'communities without community, each marked by multiple and hybrid affiliations of varying geographical reach' (Amin 2002: 16).

There are clear limits then to how far 'community cohesion' can become the basis of living with difference. Amin suggests a different vocabulary of local accommodation to difference – 'a vocabulary of rights of presence, bridging difference, getting along' (Amin 2002: 17). To adopt the language of Henri Lefebvre, this could be expressed as the right to difference, and the right to the city. The achievement of these rights depends on a politics of active local citizenship, an agonistic politics (as sketched by Donald) of broad social participation in the never completed process of making meanings, and an always emerging, negotiated common culture. But it also depends on an intercultural political culture, that is, one with effective antiracism policies, with strong legal, institutional and informal sanctions against racial and cultural hatred, a public culture that no longer treats immigrants as 'guests', and a truly inclusive political system at all levels of governance. This is the subject of the third section of this chapter. In the second section I take up the issue of difference and identity in relation to national belonging and question the adequacy of framing the issues of an intercultural society through the language of race and minority ethnicity. A significant dimension of the civil disturbances in Britain in 2001 was this aspect of identity and belonging, and this spills over into the next section.

Thinking through identity/difference

'We have norms of acceptability and those who come into our home – for that is what it is – should accept those norms.'
(David Blunkett, quoted in Alibhai-Brown 2001)

. . . seven years ago I finally decided this place was my place, and that was because I had a daughter whose father was of these islands. This did not make me any less black, Asian or Muslim – those identities are in my blood, thick and forever. But it made me kick more vigorously at those stern, steely gates that keep people of color outside the heart of the nation, then blame them for fighting each other in the multicultural wastelands into which the establishment has pushed them. A number of us broke through. The going was (and still is) incredibly hard but we are in now and, bit by bit, the very essence of Britishness is being transformed.

(Alibhai-Brown 2001)

The above remarks of David Blunkett were made in December of 2001, after Britain's summer of 'race riots'. It was a time of questioning of the previous half-century of immigration, the race relations problems that had emerged, and the policy response of multiculturalism. At the heart of this questioning was a perturbation over what it meant/means to be British (an agonizing which has only heightened since the terrorist bombings in London in the summer of 2005). Notions of identity were being unsettled. The response of Blunkett, the Home Secretary in the Blair government, was a rather crude reassertion of us-and-them thinking. His words epitomize a long-standing but much-contested view that immigrants are guests in the home of the host nation and must behave the way their hosts want them to behave: adopt the norms of 'Britishness', or get out. Implicit in this view is that there is only one correct way to be British and that it is the responsibility of newcomers to learn how to fit in with that way. Yasmin Alibhai-Brown, herself an immigrant of three decades' standing, contests this pure and static notion of national identity, counterposing it with a notion of a more inclusive, dynamic and evolving identity which can accommodate the new hybrid realities of a changing culture. She urges 'a national conversation about our collective identity' (Alibhai-Brown 2000: 10)

At stake here, and across European (or any of the large number of globalizing) cities today, are contested notions of identity and understandings of difference, and conflicting ways of belonging and feeling at home in the world. The Home Secretary expresses the view that there is a historic Britishness that must be protected from impurity. (Sections of the Austrian, Danish, French, Italian and Dutch populations have expressed similar antagonisms in recent years.) In this view, what it means to be British, to be 'at home' in Britain, is being threatened by immigrants who bring a different cultural baggage with them. Interestingly, the (fragile) notion of identity at the heart of this view is one that is both afraid of and yet dependent on difference. How does this apparent psychological paradox work?

When a person's self-identity is insecure or fragile, doubts about that identity (and how it relates to national identity may be part of the insecurity), are posed and resolved by the constitution of an Other against which that identity may define itself, and assert its superiority. In order to feel 'at home' in the nation and in the wider world, this fragile sense of identity seeks to subdue or erase from consciousness (or worse) that which is strange, those who are 'not like us'. Attempts to protect the purity and certainty of a hegemonic identity – Britishness, Danishness, and so on – by defining certain differences as independent sites of evil, or disloyalty, or disorder, have a long history.[2] There are diverse political tactics through which doubts about self-identity are posed and resolved, but the general strategy is the establishing of a system of identity and difference which is given legal sanctions, which defines who belongs and who does not. Over long periods of time, these systems of identity and difference become congealed as cultural norms and beliefs, entrenching themselves as the hegemonic status quo. Evil infiltrates the public domain, Connolly (1991) argues, when attempts are made to secure the surety of self- and national identity – and the powers and privileges that accompany it – with spatial and social and economic policies that demand conformity with a previously scripted identity, while defining the outsider as an outsider (a polluter of pure identities), in perpetuity.

There is a fascinating paradox in the relationship between identity and difference. The quest for a pure and unchanging identity (an undiluted Britishness, or Brummie-ness, or Danishness . . .) is at once framed by and yet seeks to eliminate difference; it seeks the conformity, disappearance, or invisibility of the Other. That is the paradox of identity. But what of difference and its political strategies? Surely difference, too, is constituted by its Other – as woman is in patriarchal societies, or to be gay and lesbian in heterosexual societies, or to be Black in white societies – and so is constituted by the hegemonic identity which it resists and seeks to change. Difference, defined as that which is outside, in opposition to the congealed norms of any society, is constituted by/against hegemonic identity. Identity and difference, then, are an intertwined and always historically specific system of dialectical relations, fundamental to which is inclusion (belonging) and its opposite, exclusion (not belonging). Here then is a double paradox. Some notion of identity is, arguably, indispensable to life itself (Connolly 1991), and some sense of culturally based identity would seem to be inescapable, in that all human beings are culturally embedded (Parekh 2000: 336).[3] But while the politics of pure identity seeks to eliminate the Other, the politics of difference seeks recognition and inclusion.

A more robust sense of identity must be able to embrace cultural autonomy and, at the same time, work to strengthen intercultural solidarity. If one dimension of such a cultural pluralism is a concern with reconciling old and new identities by accepting the inevitability of 'hybridity', or 'mongrelization', then another is the commitment to actively contest what is to be valued across diverse cultures. Thus Alibhai-Brown feels 'under no obligation to bring my daughter and son up to drink themselves to death in a pub for a laugh', nor does she want to see young Asian and Muslim women imprisoned in 'high-pressure ghettoes . . . in the name of "culture"', a culture that forces obedience to patriarchal authority and arranged marriages (Alibhai-Brown 2001). Negotiating new identities, then, becomes central to daily social and spatial practices, as newcomers assert their rights to the city, to make a home for themselves, to occupy and transform space.[4]

What now seems insidious in terms of debates about belonging in relation to the nation is the way in which the identities of minorities have been essentialized on the grounds of culture and ethnicity. The ethnicization and racialization of the identities of non-white or non-Anglo people in western liberal democracies, even the most officially multicultural among them (Canada and Australia), has had the effect of bracketing them as minorities, as people whose claims can only ever be minor within a national culture and frame of national belonging defined by others and their majority histories, usually read as histories of white belonging and white supremacy (Amin 2002: 21; Hage 1998). But the claims of the Asian youths in Britain's northern mill towns, just as those of Black Britons or 'Lebanese Australians' or 'Chinese Canadians', are claims for more than minority recognition and minority rights. Theirs is a claim for the mainstream (or perhaps it is a claim for 'the end of mainstream' (Dang 2002)), for a metaphorical shift from the margins to the centre, both in terms of the right to visibility and the right to reshape that mainstream. It is nothing less than a claim to full citizenship and a public naming of what has hitherto prevented that full citizenship – the assumption that to be British, Canadian, Danish, Dutch, and so on, is to be white, and part of white culture. As long as that assumption remains intact, the status of minority ethnic groups in all the western democracies will remain of a different order to that of whites, always under question, always at the mercy of the 'tolerance' of the dominant culture, a tolerance built on an unequal power relationship (Hage 1998).

The crucial implication of this discussion is that in order to enable all citizens, regardless of 'race' or ethnicity or any other cultural criteria, to become equal members of the nation and contribute to an evolving national identity, 'the ethnic moorings of national belonging need to be exposed and replaced by criteria that have nothing to do with whiteness' (Amin 2002: 22). Or as Gilroy (2000: 328) puts it, 'the racial ontology of sovereign territory' needs to be recognized and contested. This requires an imagination of the nation as something other than a racial or ethnic territorial space, perhaps an imagination that conceives the nation as a space of traveling cultures and peoples with varying degrees and geographies of attachment. Such a move must insist that race and ethnicity are taken out of the definition of national identity and national belonging 'and replaced by ideals of citizenship, democracy and political community' (Amin 2002: 23). This brings me to the necessity of rethinking 20th century notions of multiculturalism (based on ethno-cultural recognition), and that is the subject of the third section of this chapter.

Reconsidering multiculturalism

As a fact, multiculturalism describes the increasing cultural diversity of societies in late modernity. Empirically, many societies and many cities could be described today as multicultural. But very few countries have embraced and institutionalized an ideology of multiculturalism. Australia and Canada have done so since the late 1960s, as have Singapore and Malaysia, although the latter pair of countries have a different interpretation of multiculturalism than do the former pair. During the same period, the USA has lived through its 'multicultural wars', still uneasy with the whole notion, preferring the traditional 'melting pot' metaphor and its associated politics of assimilation. France has been most adamant that there is no place for any kind of political recognition of difference in its republic. The Dutch and the Danish, once the most open to multicultural policy claims, have each begun to pull up the drawbridges since 2002. Each country has a different definition of multiculturalism, different sets of public policies to deal with/respond to cultural difference, and correspondingly different definitions of citizenship.

As an ideology, then, multiculturalism has a multiplicity of meanings. What is common in the sociological content of the term in the West – but never spoken of – is that it was formulated as a framework, a set of policies, for the national accommodation of non-white immigration. It was a liberal response that skirted the reality of the already racialized constitution of these societies and masked the existence of institutionalized racism.[5] The histories of multicultural philosophies are in fact much more complex and contested than this, and genealogical justice cannot be done without a much more contextualized discussion of each country, which is not my purpose here. So instead, drawing on the distinguished British cultural studies scholar Stuart Hall, I will simply summarize the range of meanings that have been given to multiculturalism as ideology, and some of the dangers embedded in it.

Hall (2000) theorizes the multicultural question as both a global and local terrain of political contestation with crucial implications for the West. It is contested by the conservative Right, in defence of the purity and cultural integrity of the nation. It is contested by liberals, who claim that the 'cult of ethnicity', the notion of 'group rights', and the pursuit of 'difference' threaten the universalism and neutrality of the liberal state. Multiculturalism is also contested by 'modernizers of various political persuasions'. For them, the triumph of the (alleged) universalism of western civilization over the particularisms of ethnic, religious, and racial belonging established in the Enlightenment marked an entirely worthy transition from tradition to modernity that is, and should be, irreversible. Some postmodern versions of cosmopolitanism oppose multiculturalism as imposing a too narrow, or closed, sense of identity. Some radicals argue that multiculturalism divides along ethnic lines what should be a united front of race and class against injustice and exploitation. Others point to commercialized, boutique, or consumerist multi-culturalism as celebrating difference without making a difference (Hall 2000: 211).

Clearly, multiculturalism is not a single doctrine and does not represent an already achieved state of affairs. It describes a variety of political strategies and processes that are everywhere incomplete. Just as there are different multicultural societies, so there are different multiculturalisms.

> Conservative multiculturalism insists on the assimilation of difference into the traditions and customs of the majority. Liberal multiculturalism seeks to integrate the different cultural groups as fast as possible into the 'mainstream' provided by a universal individual citizenship . . . Pluralist multiculturalism formally enfranchises the differences between groups along cultural lines and accords different group rights to different communities within a more . . . communitarian political order. Commercial multiculturalism assumes that if the diversity of individuals from different communities is recognized in the marketplace, then the problems of cultural difference will be dissolved through private consumption, without any need for a redistribution of power and resources. Corporate multiculturalism (public or private) seeks to 'manage' minority cultural differences in the interests of the center. Critical or 'revolutionary' multiculturalism foregrounds power, privilege, the hierarchy of oppressions and the movements of resistance . . . And so on.
>
> (Hall 2000: 210)

Can a concept that has so many valences and such diverse and contradictory enemies possibly have any further use value? Alternatively, is its contested status precisely its value, an indication that a radical pluralist ethos is alive and well?

Given that we live in an age of migration (Castles and Miller 1998), we are inevitably implicated in the politics of multiculturalism. This in turn demands a rethinking of traditional notions of citizenship as well as a lot of new thinking about the social integration of immigrants. Given this 21st century urban reality, we need to find a way to publicly manifest the significance of cultural diversity, and to debate the value of various identities/differences; that is, to ask which differences exist, but should not, and which do not exist, but should.[6] Far from banishing the concept to political purgatory, we need to give it as rich a substance as possible, a substance that expands political possibilities and identities rather than purifying or closing them down. This leads me to re-theorize multiculturalism, which I prefer to re-name as interculturalism, as a political and philosophical basis for thinking about how to deal with the challenge of difference in the mongrel cities of the 21st century.

My intercultural theory is composed of the following premises:

- 'Culture' cannot be understood as static, eternally given, essentialist. It is always evolving, dynamic and hybrid of necessity. All cultures, even allegedly conservative or traditional ones, contain multiple differences within themselves that are continually being re-negotiated.

- Cultural diversity is a positive thing, and intercultural dialogue is a necessary element of culturally diverse societies. No culture is perfect or can be perfected, but all cultures have something to learn from and contribute to others. Cultures grow through the everyday practices of social interaction.
- The political contestation of interculturalism is inevitable, as diverse publics debate the merits of multiple identity/difference claims for rights.
- At the core of interculturalism as a daily political practice are two rights: the right to difference and the right to the city. The right to difference means recognizing the legitimacy and specific needs of minority or subaltern cultures. The right to the city is the right to presence, to occupy public space, and to participate as an equal in public affairs.
- The 'right to difference' at the heart of interculturalism must be perpetually contested against other rights (for example, human rights) and redefined according to new formulations and considerations.
- The notion of the perpetual contestation of interculturalism implies an agonistic democratic politics that demands active citizenship and daily negotiations of difference in all of the banal sites of intercultural interaction.
- A sense of belonging in an intercultural society cannot be based on race, religion, or ethnicity but needs to be based on a shared commitment to political community. Such a commitment requires an empowered citizenry.
- Reducing fear and intolerance can only be achieved by addressing the material as well as cultural dimensions of 'recognition'. This means addressing the prevailing inequalities of political and economic power as well as developing new stories about and symbols of national and local identity and belonging.

There are (at least) two public goods embedded in a version of interculturalism based on these understandings. One is the critical freedom to question in thought, and challenge in practice, one's inherited cultural ways. The other is the recognition of the widely shared aspiration to belong to a culture and a place, and so to be at home in the world (Tully 1995). This sense of belonging would be lost if one's culture were excluded, or if it was imposed on everyone. But there can also be a sense of belonging that comes from being associated with other cultures, gaining in strength and compassion from accommodation among and interrelations with others, and it is important to recognize and nurture those spaces of accommodation and intermingling.

This understanding of interculturalism accepts the indispensability of group identity to human life (and therefore to politics), precisely because it is inseparable from belonging. But this acceptance needs to be complicated by an insistence, a vigorous struggle against the idea that one's own group identity has a claim to intrinsic truth. If we can acknowledge a drive within ourselves, and within all of our particular cultures, to naturalize the identities given to us, we can simultaneously be vigilant about the danger implicit in this drive, which is the almost irresistible desire to impose one's identity, one's way of life, one's very definition of normality and of goodness, on others. Thus we arrive at a lived conception of identity/difference that recognizes itself as historically contingent and inherently relational; and a cultivation of a care for difference through strategies of critical detachment from the identities that constitute us (Connolly 1991; Tully 1995). In this intercultural imagination, the twin goods of belonging and of freedom can be made to support rather than oppose each other.

From an intercultural perspective, the good society does not commit itself to a particular vision of the good life and then ask how much diversity it can tolerate within the limits set by this vision. To do so would be to foreclose future societal development. Rather, an intercultural perspective advocates accepting the reality and desirability of cultural diversity and then structuring political life accordingly. At the very least, this political life must be dialogically and agonistically constituted. But the dialogue requires certain institutional preconditions, such as freedom of speech, participatory public spaces, empowered citizens, agreed procedures and basic ethical norms, and the active policing of discriminatory practices. It also calls for

> such essential political virtues as mutual respect and concern, tolerance, self-restraint, willingness to enter into unfamiliar worlds of thought, love of diversity, a mind open to new ideas and a heart open to others' needs, and the ability to persuade and live with unresolved differences.
>
> (Parekh 2000: 340)

A notion of the common good is vital to any political community. From an intercultural perspective, this common good must be generated not by transcending or ignoring cultural and other differences (the liberal position), but through their interplay in a dialogical, agonistic political life. Finally, a sense of belonging, which is important in any society, cannot in multicultural societies be based on ethnicity or on shared cultural, ethnic or other characteristics. An intercultural society is too diverse for that. A sense of belonging must ultimately be political, based on a shared commitment to a political community (Parekh 2000: 341; Amin 2002: 23).

Since commitment, or belonging, must be reciprocal, citizens will not feel these things unless their political community is also committed to them and makes them feel that they belong. And here is the challenge. An intercultural political community 'cannot expect its members to develop a sense of belonging to it unless it equally values and cherishes them in all their diversity, and reflects this in its structure, policies, conduct of public affairs, self-understanding and self-definition' (Parekh 2000: 342). It would be safe to say that no existing (self-described) multicultural society can yet claim to have achieved this state of affairs, for reasons

54 Leonie Sandercock

that have already been elaborated: political and economic inequalities accompanied by an unresolved postcolonial condition that we may as well name as racism. But in recent years these issues have been identified, increasingly documented, and are becoming the focus of political activity in many countries.

Conclusions: the marriage of theory and practice

This chapter has outlined three main elements of a cosmopolitan urbanism, or intercultural political philosophy. What has emerged from the descriptions and analysis of the Collingwood Neighbourhood House (CNH) is that this local institution is a catalyst for, and a working example of, living together and bridging vast cultural differences. With many different ethno-cultural groups living in this one territorially defined neighbourhood, it is neither the existence of a common culture (ethnically defined) nor a shared sense of attachment to place that makes this neighbourhood a community. Rather, what has happened in the period of twenty years of rapid demographic change from a predominantly Anglo-European to a much more ethnically mixed population is exactly what James Donald theorized, 'a broad social participation in the never completed process of making meanings and creating values, an always emerging, negotiated common culture' (1999: 151).

But that 'common culture' is not ethno-culturally grounded, nor is it the result of one dominant culture imposing its lifeways on all the rest. Rather, it is a negotiated sharing of values, established through broad social participation. This is community redefined neither as identity nor as place, but as a productive process of social interaction. The CNH is indeed a physical place: many folks even refer to it as a blessed place, one that has helped to create a sense of belonging. But, perhaps paradoxically, that belonging is only partially to do with the actual physical place, and more profoundly to do with the lived experience of building relationships. As James Donald proposed in his normative ideal, we don't need to share cultural traditions with our neighbours in order to live alongside them, but we do need to be able to talk to them. CNH has created that space for intercultural dialogue, for exchange across cultural difference, which is the precondition for relationship building.

The secret of this remarkable achievement is in the CNH mission, which embodies Ash Amin's normative ideal of a politics of local liveability, nurtured through daily habits of 'quite banal intercultural interaction in order to establish a basis for dialogue'. At CNH, these daily habits of banal interaction occur around childcare, around the learning of English as a second language, around preparing and/or sharing meals together, and sharing a multitude of other training and learning and recreational opportunities. In these 'micro-public spaces', these sites of everyday encounter and prosaic negotiation of difference, people from different cultural backgrounds come together, initially in quite practical ways, but in these moments of coming together there is always the possibility of dialogue, of initiating new attachments. And that is what happens at and through CNH. Part of what happens through such everyday contact is the gradual overcoming of feelings of strangeness in the simple process of sharing everyday tasks and/or challenges and comparing ways of doing things.

But such initiatives do not automatically become sites of social inclusion. They need organizational and discursive strategies that are designed to build voice, to foster a sense of common benefit, to develop confidence among disempowered groups, and to arbitrate when disputes arise.

And that is precisely, and systematically, what the CNH Board and leadership have done through two decades of social and demographic change. They have consciously diversified as a Board, and in the selection of staff and nurturing of volunteers. They have consciously chosen not to provide any programs or services on an ethno-culturally specific basis. They have systematically conducted outreach to marginalized groups such as First Nations and youth. They have systematically organized anti-racism and diversity training for staff and volunteers, and empowered youth to run their own anti-bullying, anti-racism and drug counseling programs. And they have proactively developed programs for homeless people. All of which reflects the values of social justice and social inclusion embedded in the mission of CNH (Sandercock and Attili 2009: chapter 6).

CNH's vocabulary of accommodation to difference is a vocabulary of 'rights of presence, bridging difference, getting along', just as proposed in Amin's normative ideal. And an important part of this pragmatic vocabulary is the recognition of conflict as inevitable, and a commitment to work through such conflict, acknowledging whatever fears and anxieties have been triggered, and devoting time to listening, talking through and arriving at new accommodations that work for residents.

But these local, neighbourhood-based organizational and discursive strategies cannot endure, let alone thrive, in the absence of a broader intercultural political culture: that is, one with effective anti-racism policies, with strong legal, institutional and informal sanctions against racial and cultural hatred and a public culture that no longer treats immigrants as 'guests'. Such a local experiment, in Frankfurt, failed in the 1990s in the absence of this broader political culture (Sandercock and Attili 2009: chapter 3). Canada as a nation and Vancouver as a city have striven, albeit imperfectly, to create such a political culture over the past three decades (ibid: part one).

One very important aspect of Canada's evolving political culture at federal government level, especially in the past decade, through the Department of Canadian Heritage, has been the effort to create a sense of national identity and national belonging that is grounded in ideals of active citizenship, democracy and political community, rather than in notions of 'Canadianness' grounded in race or ethnicity (the latter being the case in most European countries). This very important shift is also a shift in the meaning of multiculturalism, from its earlier incarnation emphasizing recognition and support of all immigrant cultures and the celebration of ethno-cultural differences, to an intercultural position emphasizing the building of bridges between cultures. And this has been reflected in actual funding shifts, away from the support of ethno-culturally specific organizations or facilities (such as a Chinese Cultural Centre or a Vietnamese Seniors Centre) to organizations with explicit intercultural mandates, like CNH. In the process, the essential political virtues of a cosmopolitan urbanism (or an intercultural society) are being nurtured: the virtues of mutual respect and concern, tolerance, self-restraint, love of diversity, minds open to new ideas and hearts open to the needs of others.

In embodying these virtues, nurturing them, and pursuing them through relationship building in everyday life, the Collingwood Neighbourhood House is a microcosm of all that Canada aspires to be (but is not, yet). It is a marriage of the theory and practice of cosmopolitan urbanism.

Notes

1   The following description is taken from James Tully's account of the sculpture (Tully 1995: 17–18).
2   For much of my interpretation in this section I am indebted to the work of William Connolly (1991, 1995) and Julia Kristeva (1991).
3   'Culturally embedded' in the sense that we grow up and live within a culturally structured world, organize our lives and social relations within its system of meaning and significance, and place some value on our cultural identity (Parekh 2000: 336)
4   Or as previously dominated groups such as gays and lesbians, women, people with disabilities, decide to engage in a politics of identity/difference, a politics of place-claiming and place-making (Kenney 2001).
5   See Hage (1998), on Australia; Hesse (2000) and Hall (2000), on the UK; Bannerji (1995, 2000), on Canada.
6   See Chantal Mouffe's discussion of this dilemma in her case for an agonistic democratic politics in *The Democratic Paradox* (2000).

References

Albrow, M. (1997) 'Travelling Beyond Local Cultures: Socioscapes in a Global City', in J. Eade (ed.) *Living the Global City: Globalization as a Local Process*, London: Routledge.
Alibhai-Brown, Yasmin (2000) 'Diversity versus Multiculturalism', *The Daily Telegraph*, 23 May.

Alibhai-Brown, Yasmin (2001) 'Mr. Blunkett Has Insulted All of Us', *The Independent*, 10 December.

Amin, Ash (2002) *Ethnicity and the Multicultural City: Living with Diversity*. Report for the Department of Transport, Local Government and the Regions, Durham: University of Durham.

Bannerji, H. (1995) *Thinking Through*, Toronto: Women's Press.

Bannerji, H. (2000) *The Dark Side of the Nation: Essays on Multiculturalism, Nationalism and Gender*, Toronto: Canadian Scholars' Press Inc.

Castles, S. and Miller, M.J. (1998) *The Age of Immigration*, 2nd edn, East Sussex: Guilford Press.

Connolly, William (1991) *Identity/Difference*, Ithaca, NY: Cornell University Press.

Connolly, William (1995) *The Ethos of Pluralization*, Minneapolis: University of Minnesota Press.

Dang, S. (2002) 'Creating Cosmopolis: The End of Mainstream'. Unpublished Masters Thesis, School of Community and Regional Planning, University of British Columbia.

Donald, J. (1999) *Imagining the Modern City*, London: The Athlone Press.

Gilroy, P. (2000) *Between Camps*, London: Penguin.

Hage, Ghassan (1998) *White Nation: Fantasies of White Supremacy in a Multicultural Society*, Sydney: Pluto Press.

Hall, Stuart (2000) 'Conclusion: The Multi-cultural Question', in B. Hesse (ed.) *Un/settled Multiculturalisms*, London: Zed Books.

Hesse, B. (2000) 'Introduction: Un/settled Multiculturalisms', in B. Hesse (ed.) *Un/settled Multiculturalisms*, London: Zed Books.

Home Office (2001) *Building Cohesive Communities: A Report of the Ministerial Group on Public Order and Community Cohesion*, London: Home Office/Her Majesty's Government.

Kenney, Moira (2001) *Mapping Gay L.A.: The Intersection of Place and Politics*, Philadelphia: Temple University Press.

Kristeva, Julia (1991) *Strangers to Ourselves*, New York: Columbia University Press. Translated by Leon S. Roudiez.

Mouffe, Chantal (2000) *The Democratic Paradox*, London: Verso.

Parekh, B. (2000) *Rethinking Multiculturalism*, London: Macmillan.

Rocco, R. (2000) 'Associational Rights-claims, Civil Society and Place', in E. Isin (ed.) *Democracy, Citizenship and the Global City*, London: Routledge.

Sandercock, Leonie (2003) *Cosmopolis 2: Mongrel Cities of the 21st Century*, London: Continuum.

Sandercock, L. and Attili, G. (2009) *Where Strangers Become Neighbors: The Integration of Immigrants in Vancouver*, New York: Springer Publishing.

Sassen, Saskia (1996) 'Whose City Is It? Globalization and the Formation of New Claims', *Public Culture*, 8: 205–223.

Sennett, Richard (1994) *Flesh and Stone: The Body and the City in Western Civilization*. New York: Norton.

Tully, James (1995) *Strange Multiplicity: Constitutionalism in an Age of Diversity*, Cambridge: Cambridge University Press.

Urban Task Force (1999) *Towards an Urban Renaissance*, London: E & FN Spon.

Chapter 4
For the equitable city yet to come

Tanja Winkler

Civil society is widely believed to have the potential to make a positive contribution to democracy and development in South Africa. However, its transformative planning potential remains under-analyzed. This chapter offers a critical analysis of three Johannesburg-based civic associations that employ different planning strategies to reshape the public domain. It investigates the extent of their contributions to local planning policies, and their capacity to challenge and change the state's current cost recovery policies. Findings show that all three associations contribute to pluralism and democratic values, but that they fail to make significant inroads into public decision-making processes.

Introduction

> After the first post-apartheid administration, civil society organizations began to lose their special place in the language of development. They were seen as at best a nuisance and at worst a threat to the democratic government. The argument was that they would delay service delivery. But, it is indeed the re-emergence of [such] organizations that has provided a mirror for the ANC to examine itself in. Civil society must, once again, occupy a central position within the development agenda
>
> (Mangcu 2008: 123–124)

Since 1994, South Africa has been a multi-party democracy in which civil and political liberties are formally recognized. It is governed by the African National Congress (ANC), which has won three successive general elections by a wide margin.[1] While the liberal democratic features of the South African system are firmly

---

Submitted by the Association of African Planning Schools.

Originally published in *Planning Theory & Practice* 10: 1, 65–83. ©2009 By kind permission of Taylor & Francis Ltd. www.informaworld.com and the Royal Town Planning Institute.

in place, "the ANC's electoral dominance means that the parliamentary opposition is weak and a largely ineffective source of government accountability" (Robinson and Friedman 2007: 646). For this reason, a number of local scholars argue that the recent re-emergence of civic associations in South Africa has the potential to significantly improve governance (Ballard et al. 2006; Bond 2005; Buhlungu 2006; Mangcu 2008; Masiwa 2007; Pieterse 2006; Robinson and Friedman 2007). They have become invaluable agents for furthering democracy and countering the imposition of structural adjustment policies, which the state adopted in 1996 under the misnomer of "Growth, Employment and Redistribution."

This view of civil society, however, is not one that regards it as taking on state welfare functions (Alnoor 2003). Rather, it sees civic associations as advocates for disadvantaged social groups whose interests have been disregarded by the governing regime. Yet despite the often unquestioned assumption that civil society can play an important role in strengthening democracy, relatively little is known about its effectiveness and impact on local planning policies. This chapter aims, through the presentation of case studies of oppositional associations, to explore the planning strategies employed by three different Johannesburg-based civic associations seeking to reshape the public domain. Accordingly, the main research question asks: to what extent can such associations succeed in challenging and changing the local state's current cost recovery policies?

It is important to note that despite a marked decline in structured engagement by civil society groups since 1999, they retain the right to seek to influence local government policies through formal integrated development planning (IDP), and other processes (Robinson and Friedman 2007). Nonetheless, both the City of Johannesburg's and the Ekurhuleni Metropolitan Municipality's bureaucracies are large and unwieldy, and their policy contexts tend to be scattered in focus and competing in orientation (Charlton and Gotz 2007). Both of these local authorities have jurisdiction over the greater Johannesburg region, they are governed by the ruling ANC, and their IDP processes have become annual document-producing activities rather than sustained public engagement initiatives (Charlton and Gotz 2007). Greater Johannesburg's geographic area is also highly uneven, representing dramatic spatial and socioeconomic inequalities amongst its resident population. Over 45% of its population are officially unemployed (CoJ 2006). Through economic growth and cost recovery strategies, these municipalities hope to ameliorate persistent inequalities. Johannesburg-based findings will, therefore, reach similar conclusions to those found by Alfasi (2003) and Martens (2005) in their respective studies of Israeli non-government organizations (NGOs), namely that "conventional approaches to public participation have not made planning more democratic" (Alfasi 2003: 185), and that "dominant actors remain the ones who shape new [planning] practices" (Martens 2005: 2). However, Yacobi argues that "both Alfasi and Martens ignore the Israeli socio-political structure, and thus discuss planning as an isolated social realm detached from wider ideological contexts" (2007: 746). By drawing from conceptual frameworks of planning and political theory to analyze the complex configurations of associational life in relation to the state, this chapter also aims to show how the local socio-political structure shapes the activities of civic associations.

Within the wide spectrum of civic associations that promote democratic development projects, Yacobi (2007) identifies at least three general categories of associations based on different strategies employed by them to reshape the public domain. For analytical and comparative purposes, Yacobi's (2007) three categories will be applied to the Johannesburg study. The first category encompasses associations in civil society that use their professional planning expertise as a strategy to achieve spatial and socioeconomic justice for disadvantaged communities. They employ planners as full-time members of staff. Most of these associations have their roots in South Africa's anti-apartheid struggle, when segregated urban townships represented the battlefields of the physical struggle against the apartheid regime. Their common programme areas include facilitating community participation in low-income housing delivery and working in support of the institutional development of local authorities. Eight such NGOs operate throughout South Africa, but only one, namely Planact, is based in Johannesburg. For this reason, only Planact will be examined.[2]

A second grouping centres on civic associations that see planning as a vehicle to reach other claims through social mobilization. These associations do not employ planners. They employ mobilization strategies to challenge the state, and they have their roots in post-apartheid struggles. As such, they are often referred to as "new social movements." This group includes the Anti-Privatisation Forum, the Homeless People's Alliance, the Landless People's Movement, and the Western Cape Anti-Eviction Campaign. Although all engage in aspects of planning for social transformation practices, the Anti-Privatization Forum (APF) is the most active social movement in the greater Johannesburg region. Its transformative activities will, therefore, be assessed.

A third category includes associations that deal with human rights issues from a legal, rights-based, perspective. They also do not employ planners, but they work with planners in their professional teams when dealing with planning policies, which they encounter in their legal work. Accordingly, the activities of the Centre for Applied Legal Studies (CALS) will be evaluated, as this non-profit organization is actively engaged in defending the rights of Johannesburg's inner-city residents by challenging local planning policies.

This chapter will be structured as two sections. The first will establish the conceptual framework for this study. The second section will then evaluate the extent to which Planact, the APF, and CALS contribute to democratic planning practices and policies. Here, the particular circumstances or conditions which call for a response from civic associations, and the obstacles they face, will also be discussed.

Setting up the theoretical and analytical frameworks for this study

To explore the linkages and tensions between the state and associational modes of organization, focus will be placed on the institutional sector that, metaphorically speaking, lies "in between" the state and the economy (Emirbayer and Sheller 1999). This "in between" space is occupied by NGOs, social movements, and other associational modes of organization, but it excludes the state and the economy. Such a conceptualization of civil society is certainly controversial, given the lengthy history of the term found in the theories of, for example, John Locke, Georg Hegel, Antonio Gramsci, Jürgen Habermas, and Manuel Castells. But rather than embark on an archaeological expedition into these theoretical differences, and given its extensive treatment elsewhere,[3] what is important for this "in between" conceptualization is civil society's legitimacy to act in the public domain (to speak on behalf of "planning" or marginalized residents).

A legitimacy to act on behalf of something or someone serves to reaffirm and stabilize the relative autonomy gained by civil society vis-à-vis the state, as well as to safeguard democratic advances achieved within civil society itself (Friedmann 2005). Moreover, theorists of the "Just City" and planning for social transformation argue that the legitimacy to act in the public domain is "created and defended from below," and that civic associations owe their legitimacy to act on behalf of something or someone ultimately to civil society itself (Beard 2003; Castells 1983; Fainstein 2000; Friedmann 2005). By implication, then, the state is not the only legitimate "planner." However, as theorists of collaborative planning argue, the state is no more an evil to be kept at bay than civil society is itself necessarily a good (Forester 1999; Healey 1997). Often, in fact, civil society, if left to itself, "generates radically unequal power relationships which only state power can challenge. [T]he state is an indispensable agent, even if associational networks also, always, resist the organizing impulses of state bureaucrats" (Walzer 1992: 104). Assessing Planact's, the APF's, and CALS's legitimacy to act in the planning field, either with or without the state, to reshape the public domain therefore involves dealing with opposing theoretical standpoints, which need to be clarified before the analytical framework used for such an assessment can be presented.

Establishing the theoretical framework by drawing from planning theories

A range of theoretical positions exist to explain planning as a phenomenon and to provide ideas for how planning should be conducted, and this chapter will draw from aspects of collaborative planning, social transformation, and Just City theories to explain the different approaches adopted by civic associations in their quest to reshape the public domain. All three theoretical positions value a role for civil society in democratic planning processes, and associated empirical approaches with their concern for the empowerment of social groups that are located outside of (and sometimes against) the state (Watson 2002).

Collaborative planning theory draws inspiration from Habermas to posit communication as an important element of planning practice (Forester 1999; Healey 1997). In the Habermasian perspective, "the public sphere" enables "citizens [to] behave as a public body when they confer about matters of general [public] interest" (Habermas 1974: 49). Such a conference entails communicating ideas with different interest groups, arguing about these ideas, debating differences, and, eventually, reaching some kind of consensus on an appropriate course of action. While this approach recognizes that communication can be distorted in various ways, it argues that if public decision-making processes are inclusive and transparent, and if power differences between participants can be defused, then the outcome of such a process can be considered valid (Habermas 1974). In other words, "the power of dominant discourses can be challenged through the transformations that come as people learn to understand and respect each other across their differences and conflicts" (Healey 1999: 119). The overarching aim of a collaborative approach for planning is to facilitate a just process. If the planning process is just, then the outcome will also be just. Collaborative planning theorists also embrace Habermas's faith in civil society as a source of democracy, and as a vehicle for placing pressure on the state to act more responsively (Watson 2002). Collaboration, therefore, "seeks ways of recovering a new participatory realisation of democracy [to] focus the activity of governance according to the concerns of civil society" (Healey 1999: 119).

Planning for social transformation, on the other hand, begins with a critique of the present situation that is creating systemic inequalities, and it then provides an alternative response to the critique through policies for structural transformation and the potential empowerment of those who have been systematically disempowered (Beard 2003; Friedmann 1998; Holston 1998; Reardon 1998; Sandercock 1998). However, a critical analysis of the status quo is not sufficient. Identifying possible policy actions to ameliorate systemic inequalities is equally important. The role of the planner in this transformative model is thus to assist civil society in creating practical solutions which are not the monopoly of professional planners but are collectively formulated through a respectful process of "social/ mutual learning" (Friedmann 1998). Here, civil society, and not the state, is the change agent in an ongoing struggle for a more just and socially inclusive public domain. As such, local agents are viewed as active planners for themselves and not as passive recipients of determining ecological factors. Planning for social transformation enlarges the traditional planning field from professional practitioners alone to include civic associations, activists, and citizens as "planners." Planning is, therefore, no longer "only that professional domain that constitutes the field of city-building, but [it is] also that form of collective action which we might call community-building" (Sandercock 1998: 39). For Sandercock, socially transformative practices do not necessarily need to begin with large-scale interventions, but can instead be initiated through smaller actions or what she calls "a thousand tiny empowerments" (1998: 157). She outlines two different approaches to planning for social transformation: "insurgent" and "radical" planning (Sandercock 1998). Insurgency implies something oppositional: a mobilization against the state, the market, or both. Its aim is to challenge and transform existing power relationships through mobilized community actions. By contrast, "radical planning is not always, or necessarily, oppositional" (Sandercock 1998: 41). However, neither radical nor insurgent planning are mainstream practices, but this type of planning can lead to a change in mainstream cultures.

A form of planning that seeks to promote redistribution, equity, and justice informs Fainstein's alternative Just City position. Fainstein, along with communicative planning and social transformation theorists, is concerned with planning processes and participation (Watson 2002). However, the primary audience for Just City theorists are the leaders of urban social movements, rather than the state, which may be neither neutral nor benevolent (Fainstein 2000). As such, the Just City distances itself from communicative planning theorists who "primarily speak to planners employed by the state, calling on them to mediate among diverse interests" (Fainstein 2000: 468). Instead, proponents of the Just City aim to capacitate civil society in public decision-making processes and actions, and to be attentive to social inequalities and asymmetrical power rationalities that exist not only between the state and civil society, but also within civil society itself. For this reason, Fainstein is cautious of accepting the validity of all the claims made by different actors. Claims cannot be judged by procedural rules alone, and just processes do not necessarily result in just outcomes, as Habermas argued. Instead, both the substantive content of planning policies and the process of planning need to be equally judged for their impact on equity and democracy (Fainstein 2005). Whereas in collaborative planning the role of both the state and the planner is to facilitate an inclusive and transparent planning approach, in Just City theories the role of the state is to facilitate the equitable redistribution of resources, while the role of the planner in this model is similar to the role envisaged by social transformation theory, producing practical solutions to broad-based issues. A Just City should benefit from a welfare state, an active and democratic civil society, and good governance that is attentive to the needs of as many citizens as possible (Fainstein 2005). This conceptualization of planning is thus inspired by goals of spatial, political, and socioeconomic justice and equity.

Establishing the analytical framework by drawing from political theory

Analytical frameworks developed by Emirbayer and Sheller (1999) and Hadenius and Uggla (1996) offer a comprehensive approach for assessing the extent to which civil society is able to challenge and change the state's cost recovery policies. Accordingly, the democratizing potential of civil society is assessed along three main axes, namely: contribution to pluralism, promotion of democratic values, and facilitation of public participation in policy decision-making processes. Contributions to pluralism, in turn, are evaluated on the basis of three criteria: a multiplicity of popular associations to counter political oppression; a high degree of autonomy from the state; and associational diversity to provide a balance between contending power centres. The issue of power is, therefore, central to this analytical strategy, and to the three planning positions discussed earlier. As such, it is important to identify nodes of power, because actors occupying such nodes are uniquely positioned to extend their influence to others within their environments. Power also depends on an actor's location within networks of trust, as "those who hold trust hold power" (Lewis and Weigert 1985: 459).

Promoting democratic values is the second attribute of civil society in this framework. To this end, the structure, agency, and actions of civil associations need to be assessed. If they facilitate structures, agencies, and actions for inculcating democratic norms and consensus building, provided they have a leadership that is accountable and responsive, they will be able to promote democratic values and to speak on behalf of citizens' concerns (Hadenius and Uggla 1996). A multi-layered organizational structure, characterized by small homogenous groups sharing similar problems and resources ensures transparency and accountability in associational life (Hadenius and Uggla: 1996). This analytical framework, therefore, aims to identify the presence of "cohesive subgroups" that comprise a subset of actors among whom there are relatively strong, direct, and frequent structural ties (Emirbayer and Sheller 1999). A lack of structural cohesion between subgroups, on the other hand, may reveal the ways in which barriers develop between civic associations that discourage the open-endedness required for effective engagement in public policy (Emirbayer and Sheller 1999). Furthermore, it is important to identify the possible existence of "structural holes" between the state and civil society. Originally conceived by Burt, "structural holes" represent "the separation between contacts [that are] disconnected either directly, in the sense that they have no direct contact with one another, or indirectly, in the sense that one has contacts that exclude others" (Burt 1992: 18). A sensitivity to the absences (and not only the presences) of structural ties among civic associations, and between the state and civil society, may then suggest reasons for minimal policy impacts. Adopted agencies and related actions, in turn, are informed by associations' "self-identified role" (Emirbayer and Sheller

1999). Distinct roles elicit equally distinct attitudes, values, ethics, and discourses that may either enable or constrain action in different ways (Emirbayer and Sheller 1999).

A capacity to foster public participation in policy-making processes is the third main attribute of civil society in this framework. Here, parallels may be drawn with collaborative planning. A participatory approach to decision making and an open and accountable leadership are positively associated with political efficacy, namely, the ability of civic associations to influence state policies (Hadenius and Uggla 1996). However, if civic associations are too densely woven around a single issue, they may fail to influence targeted institutions. "Even with high degrees of participation [among civic associations and/or between the state and civil society], influence may actually decrease under conditions of excess centralization" (Emirbayer and Sheller 1999: 169).

These theoretical considerations posit that if civic associations are internally democratic and motivated by broader societal concerns (rather than by narrow self-interests) they can make a positive contribution to the process of democratization by fostering pluralism, promoting democratic values, and enhancing political participation. For the purpose of this chapter, Planact's, the APF's, and CALS's contribution to planning policies will therefore be judged on the extent to which they are able to influence public policy and make office holders accountable to the needs of disadvantaged social groups living in the greater Johannesburg region. In the process, this analysis will attempt neither to "bracket" existing power asymmetries, nor to elicit unrealistic normative planning solutions that are detached from wider socio-political realities (Campbell 2005; Yacobi 2007).

This discussion will begin by introducing Planact's "community development" agency to highlight how collaborative planning practices are used as an approach to achieve change. The APF, on the other hand, embraces "insurgency" as an agency to facilitate social transformation. Finally, CALS makes use of a "rights-based" agency to challenge the state and to promote aspects of the Just City.

Collaborating with Planact

Planact was established in 1985 as a voluntary association of planners who came together to assist community organizations in advancing alternative development strategies to those implemented by the apartheid regime. It is now a formally registered NGO, and it employs full-time planners. After the establishment of democracy in 1994, Planact began to elicit new genres of engagement, favouring collaborations with the state as opposed to its former oppositional approach. Its agency, nevertheless, remains rooted in disadvantaged communities, and its self-identified role aims to promote "strategic interventions in the areas of local government transformation and community development" (Planact: www.planact. org.za). To this end, Planact facilitates "community participation processes to enhance good governance at the local level" (Planact: www.planact.org.za). Strategic intervention, community participation, and good governance values, derived from professional planning ethics and discourses, are, in turn, realized "through networking with likeminded organisations" (Planact: www.planact. org.za).[4] In other words, Planact's community development activities are supported by a cohesive subgroup of civic actors among whom there are relatively strong, direct, and frequent structural ties (participant interview 2008). Planact and its cohesive subgroup, therefore, represent an example of Hadenius and Uggla's (1996) multi-layered organization, characterized by autonomous groups sharing similar problems and resources. Planact's organizational structure and its actions also include democratic norms, and its leadership is accountable and responsive. As such, Planact promotes transparency and accountability in associational life, and it values a participatory approach to decision making, which, according to Hadenius and Uggla (1996), is essential if civic associations hope to influence state policies. Attention will now turn to how Planact challenges the state's cost recovery policies by assessing the implementation of a People's Housing Process (PHP)[5] in Vosloorus, located 27 km to the south-east of downtown Johannesburg within the jurisdiction of the Ekurhuleni Metropolitan Municipality.

Within the PHP policy remit, local authorities are obliged to collaborate with established NGOs and community leaders to facilitate housing delivery in poor urban districts. It was on this basis that the Ekurhuleni Metropolitan Municipality appointed Planact, in 2000, as the project manager for the Vosloorus PHP project (Robbins and Aiello 2005). Planact's legitimacy to act in a poor community was thus bestowed by the local government. The overarching aim of this project was to facilitate the in-situ upgrading of 1,000 owner-occupied residential units while simultaneously facilitating capacity-building programmes for the Vosloorus Steering Committee (VSC), a body formed by community leaders.

Proposals were prepared by Planact, in collaboration with the VSC, and presented to senior provincial and local government officials with the hope of securing the provincial government's PHP facilitation and establishment grants. Once the first subsidies were approved, management agreements were signed between all four stakeholders, namely, the provincial and local state, the VSC, and Planact. The local authority, as the account administrator, also provided Planact with additional funds so that it could pay community-based contractors (Planact 2003). Construction began in November 2002 and by September 2003 all 250 units of Phase I were completed. Additionally, Planact facilitated an on-site building training programme, and 16 community-based contractors completed skills training initiatives during the first construction phase. Phases II and III of the project were completed in May 2005 and in March 2006 respectively. But these phases were hindered by difficulties in procuring local (community-based) contractors, the discovery of geotechnical constraints that increased the cost of construction, and payment delays from the provincial and local state. While training programmes facilitated during both construction periods enjoyed relatively high levels of participation, Planact had to pre-finance these initiatives (Planact 2005). Overall, 750 new housing units were built, 577 beneficiaries of housing units were trained in construction skills, and 267 residents managed to secure temporary employment during the five-year project period (Robbins and Aiello 2005). Government officials, therefore, claimed that the success of this project was a result of their collaborative approach to planning (participant interview 2008).

However, collaborative commitments to this PHP were lacking. The Metropolitan Municipality refused to provide the necessary bridging finance for construction after the discovery of geotechnical constraints, and, as the financial administrator of the project, they continually delayed the release of working capital to Planact so that contractors could be paid on time (participant interview 2008). Financial delays, in turn, began to erode the VSC's trust in Planact as a credible service provider, particularly during the third phase of the project when Planact could no longer afford to supply requisite capital outlays from its own coffers. According to Gladys Macala, a member of the VSC:

> Initially the subsidy was $1,450 for each unit, but the amount [was] increased to $2,400. Our biggest problem with Planact is that we do not know what [happened] to the balance [of the money] after the completion of each unit.
> (Macala, cited in Seokoma 2006: 407)

Problems resulting from financial delays were not communicated to the VSC, and a consensus on how to deal with identified problems was not reached (Seokoma 2006). Stakeholders, therefore, failed to act as "a public body" (Habermas 1974) and a breakdown of trust resulted in a breakdown of Planact's power to plan on behalf of the VSC. Without community members' explicit support, Planact was also no longer in a position of power to challenge the local state's myopic view of the PHP. Rather, the Ekurhuleni Municipality was now in a position of power to dictate their preferred contractors to hasten delivery processes. Preferred contractors were not based in the community, as originally envisaged, instead they were professional outsiders who "routinely discredited the PHP process by usurping decision-making processes" (Planact, cited in USN 2003: 63). Further, the VSC and community-based contractors were excluded from decision-making processes, thereby nullifying the local state's commitment to collaborative planning (Planact, cited in USN 2003: 63). As a consequence of the local state's power to accelerate the delivery process, more than 50% of housing recipients interviewed by Robbins and Aiello (2005) complained about the poor construction quality of their homes.

The provincial government also exerted its power by electing not to reappoint Planact to complete the fourth and final phase of this project, and by shifting the financial administration of the Vosloorus PHP away from the Ekurhuleni Municipality. Instead, the Provincial Department of Housing created a non-profit organization, the Xhasa Accounting and Technical Centre, to act as an implementing and financial management agent for the final phase (Mokonyane 2005). In the process, the state legitimized its own non-profit organization as an agent to act on behalf of poor communities. However, Planact's official retreat from the Vosloorus PHP at the end of March 2006 left the VSC anxious about the future of the project: "we are empowered to take the project forward, but we still need Planact to play a watchdog role" (Macala, cited in Seokoma 2006: 407). By creating Xhasa, a "watchdog role" was, officially, relinquished. From a collaborative planning standpoint, this PHP was unjust, and, as a result, the outcome was unjust. Public decision-making processes were not inclusive and transparent, and power differences between the four stakeholders were not resolved.

Still, the creation of Xhasa, according to the provincial government, will fulfil the National Department of Housing's policy mandate to abolish PHP programmes in favour of Community-Driven Housing Initiatives (CDHIs) (Mokonyane 2005). Consequently, the People's Housing Partnership Trust (PHPT), the institutional home of the PHP, is in the process of being dismantled. While NGOs welcome the state's move to dismantle the PHPT, their reflexive PHP insights, based on local knowledge and experience, have been silenced (participant interview 2008). Development orientated NGOs, including Planact, collectively lobbied the state for more holistic and context-specific approaches to affordable housing delivery than the state's envisaged capital-saving and generic programmes (Masiwa 2007). However, further holistic approaches will require greater state funding and time for effective skills training (Masiwa 2007). For the state, such requirements are seemingly counterproductive as they continue to view future CDHIs in terms of finances alone, as:

> a way of saving $331 per unit by using voluntary labour. Beneficiaries of low cost housing units will thus become nothing more than cheap labour in projects planned by outsiders.
>
> (Eglin 2006: 95)

Planact's collaboration with other development orientated NGOs around the future of CDHIs is demonstrated here, but, for now, their collective attempts to change housing policies have been largely suppressed by the state's dominant discourses. The state's cost recovery agenda remains unaltered, and an "indirect structural hole" (Burt 1992) has been created between the state and civil society through the replacement of Planact (and other NGOs) with Xhasa. This case study therefore illuminates Emirbayer and Sheller's (1999) concern that if civic associations are too densely woven around a single issue, they may fail to influence targeted institutions.

Mobilizing insurgency

By the late 1990s the City of Johannesburg had accumulated a $53 million overdraft, due to a decade of bad management. In an attempt to ameliorate this financial crisis, the City adopted a World Bank model that included, amid other structural adjustment policies, "corporatizing" the municipality's utility services (Khan 2000). In response, the Anti-Privatisation Forum (APF) was established in July 2000 by activists involved in the struggle against the municipality's cost recovery model (Ngwane 2003). These activists constituted a disparate group of individuals who had been cast aside by the ANC, the South African Communist Party (SACP), and the Congress of South African Trade Unions (COSATU), and who, up to then, had not found a loud enough voice to publicize their plight (Buhlungu 2006). With the support of the Municipal Workers' Union and the National Council of Trade Unions, the APF's struggle against the City of Johannesburg eventually forced the municipality to reconceptualize its cost recovery policy, which symbolized a decisive and symbolic, albeit short-lived, victory for the fledgling Forum. However, the City's utility services remained "corporatized," a significant number of public sector employees were retrenched, and the cost recovery policy survived in a new incarnation, as a long-term economic growth policy known as "Jo'burg 2030" (CoJ 2002).

By embracing "insurgency" as an oppositional strategy to facilitate social transformation, the APF hopes to "unite struggles against privatisation by linking workers' struggles for a living wage with community struggles for housing, water, electricity, and fair rates and taxes" (APF: www.apf.org.za). Its campaigns for the provision of public services, affordable housing, and the prevention of job losses provide the Forum with "perhaps the most powerful weapon against the ANC, as it enables activists to expose the failure of the government on issues that are closest to people's hearts" (Buhlungu 2006: 75). The APF is highly critical of the present political and socioeconomic situation that is creating systemic inequalities. It, therefore, fights against the perceived "sell-out" leadership of the ANC/SACP/COSATU Tripartite Alliance fold.[6] The APF also serves as one of the few civic associations available to the most vulnerable residents of the greater Johannesburg region, and it sees itself as "based in communities and accountable to this class" (Ntuli and Ngwane 2006: 116). Its legitimacy to act on behalf of marginalized communities is thus "created and defended from below." It also perceives itself, and not the state, as an agent for change in the ongoing struggle for an equitable public domain.

The APF is structured as an autonomous umbrella forum constituting four political (mass-based) groups and 21 (community-based) affiliates. Collectively, they represent what social transformation theorists would call "active planners for themselves." The APF and its affiliates are also examples of Hadenius and Uggla's (1996) multi-layered organizations, creating a sense of solidarity and trust between affiliated networks. Pensioners, students, and women from apartheid-segregated black townships, informal settlements, and poor inner-city neighbourhoods have a significant presence within the APF's affiliated networks (participant interview 2008). Representatives from these affiliates constitute the Forum's governing council, whereas one delegate from each mass-based organization is a member of its coordinating committee (participant interview 2008). In accordance with Hadenius and Uggla's (1996) analytical prescriptions, the APF also facilitates actions for including democratic norms and its leadership is accountable. It, therefore, promotes transparency in associational life.

The Forum's largest supported demonstration to date was held between 26 August and 4 September 2002 in protest of the United Nations' sponsored World Summit on Sustainable Development. Hailed by the Forum as "a watershed event in post-apartheid South Africa," demonstrations at the Johannesburg World Summit undermined the official Tripartite Alliance (APF 2003: 1).

The APF is, however, not an organization. Rather, it styles itself as a "'new workers' movement" that deliberately operates outside of the auspices of former liberation movements and other civic associations of the struggle era (Buhlungu 2006). To survive, the APF must "continually engage in attracting and holding supporters by reframing [its] undertakings and actions" (Drakeford 1997: 74). It also has to allow its affiliates to operate in an autonomous manner. Its self-identified position is one that represents new constituencies facing new struggles in post-apartheid South Africa. It is, therefore, deeply suspicious of NGOs that do not overtly fight against the state's structural adjustment policies. Instead, "these NGOs are in the service of the state, and, as a result, they simply humanize capitalism" (Ntuli and Ngwane 2006: 116). While these claims may be valid, the Forum may equally view the presence of other NGOs as a threat to their power in local settings, as Ntuli and Ngwane go on to argue that "NGOs compete with socio-political movements for influence among the poor, women, youth, and workers" (Ntuli and Ngwane 2006: 116).

A struggle for power to speak on behalf of communities was evident at a Bekkersdal mass meeting held in August 2006, and Johannesburg branch of the SACP. The aim of this meeting was to raise concerns around residents' access to basic public services. The SACP perceived its role as leading the struggle, whereas the APF was invited to the meeting to introduce itself to members of the Bekkersdal community. However, the APF objected to the SACP, as a political party, taking the struggle forward on the grounds that such an approach would limit and divide resident participation. As an alternative, the APF offered its services to facilitate "a unified struggle so that full participation from all members of the community may be secured" (APF 2007: 6). This community-building offer was seemingly welcomed by community members as they "vowed to tackle the problem of non delivery through a structure that is more open to all members of the public" (APF 2007: 6). The meeting was thus deemed "a resounding success" by the APF, since the power to speak on behalf of community members was snatched away from the SACP and was now vested in the APF.

Besides creating a continual need to reaffirm and secure its power in local settings, the APF's flexible structure is also a weakness in that it limits the ability of the Forum to be decisive (Buhlungu 2006). For example, some members question the legitimacy of the state and opt to boycott formal state election processes, whilst others lobby for involvement in these processes. Consequently, the APF did not contest the 2006 local government elections, but individual affiliates, like the Soweto Electricity Crisis Committee, publicly rejected election outcomes (participant interview 2008). A flexible structure means that the Forum has to contend with contradictions within its ranks which may, in the future, erode structural ties between the APF and its affiliated subgroups. Similarly, the size of and active contributions made by affiliated subgroups are extremely uneven, thereby diminishing structural ties between subgroups (participant interview 2008).

Campaigns against the City of Johannesburg in 2000 and at the World Summit are two examples of how the APF mobilizes against the state, whereas a staged picket that lasted for three months in front of BHP Billiton's head office, for the release of pension funds from retrenched employees, represents the APF's "insurgency" against the market (participant interview 2008). Still, the Forum's capacity to change state policies is limited. Buhlungu (2006) and Hamilton (2002) propose that one reason for this limitation stems from a combination of naivety, arrogance, and poor political judgement on the part of some leading activists in the APF. "There is a notion that because their cause is a just one, the masses [will] come by their thousands to join the APF" (Buhlungu 2006: 78). Ephemeral victories achieved in 2000 and at the World Summit reinforce this view. However, at these, and other, campaigns sustainable alliance building is neglected in favour of political posturing that antagonizes not only the leadership of the Tripartite Alliance but also other civic associations with whom bridging initiatives may be fostered. The APF, therefore, fails to facilitate Friedmann's (1998) respectful process of "social/mutual learning." Instead, barriers, or "direct structural holes" (Burt 1992), tend to develop between the APF and other associations, and these discourage the open-endedness required for collective action and for effective engagement in public policy (Emirbayer and Sheller 1999). Additionally, regardless of the fact that the APF styles itself specifically as a "'new workers' movement," it has no relations with established workers' unions (Buhlungu

72 Tanja Winkler

2006). Consequently, it fails to make any serious inroads in the public domain by organizing employed workers. Workers' unions (like the Municipal Workers' Union and the National Council of Trade Unions) that once supported the APF, no longer do so as they believe this movement to be "hijacked" by a few activists (Buhlungu 2006).

The task of building a counter-hegemonic movement in a post-apartheid society will, therefore, entail more than pointing out and fighting against the shortcomings of the existing order. Mobilized actions need to be turned into equitable and redistributive planning policies and implementable outcomes that move beyond mass demonstrations and pickets alone. This is not to devalue the APF's contributions to "a thousand tiny empowerments" (Sandercock 1998), or to suggest that the APF fails to have any impact on the state. Rather, and in accordance with the social transformation position, equitable outcomes will require alternative responses to the status quo by establishing policies for structural transformation.

Pursuit of a Just City by the Centre for Applied Legal Studies

The Centre for Applied Legal Studies (CALS) was founded in 1978, within the Faculty of Law at the University of the Witwatersrand, at a time when public interest law groups did not exist in South Africa. Its initial aim of promoting human rights through research and education soon expanded to include a wide range of public impact litigations and extra-curial mediations. It operates as "an independent, non-profit organisation [that is] committed to promoting democracy, justice, and equality in South Africa" (CALS: http://web.wits.ac.za/Academic/Centres/CALS). To achieve this, CALS's self-identified position aims to "lever intellectual, legal, and political skills in pursuit of human rights by building relationships with diverse communities most affected by the deprivation of rights" (CALS: http://web.wits.ac.za/Academic/Centres/CALS).

Only one of the Centre's human rights pursuits will be evaluated: its defence of evicted tenants from inner-city buildings. Ultimately, CALS hopes to capacitate inner-city residents in public decision-making processes, so that the local state may become more attentive to their needs. It makes use of the law to challenge and change the City's eviction and regeneration policies, and to politicize tenants to campaign against the City when conventional approaches to public participation (through formal IDP processes) seem to fail. The Centre does, however, value participatory approaches to decision making, provided that processes are inclusive and transparent (http://web.wits.ac.za/Academic/Centres/CALS). This suggests that CALS's rights-based approach sometimes entails mobilized action, while, at other times it entails collaborative action. Different approaches used by civic associations to reshape the public domain, therefore, interact and reinforce each other. An alleged violation of human rights legitimizes the Centre to act on behalf of evictees, but tenants also trust CALS to act in their interests, which, in turn, empowers CALS's rights-based approach. The Centre's organizational structure and actions, therefore, include democratic norms, as envisaged by Hadenius and Uggla (1996). Before assessing CALS's policy impacts, reasons for why tenants are being evicted from inner-city buildings need to be contextualized.

Following the first local democratic government elections held in December 2000, "inner city regeneration" was declared to be one of six mayoral priorities by the newly appointed executive mayor of Johannesburg. At the same time, policy makers formulated a competitive economic growth policy with a thirty-year horizon known as "Jo'burg 2030." This policy laid the foundation for the first Inner City Regeneration Strategy with an overarching goal "to raise and sustain private investment leading to a steady rise in property values" (CoJ 2003: 2). Achieving this goal required an overt demonstration by the municipality to accommodate investor needs through, among a host of other incentives, facilitating the Better Buildings Programme (BBP). The BBP allowed the City to write off arrears on identified "bad buildings" and to transfer the ownership of these buildings to private-sector developers for renovation. "Bad buildings" were abandoned by their owners, but they were occupied by residents who were unable to find affordable accommodation through the private housing market. While conditions in these buildings were abysmal, Wilson and du Plessis from CALS argued that "bad buildings housed the poorest and most vulnerable residents of the inner city" (Wilson and du Plessis 2005: 3). Conversely, the City's BBP manager was of the opinion that:

> Developers want empty occupation because they can't fix a building unless we get rid of the people. For us the big issue is to decant existing tenants to other buildings, because judges often only grant eviction notices [based on] alternative [tenant accommodation]. And that's a tough one because the City doesn't always have alternatives.
>
> (Interview, BBP manager, 2004)

At least 250 "bad buildings" have been identified by the City for its BBP, and "getting rid" of current occupiers means evicting them: "In doing so, the City exercises its power in terms of the National Building Regulations and Building Standards Act, of 1977, which empowers a local authority to order the evacuation of a property that poses a threat to the health and safety of those occupying it" (Wilson and du Plessis 2005: 4). However, these buildings are currently occupied by approximately 60,000 residents, and capital investments required to repair "bad

74 Tanja Winkler

buildings" exclude many evictees from being able to afford rehabilitated building rentals (Wilson and du Plessis 2005: 4). Since 2002, 125 inner-city buildings have been cleared, resulting in the eviction of thousands of residents without the City providing suitable alternative accommodation for evictees (participant interview 2008):

> This is immoral, because it further victimises people who, through no fault of their own, live in unsafe and unhealthy conditions. It is impractical, because [residents], once evicted, invariably occupy other slum properties in the inner city. It is also unconstitutional [in terms of] Section 26 of the Constitution
> (Wilson and du Plessis 2005: 6)

Wilson and du Plessis go on to argue that the City employs legislation passed under apartheid to secure eviction orders from the High Court, and that it acts in contravention of the Promotion of Administrative Justice Act, 2000/2002, by not convening hearings with tenants of "bad buildings" before it takes the decision to evict: "Assuming that evictions are absolutely necessary, the City should instead use the Prevention of Illegal Eviction and Unlawful Occupation of Land Act (the 'PIE' Act), which was passed in 1998 specifically to ensure that mediation takes place before evictions take place" (Wilson and du Plessis 2005: 6). The PIE Act equally requires the state to provide tenants with suitable alternative accommodation. Armed with post-apartheid legislation, CALS seeks not only to defend the rights of residents to live in the inner city, but also to pressure the municipality to change its market-driven regeneration policies that fail to promote the equitable redistribution of resources. In the process, CALS actively engages in the planning field by fighting for aspects of Fainstein's Just City, where "the concept of justice [is] situated, and theorizing about the just city actually means theorizing about justice within [a] particular urban milieu" (Fainstein 2005: 126). Accordingly, "the City of Johannesburg's responsibility for fulfilling spatial and socioeconomic rights [needs to] begin where the market fails" (Wilson 2006: 8). In this Just City-style pursuit for spatial, political, and socioeconomic equity, CALS is supported by a small but highly cohesive subgroup of planners including a planning consultancy, Development Works, and an international planning and human rights NGO, the Centre on Housing Rights and Evictions (COHRE). CALS and its cohesive subgroup, therefore, represent Hadenius and Uggla's (1996) autonomous and multi-layered organizations that promote transparency and accountability in civil society. Strong structural ties between members of this cohesive subgroup allow CALS to engage in a host of litigations against the local state (participant interview 2008).

One of these drawn-out litigations began in April 2005, when the City of Johannesburg (the applicant) sought to evict 872 residents from five inner-city buildings (the respondents). The municipality, however, "refused to offer the respondents alternative accommodation, as the apartheid Building Standards Act contains no such provision" (CALS 2005: 1). The Centre defended respondents against the eviction on the grounds that they were entitled to the PIE Act, and that procedures of the Promotion of Administrative Justice Act were not followed. They also argued that the Building Standards Act was in contravention of Section 26(3) of the Constitution and Chapter 12 of the National Housing Code, which stipulates that municipalities are responsible for implementing an Emergency Housing Programme (EHP) for evictees of unsafe buildings. Additionally, CALS made use of this opportunity to have the Inner City Regeneration Strategy declared unconstitutional (participant interview 2008).

Almost a year later, in March 2006, the Johannesburg High Court effectively banned the municipality from evicting residents of "bad buildings" without implementing Chapter 12 of the National Housing Code (JHC 2006, Court Order 3). This ruling forced the municipality to promulgate a Human Development Strategy (CoJ 2006) to tame its Jo'burg 2030 economic growth policy (participant interview 2008). Nonetheless, the City appealed the High Court judgment in the Supreme Court of Appeals with the aim of protecting its Inner City Regeneration Strategy, and the Supreme Court ruled in favour of the municipality in April 2007. Residents were ordered to vacate premises (SCA 2007). The Supreme Court held that occupiers of "bad buildings" did not have a constitutional right to alternative housing in the inner city (SCA 2007). However, the Supreme Court also ordered the City "to provide those residents who needed it with alternative shelter [on the urban edge]" (SCA 2007: 32).

> CALS' in-depth research with Development Works and COHRE shows that affected residents are too poor to travel from far-flung settlements to their work places in the inner city. To relocate them to places far away from the city centre will have disastrous implications for the survival strategies of many families. This judgment appears to condone the City's decision to exclude the poor from its Inner City Regeneration Strategy. We are studying the judgment carefully, and we will appeal [this ruling] to the Constitutional Court
>
> (Wilson 2007: 2)

With CALS's legal support, residents lodged a suspension and repeal of the Supreme Court's eviction orders in the Constitutional Court, the highest court in South Africa (Wilson 2007). In anticipation of the Constitutional Court's likely ruling in favour of poor inner-city residents, in May 2007 the municipality swiftly formulated and publicly launched its most recent regeneration framework: the Inner City Regeneration Charter (CoJ 2007b), inclusive of Chapter 12 of the National Housing Code. The City's affidavit submitted to the Constitutional Court

76 Tanja Winkler

strategically outlined the aim of its new regeneration policy, which, for the first time, entails converting some BBP buildings into emergency shelters for evictees from "bad buildings" (CoJ 2007a). Despite these mitigation attempts by the municipality, the Constitutional Court overturned the Supreme Court ruling in February 2008, and the municipality was ordered to provide "the occupiers of ['bad buildings'] with affordable and safe accommodation in the inner city where they may live secure against eviction" (RSACC 2008: 2).

In the eyes of many, CALS had won: "For the applicants, as well as for poor inner city residents more generally, the judgement represents a victory" (Dugard, 2008: 1). However, while CALS's rights-based approach forced the City to promulgate a Human Development Strategy and a new Regeneration Charter, this approach failed to secure residents' involvement in policy making. Residents' voices remain excluded from these new policies. Nonetheless, CALS, in collaboration with Development Works and COHRE, will continue to press the City to formulate and implement equitable policies for the inner city (participant interview 2008). But equitable policies will need to be judged on their substantive content and the process employed by the City in formulating them if a Just City is sought.

Concluding remarks

While the original aim of this chapter was to establish the extent of civic associations' contributions to local planning policies, and their capacity to change the state's cost recovery strategies, the three cases under study also highlight the particular circumstances or conditions which call for a response from associations of civil society. Accordingly, research findings demonstrate that planning in Johannesburg is embedded in political processes based on asymmetrical power relations that raise serious constraints for civil society groups to change this field. Successes achieved in the Vosloorus PHP (the construction of 750 new housing units, training programmes, temporary employment opportunities) are undermined by an unjust process, and by the state's power to curtail associations' contributions to equitable planning policies. Similarly, while the actions adopted by the APF and CALS forced the City of Johannesburg to reconceptualize its cost recovery and urban regeneration policies, the new documents fail to include civil society in the formulation of future policies. Civic associations remain outside of public decision-making processes, which hampers the extent of their contributions to local planning policies and their capacity to change the state's dominant discourse. Such findings, however, should not diminish civil society's role as a necessary and valuable agent of change. Rather, findings spotlight that change is a slow process, and that an equitable policy context requires sustained pressure on the state to be more responsive to the concerns of disadvantaged social groups. Findings also demonstrate how the activities of different associational groups are shaped by the local socio-political context, which, in turn, elicits equally different agencies and actions in response to this context.

Such findings, then, raise questions about certain aspects of both Emirbayer and Sheller's (1999) and Hadenius and Uggla's (1996) analytical frameworks. For example, why is a high degree of autonomy necessary to influence state actors and actions? Planact, the APF, and CALS all operate as autonomous entities. Nonetheless, this attribute translates neither into policy-making efficacy, nor into a balance between contending power centres. The creation of Xhasa to suppress civic associations' "watchdog" roles, and the City of Johannesburg's reactionary, but exclusionary, response to the mobilized and legal actions of the APF and CALS, undermines the idea of an open exchange between the state and civil society which can influence state actors and actions. Similarly, why is a multi-layered organizational structure, characterized by small homogenous groups, considered a determinant of successful policy impact? Regardless of the fact that all three associations conform to Hadenius and Uggla's "multi-layered" prescription, and that they facilitate organizational structures for including democratic norms and accountable leadership, all of them remain excluded from public decision-making processes. Conventional policy-making approaches have not made planning more democratic, and state actors continue to shape new planning policies and practices. It is, therefore, unclear why a multiplicity of associations is considered necessary but a diversity of strategies and actions to challenge the state is not. The framework seems to provide intellectual support for this approach but without building a compelling argument in its defence. Findings also suggest that civil society's legitimacy to act in the public domain serves mainly to affirm its relative autonomy vis-à-vis the state and not as a policy transformation determinant. The fact that legitimacy was bestowed by the state in the Vosloorus case, as opposed to by residents in the APF's and CALS's campaigns, seems irrelevant to policy contributions.

However, the case studies show that the contribution made by civic associations to democracy is not only manifest in the extent of their ability to influence policy. If measured on the basis of this criterion alone, their impact is judged to be very minimal, as demonstrated. But the evidence also suggests that their contribution to democracy extends to their ability to foster democratic values and to build leadership capacity. In other words, their capacities to enhance pluralism and democratic values are as important as their ability to influence decision making and demand accountability from state actors. Such findings question inflated expectations of the policy-influencing potential of civil society and highlight the mutually reinforcing dimensions of its contribution to democracy (Robinson and Friedman 2007).

Civic associations also generate effective awareness of persistent inequalities in the greater Johannesburg region, and perform the prime democratic function of giving citizens a voice, even if this voice is currently not captured in public policy (Robinson and Friedman 2007). In the future, if civil society sustains its pressure on the state to act more responsively to citizens' concerns, a more inclusive and equitable planning approach may be realized. In accordance, then, with a collaborative, a transformative, and a Just City position, civil society is valuable for its ability to respond to a situated context. Said differently, civil society is generally shaped by a need to respond to local socio-political realities, and this response calls for flexible and pluralistic strategies and actions to counter inequalities. Similarly, all three theoretical positions aim to promote greater spatial, political, and socioeconomic equity in the public domain, even if the means of achieving equity differs from one theoretical position to the next. An equitable city is then "as likely to [be] derived from court decisions as from deliberative democracy [or mobilized actions]" (Fainstein 2005: 126).

Acknowledgements

Research findings are based on an ongoing and wider study of local civil society organizations. This wider study is financially supported by the National Research Foundation. The author would also like to thank Claire Bénit-Gbaffou, Kiera Chapman and all four anonymous referees for their insightful and invaluable comments.

Notes

1   The next general election will take place in April 2009.
2   The other seven development orientated NGOs include the Development Action Group, the Foundation for Contemporary Research and the Isandla Institute, which are all based in Cape Town, Afesis-Corplan, based in East London, the Built Environment Support Group, based in Pietermaritzburg, Habitat for Humanity South Africa, based in Tshwane, and the Urban Services Group, based in Port Elisabeth.
3   See, for example, S. Chambers and W. Kymlicka (2002); J. Cohen and A. Arato (1992); S. Kaviraj and S. Khilani (2001); J. Keane (1988); K. McCarthy (2001); C. Mouffe (1992).
4   These "like-minded" organizations include Afesis-Corplan, the Centre for Civil Society, Civicus, COPE Housing, the Foundation for Contemporary Research, the Institute for Democracy in South Africa, and the South African Cities Network (participant interview 2008).
5   The National Department of Housing introduced the PHP policy in 1998 as one of a number of affordable housing delivery mechanisms. It specifically involves "beneficiary communities," as contractors or in associated services, in the housing

delivery process. By March 2004, subsidies approved for the delivery of houses via the PHP policy totalled 272,165. This represented 11.17% of all of the 2,436,404 housing subsidies approved between 1994 and 2004 (Robbins and Aiello, 2005).

6 The Tripartite Alliance refers to the South African state's current parliamentary wing of a three-part alliance between the African National Congress (ANC), the South African Communist Party (SACP) and the Congress of South African Trade Unions (COSATU).

References

Alfasi, N. (2003) "Is public participation making urban planning more democratic? The Israeli experience," *Planning Theory and Practice*, 4: 2, 185–202.

Alnoor, E. (2003) *NGOs and Organizational Change*, Cambridge: Cambridge University Press.

APF. Available at http://apf.org.za/.

APF (Anti-Privatisation Forum) (2003) Narrative Report, Johannesburg, in-house publication.

APF (2007) Narrative Report, Johannesburg, in-house publication.

Ballard, R., Habib, A. and Valodia, I. (2006) *Voices of Protest: Social Movements in Post-Apartheid South Africa*, Scottsville: University of KwaZulu-Natal Press.

Beard, V.A. (2003) "Learning radical planning: the power of collective action," *Planning Theory*, 2: 1, 13–35.

Bond, P. (2005) *Fanon's Warning: A Civil Society Reader on the New Partnership for Africa's Development*, Trenton, NJ: Africa World Press.

Buhlungu, S. (2006) "Upstarts or bearers of tradition? The anti-privatisation forum of Gauteng," in: R. Ballard, A. Habib and I. Valodia (eds) *Voices of Protest: Social Movements in Post-Apartheid South Africa*, pp. 67–87, Scottsville: University of KwaZulu-Natal Press.

Burt, R. (1992) *Structural Holes*, Cambridge, MA and London: Harvard University Press.

CALS. Available at http://web.wits.ac.za/Academic/Centres/CALS/.

CALS (Centre for Applied Legal Studies) (2005) City of Johannesburg (applicant) and the occupiers of San Jose; and the occupiers of ERF 378 to 381, Berea Township (respondents), Litigation News, April. Available at http://web.wits.ac.za/Academic/Centres/CALS (accessed 22 February 2008).

Campbell, H. (2005) "Planning and institutionalism: what value in the new institutionalism?" Paper presented to the *Association of Collegiate Schools of Planning* (ACSP) conference, October 27–30, Kansas City, Missouri.

Castells, M. (1983) *The City and the Grass-Roots: A Cross-Cultural Theory of Urban Social Movements*, London: Edward Arnold.

Chambers, S. and Kymlicka, W. (2002) *Alternative Conceptions of Civil Society*, Princeton, NJ: Princeton University Press.

Charlton, S. and Gotz, G. (2007) "Thinking about Johannesburg," *Planning Theory and Practice*, 8: 3, 383–412.

Cohen, J. and Arato, A. (1992) *Civil Society and Political Theory*, Cambridge, MA: MIT Press.

CoJ (City of Johannesburg) (2002) *Jo'burg 2030 Vision*, Johannesburg: City of Johannesburg Publication.

CoJ (2003) *Inner-City Regeneration Strategy* (ICRS), Johannesburg: City of Johannesburg Publication.

CoJ (2006) *Human Development Strategy* (HDS), Johannesburg: City of Johannesburg Publication.

CoJ (2007a) Affidavit submitted to the Constitutional Court of South Africa, Constitutional Court Records, CCT 24/07, ZACC 1, October.

CoJ (2007b) *Inner City Regeneration Charter*, Johannesburg: City of Johannesburg Publication.

Drakeford, M. (1997) *Social Movements and Their Supporters: The Green Shirts in England*, London: Macmillan.

Dugard, J. (2008) "Constitutional court overturns Supreme Court of Appeal decision to grant an eviction order," CALS Media Statement, 19 February.

Eglin, R. (2006) "The evolution of the people's housing process," *Local Government Transformer*, p. 95. Available at www.afesis.org.za/index.php?option=com_content&task=view&id=240&Itemid=95 (accessed 16 February 2008).

Emirbayer, M. and Sheller, M. (1999) "Publics in history," *Theory and Society*, 28: 1, 145–197.

Fainstein, S. (2000) "New directions in planning theory," *Urban Affairs Review*, 35: 4, 451–478.

Fainstein, S. (2005) "Planning theory and the city," *Journal of Planning Education and Research*, 25: 121–130.

Forester, J. (1999) *The Deliberative Practitioner*, Cambridge, MA: MIT Press.

Friedmann, J. (1998) "Claiming rights: citizenship and the spaces of democracy," *Plurimondi*, 2, 287–303.

Friedmann, J. (2005) "Civil society revisited: travels in Latin America and China," in: M. Keiner (ed.) *Managing Urban Futures: Sustainability and Urban Growth in Developing Countries*, pp. 127–142, Burlington, VT: Ashgate.

Habermas, J. (1974) "The public sphere," *New German Critique*, 3, 49–55.

Hadenius, A. and Uggla, F. (1996) "Making civil society work, promoting democratic development: what can states and donors do?" *World Development*, 24: 10, 1621–1639.

Hamilton, W. (2002) "The challenges facing the working class in South Africa, Khanya," *A Journal for Activists*, 2, 15–18.

Healey, P. (1997) *Collaborative Planning*, London: Macmillan.

Healey, P. (1999) "Institutional analysis, communicative planning, and shaping places," *Journal of Planning Education and Research*, 18: 2, 111–121.

Holston, J. (1998) "Spaces of insurgent citizenship," in: L. Sandercock (ed.) *Making the Invisible Visible*, pp. 37–56, Berkeley, Los Angeles & London: University of California Press.

JHC (Johannesburg High Court) (2006) City of Johannesburg versus Rand Properties (Pty) Limited and others, the High Court of South Africa (Witwatersrand Local Division), Case Numbers: 04/10330; 04/10331; 04/10332; 04/10333; 03/24101; 04/13835, 3 March, Johannesburg. Available at www.saflii.org/za/cases/ZAGPHC/2006/21. html (accessed 1 March 2008).

Kaviraj, S. and Khilani, S. (2001) *Civil Society: History and Possibilities*, Cambridge: Cambridge University Press.

Keane, J. (1988) *Democracy and Civil Society*, Cambridge: Cambridge University Press.

Khan, F. (2000) *iGoli 2002: is the future private? Green Left online*. Available at www.greenleft.org.au/2000/ 415/23093 (accessed 16 February 2008).

Lewis, J. and Weigert, A. (1985) "Social atomism, holism, and trust," *Sociological Quarterly*, 26: 4, 455–471.

Mangcu, X. (2008) *To the Brink: The State of Democracy in South Africa*, Scottsville: University of KwaZulu-Natal Press.

Martens, K. (2005) "Participatory experiments from the bottom up: the role of environmental NGOs and citizen groups," *European Journal of Spatial Development*, 8, 1–20.

Masiwa, B. (2007) "Community-driven housing initiatives," *Local Government Transformer*, August/September, 52.

McCarthy, K. (2001) *Women, Philanthropy and Civil Society*, Bloomington, IN: Indiana University Press.

Mokonyane, N. (2005) Budget speech for the 2005/06 financial year by the Gauteng MEC for housing, Gauteng Online, 13 June. Available at www.gautengonline. gov.za/portal/dt?serviceAction=speechDetails&smID=GPGSPEECHANDMEDIA_ 25069 (accessed 16 February 2008).

Mouffe, C. (1992) *Dimensions of Radical Democracy*, London: Verso.

Ngwane, T. (2003) "Sparks in the Township," *New Left Review*, 22, 37–56.

Ntuli, L. and Ngwane, T. (2006) NGOs and the struggle of the working class: a socialist perspective, unpublished research paper, the Anti-Privatisation Forum, 29 June. Available at http://up191.apf.m2014.net/article.php3?id_article=116 (accessed 22 December 2007).

Pieterse, E. (2006) "Building with ruins and dreams: some thoughts on realising integrated urban development in South Africa through crisis," *Urban Studies*, 43: 2, 285–304.

Planact. Available at www.planact.org.za

Planact (2003) Annual Review, Johannesburg: in-house publication.

Planact (2005) Annual Review, Johannesburg: in-house publication.

Reardon, K. (1998) "Enhancing the capacity of community-based organizations in east St Louis," *Journal of Planning Education and Research*, 17: 4, 323–333.

Robbins, G. and Aiello, A. (2005) *Employment Aspects of Slum Upgrading: Practices and Opportunities Identified in Two South African Case Studies, Project Report*, pp. 1–72. Geneva: International Labour Office.

Robinson, M. and Friedman, S. (2007) "Civil society, democratization, and foreign aid: civic engagement and public policy in South Africa and Uganda," *Democratization*, 14: 4, 643–668.

RSACC (Republic of South Africa, Constitutional Court) (2008) Occupiers of 51 Olivia Road, Berea Township and 197 Main Street, Johannesburg versus City of Johannesburg, Rand Properties (Pty) Ltd, CCT 24/07, ZACC 1, 19 February 2008.

Sandercock, L. (1998) *Making the Invisible Visible*, Berkeley, Los Angeles: University of California Press.

Seokoma, B. (2006) A community's journey to housing and independence, *SANGONeT website*, pp. 407. Available at http://www.sangonet.org.za/portal/ index.php?option=com_content&task=view&id=4109&Itemid=407 (accessed 14 February 2008).

SCA (Supreme Court of Appeal of South Africa) (2007) City of Johannesburg v. Rand Properties (Pty) Ltd and others, SAC, Case No. 253/ 06, 2007(6), SA 417. Available at www.law.wits.ac.za/cals/Rand%20Properties/innercityjudgment_SCA.pdf (accessed 22 February 2008).

Urban Sector Network (USN) (2003) Planact self-assessment report, Experience of the Peoples Housing Process (PHP), USN in-house publication, Johannesburg, pp. 23–45.

Walzer, M. (1992) "The civil society argument," in: C. Mouffe (ed.) *Dimensions of Radical Democracy*, pp. 87–107, London & New York: Verso.

Watson, V. (2002) "The usefulness of normative planning theories in the context of sub-Saharan Africa," *Planning Theory*, 1: 1, 27–52.

Wilson, S. (2006) "Human rights and market values: affirming South Africa's commitment to socio-economic rights," *CALS News Letter*, September, 1: 1, 1, 6–8.

Wilson, S. (2007) "Inner city residents appeal Jo'burg inner city evictions decision," media release, 18 April.

Wilson, S. and du Plessis, J. (2005) "Housing and evictions in Johannesburg's Inner City." Paper presented at the Cities and Slums Workshop: *The World Housing Congress*, 25 September. Available at www.law.wits.ac.za/cals/WHC%20Cities%20and%20Slums%20Statement%5B1%5D.pdf (accessed 13 February 2008).

Yacobi, H. (2007) "The NGOization of space: dilemmas of social change, planning policy, and the Israeli public sphere," *Environment and Planning D: Society and Space*, 25, 745–758.

Chapter 5

Shocking the suburbs

Urban location, homeownership and oil vulnerability
in the Australian city

Jago Dodson and Neil Sipe

Energy security is receiving increasing attention from governments and scholars at the global and national scale. Petroleum security and rising fuel prices are a challenge for cities whose housing systems are highly dependent on automobile transport. This study assesses transport and socio-tenurial patterns within Australian cities to identify how the combined present and future effects of rising fuel costs, mortgage interest rates and general inflation will be spatially distributed. Using an 'oil vulnerability' assessment methodology based on Australian Census data, the study reveals broad-scale mortgage and oil vulnerability across the outer suburbs of Australian cities. The chapter concludes with some observations about spatially equitable policy responses to ameliorate the housing and urban impacts of rising petroleum costs.

Introduction

One of the most publicly discussed economic phenomena since early 2005 has been the marked rise in the global price of oil. Rising oil prices have translated into higher domestic petroleum fuel costs in most developed nations. In turn, these rising fuel costs have raised questions about their impact on urban areas, especially the heavily car reliant metropolitan regions of North America and Australia which depend on petroleum fuel for transportation.

Governments in these regions are increasingly recognising growing uncertainty over future petroleum production and supply as a major strategic issue (Australian Senate 2007; Government Accountability Office 2007). A new and uncertain geopolitics based on energy is emerging, between oil producing and oil consuming nations (Klare 2005; Wesley 2007). A flourishing body of commentary and popular

---

Submitted by the Australia and New Zealand Association of Planning Schools.

Originally published in *Housing Studies*, 23: 3, 337–410. © 2008 By kind permission of Taylor & Francis Ltd. www.informaworld.com.

discussion of the recent oil price increases and their potential future trajectory has begun to predict dire consequences for petroleum reliant cities, particularly those with extensive suburban areas (Heinberg 2004; Kunstler 2005). However, to date, the discussion about rising petroleum prices and their impacts on the urban and housing systems of developed nations has not been accompanied or buttressed by substantial empirical or scholarly analysis. Socio-economic impacts of rising fuel costs so far have received little attention from researchers; few housing scholars have begun to address these issues. The result is that the research community has limited understanding of how the socio-economic and housing effects of rising fuel costs may be distributed across cities, especially the dispersed and car-dependent metropolitan regions of North America and Australia.

There is a need for new research that begins the task of assessing urban socio-economic and housing vulnerability to rising fuel costs so that we can both understand how future impacts from insecure global energy markets may be distributed and begin to identify what research is necessary to better comprehend this problem. Such concerns imply a new front for research into socio-spatial differentiation in cities and potential housing impacts. Processes of urban socio-spatial differentiation have received a great deal of attention internationally over the past two decades, largely in response to global economic restructuring (Badcock 1984; Fainstein et al. 1992; Van Kempen & Marcuse 1997). These studies have typically focused on such phenomena as socio-economic polarisation (Hamnett 1996), socio-spatial polarisation (Wessel 2000), social segregation (Musterd & Ostendorf 1998) and social marginalisation (Speak & Graham 1999) as well as notions of socio-economic opportunity and vulnerability (Stimson et al. 2001). Housing has been the subject of much of this work (Somerville 1998) with tenure a central concern, whether in the form of public housing (Taylor 1998) or owner occupation (Randolph & Holloway 2005; Winter & Stone 1998). Meanwhile Graham and Marvin (2001) have demonstrated the centrality of infrastructure to social exclusion processes while a later but growing literature has emerged on the exclusionary dimensions of transport systems (Dodson et al. 2006; Lucas 2004).

The role of energy use in urban social processes has so far received little attention from scholars. Rising global oil prices and uncertainties over natural gas supplies in some regions such as Europe suggest that urban energy dependence deserves greater research attention. To date there has been little consideration of how urban energy consumption and transport systems may interact with housing systems and tenure structures to generate new or previously unappreciated forms of socio-economic vulnerability. Scholarly understanding of the urban spatial distribution of housing tenure and the vulnerability that differential distribution of housing debt may imply is also limited at present. There is scant research that examines directly the links between transport, energy use, tenure and socio-spatial vulnerability, particularly at the local scale.

Most current attempts to comprehend urban energy dependence focus on technological design or consumption at the individual household scale (Lenzen et al. 2004; Troy et al. 2003) and almost completely ignore the interaction of energy use patterns with broader urban socio-spatial processes, such as housing markets or socio-economic differences. Some research has used the notion of urban socio-economic vulnerability to assess prospective exposure to adverse social or economic changes in cities (Baum et al. 1999; Stimson et al. 2001), but most current vulnerability research is bio-physically focused and oriented towards developing nations (Bankoff et al. 2004). There is almost no literature on the links between socio-economic and energy vulnerability in developed nations. Achieving socially sensitised understanding of urban energy vulnerability is essential, given the mounting policy imperatives towards achieving urban sustainability.

Concerns about metropolitan and suburban sustainability must also attempt to intersect with questions about housing tenure because the structure of some cities is associated with particular tenure forms. This is especially so in societies where homeownership (including households with mortgages) comprises a dominant tenure, such as in Canada, the United States and Australia and where this is typically expressed through dispersed suburban dwellings (Randolph 2006). Weakness in US sub-prime mortgage finance markets recently has been the subject of economic concern there. What might be the consequences of rising household fuel costs for highly petroleum dependent dispersed suburban areas where income-based mortgage finance is the predominant method by which households achieve homeownership?

This chapter suggests that knowledge about current urban housing market patterns and conditions can assist us to assess the socio-spatial prospects for cities under conditions of increasingly insecure energy supplies. Scholarship has so far given little attention to the socio-economic impacts that rising fuel, mortgage and general consumption costs might have on households. This chapter begins the task of responding to these internationally important questions concerning the interaction of rising urban fuel costs with housing markets through an examination of the Australian case. In Australia rising fuel prices have produced upward inflationary pressures on mortgage interest rates and concerns have been raised about the level of socio-economic stress these twin pressures have generated. Given the socio-spatial patterning apparent in Australian cities, any impacts are likely to be felt differentially by different households in different urban locations.

The chapter assesses the distribution of mortgage and oil vulnerability in Australian cities in three ways. First, the chapter reviews the literature that reveals patterns of car dependence, mortgage tenure and household housing debt. Next, the chapter develops and deploys a methodology to assess household spatial vulnerability to longer-term rising fuel, mortgage and general consumption costs, using data for Australia's four largest cities. This assessment builds on a previous oil vulnerability index developed by the authors (Dodson & Sipe 2007). Third, the chapter discusses the empirical insights provided by this assessment and identifies a number of new imperatives for governments' urban and housing policies within the new context of urban oil vulnerability. The chapter concludes by discussing the international transferability of the study methods and findings to other jurisdictions and sets out the imperatives and opportunities for more research into this issue. The research methods and results that we present are highly relevant to other jurisdictions, particularly in North America where transport and tenure patterns are broadly comparable to those in Australian cities. The approach and insights also apply to Europe where governments in recent years have pursued policies of encouraging suburban homeownership, including deregulation of the mortgage finance sector. How energy insecurity might influence housing patterns in European cities is presently unknown and deserves attention from researchers.

Australian cities in an international context

Australian cities provide a useful set of cases for examining questions about urban oil and mortgage vulnerability. Australia's major cities are among the most car dependent in the developed world, ranking only behind US cities (Newman & Kenworthy 1999). Large urban spatial and structural disparities in levels of car dependence mean that residents of dispersed outer and fringe suburban locations are typically more reliant on motor vehicles than those in more compact inner and central zones (Mees 2000b; Morris et al. 2002; Newman et al. 1985), a phenomenon which is closely related to the adequacy of public transport (Newman 2006).

As in Canada, the USA and the UK, homeownership has been a favoured and dominant tenure in Australia for many decades (Kemeny 1983; Productivity Commission 2004). Many households aspire to purchase their housing and typically achieve this via mortgage finance. Urban housing market structures and spatial price gradients constrain home purchasers' locational opportunities such that modest income households seeking the preferred housing form of single-family dwellings are more likely to be allocated to lower-priced outer and fringe areas compared to higher income households (Burke & Hayward 2001). Such housing is generally located away from public transit nodes and activity centres (Wood et al. 2007). The dispersion of land uses and relatively poor public transport in Australian cities' outer suburbs means that these households are likely to require long journeys by private car to access services and employment, when compared to households in more central locations (Dodson et al. 2004; Maher et al. 1992; Mees 2000b).

Researchers have so far dedicated very little attention to investigating the implications of the intertwined geography of car dependence and mortgage tenure in Australian cities, and research that investigates such urban relationships is rare internationally. Yu's (2005) study of housing market patterns in Sydney examined mortgage, income and urban structural dimensions of homeownership, but not transport pressures. Newman, Kenworthy and Lyon's (1985; Newman et al. 1990) groundbreaking work on urban car dependence examined the energy consumption of suburban transportation, but not social or housing tenurial aspects of suburban transport and housing geography. Burnley et al. (1997) found that modest income households in Sydney who sought homeownership by moving to cheaper outer suburban locations experienced longer commuting times and became more car dependent after their move. These authors did not consider transport costs directly, but noted that those relocating "sacrificed much" (p. 1124) in the trade-offs they made to achieve homeownership.

In the UK, Burrows (2003) investigated the relationship between mortgage tenure and poverty, but did not include transport or urban structural variables. Some international research has partially investigated the implications for households of the combined costs of transport and home purchase. Krizek (2003) considered the notion of 'locational efficiency' in mortgage lending in the USA as a method of promoting 'smart growth', but so far this has received only limited attention from home finance lending institutions. Attempts in the US to limit urban sprawl and encourage greater use of public transport through 'smart growth' strategies have been controversial because of their potential impacts on housing affordability (Danielsen et al. 1999).

The debate over smart growth or compact city outcomes in Australia and North America has displayed limited sensitivity to tenurial differences, especially in terms of the impact on homeownership. Randolph (2006) has demonstrated that higher-density residential development in Australian cities is associated with markedly lower levels of homeownership than conventional lower-density suburban development. This has raised highly contentious questions for the dominance of 'urban consolidation' in Australian metropolitan policies. Given these dynamics, there is a major research imperative to begin to investigate transport and mortgage cost interactions, and to understand potential impacts and solutions both in Australia and overseas. The authors of this chapter are not aware of any recent Australian or international research that has specifically addressed the implications for households of the combined costs of transport and home purchase—a research void that is surprising, given the prominence of suburban homeownership and dependence on private motor vehicles.

The remainder of this chapter responds to these concerns by considering the distribution of mortgage and oil vulnerability in Australian cities in three ways. First, the chapter reviews the literature that reveals patterns of car dependence, mortgage tenure and household housing debt. Next, the chapter develops and deploys a methodology to assess household spatial vulnerability to longer-term rising fuel, mortgage and general consumption costs, using data for Australia's four largest cities. This assessment builds on a previous oil vulnerability index developed by the authors (Dodson & Sipe 2007). Then, the chapter discusses the empirical insights provided by this assessment and identifies a number of new imperatives for governments' urban and housing policies within the new context of urban oil vulnerability.

The chapter concludes by discussing the international transferability of the study methods and findings to other jurisdictions and sets out the imperatives and opportunities for more research into this issue. The research methods and results that have been produced are highly relevant to other jurisdictions, particularly in North America where transport and tenure patterns are broadly comparable to those in Australian cities. The approach and insights also apply to Europe where governments in recent years have pursued policies of encouraging suburban home-ownership, including deregulation of the mortgage finance sector. How energy insecurity might influence housing patterns in European cities is presently unknown.

Rising oil prices

The problem addressed here begins with rising global oil prices. The global price of oil has increased markedly since late 2004 (Figure 5.1). In mid-2006 crude oil cost approximately US$70 per barrel, which represents an increase of approximately 40 per cent from the May 2005 price of approximately US$50 per barrel and is higher in real terms than most of the period since 1950, except for the oil crisis of 1979–83. The current (2006) high prices appear to be the result of a combination of growing global demand relative to production, geopolitical insecurity and longer-term supply uncertainty.

5.1 Real and nominal global prices 1950–2006 in US dollars.
Source: US Government Accountability Office 2007: 11

Australian petrol prices are the third lowest in the OECD (in Australian dollars) and receive the fourth lowest levels of fuel excise tax (Australian Treasury 2006: 252–253). The rising global oil price has translated into similar petrol price increases in Australian cities, which exhibit little difference in average at-pump costs. In early 2005 petrol cost approximately AU$0.95/L in Sydney, but by late 2006 it had settled at around AU$1.38/L, equating to an increase of approximately 45 per cent (Motormouth 2006). Similar price escalation has been observed in other Australian cities as well as overseas (see Dodson & Sipe 2007: 39, Table 1).

Longer-term concerns surround the security of global petroleum supplies. Growing demand from developing economies and increasing concentration of petroleum reserves in politically unstable regions have put pressure on oil prices (Klare 2005; Wesley 2007). Longer-term anxieties concern the sustainability of global oil supplies and the potential for a peak in global oil production followed by a period of production decline (Australian Senate 2007; Government Accountability Office 2007). Given this context, the imperatives to begin assessing the implications for cities seem strong.

The marked recent growth in fuel costs between 2004 and 2006 contrasts with annual change in general consumer prices in Australia, which increased from around 2.5 per cent to just below 4 per cent over the period. Wages growth has also been relatively modest compared to fuel price increases. The short-term volatility of fuel prices makes direct comparisons with CPI and wages difficult. Australian general transportation costs, of which petroleum costs are a major component, grew much faster over the period 2004–06 than in the preceding three years and largely exceeded both wages and general price growth from late 2005. W[J1]hile transport costs have comprised a modest proportion of household expenses over the past two decades, these appear to have increased since 2004. Dodson and Sipe (2007) examined the urban spatial distribution of fuel price pressures. Recent assessments of global energy security suggest the cost of petroleum will continue to rise over the medium and long term. Such growth in fuel prices, in combination with other cost increases such as mortgage expenses, implies increasing pressure on household budgets. Future household fuel and mortgage exposure, which is the focus of this chapter, is thus also likely to be strongly spatially expressed.

Rising fuel prices have 'contributed significantly' to consumer price inflation, which rose from just over 2 per cent in mid-2004 to almost 4 per cent in late 2006 (Reserve Bank of Australia 2006) (RBA). Between March 2005 and November 2006 the RBA made four 0.25 percentage point increases to the official interest rate in response to rising inflation, which brought the official interest rate to 6.25 per cent. In turn, these shifts propelled average mortgage interest rates from 7.05 per cent up to 8.05 per cent over the period from February 2005 to November 2006.

The past two years have seen a proliferation of media reports in Australia suggesting that rising fuel prices and mortgage interest costs have begun to impact on household financial circumstances (Brown 2006; Fishman 2006; Garnaut and Baker 2006; Gittens 2006; Gordon 2006; McMahon 2006; Smith 2006). Other reports suggest changes to household consumption patterns. Lower fuel consumption and fewer car trips combined with reductions in discretionary spending such as 'going out' and entertainment appear to be the main behavioural responses (AC Nielsen 2006; Commonwealth Bank Research 2006; Sensis 2005). However, these reports are unsupported by research-based evidence, at least in part because scholars so far have been slow at investigating these issues. Media reporting indicates the issue is of immense public concern. However, what is clear is the lack of a substantive research base that can better inform public understanding of the problem and its urban dynamics. Differences in spatial car dependence within Australian cities that are central to this issue have received almost no attention so far.

Spatial car dependence

Australian cities are highly car-dependent and thus highly oil-dependent when compared to Canadian, UK or European cities (Newman & Kenworthy 1999). Only the US and New Zealand appear to exhibit greater reliance on motor vehicles for urban travel than Australia. The private car is used for most trips in Australian cities, both for work and other purposes. In Sydney, for example, 70 per cent of all trips are made by private car (Table 5.1) while in Melbourne the figure is somewhat greater at 75 per cent (DOI 2000: 27). High levels of car dependence leave Australian cities economically and socially exposed to rising global oil prices.

Levels of car dependence are unevenly spatially distributed in Australian cities. In general, households located closer to the central business districts (CBD) demonstrate less dependence on automobiles than those in middle and outer locations (Newman et al. 1985). Sydney exemplifies this pattern, as demonstrated by data describing the average level of individuals' vehicle kilometres travelled (VKT) (DIPNR 2003). The average daily VKT for residents of inner eastern Sydney (including and surrounding the CBD) was 10.1 km in 2003, compared to approximately 18 km for those in the further out north-east areas and up to 33 km for those in the distant outer west (Table 5.1). The spatial trends in daily VKT growth are also uneven. Daily per capita VKT declined almost 10 per cent for residents of inner eastern Sydney during the period 1991–2001, but increased by approximately 23.6 per cent and 22.8 per cent for those in Sydney's south and outer west, respectively.

The above data demonstrate that spatial differences in car dependence are widening over time in Sydney as car use increases in the outer suburbs and declines in areas nearer the CBD. Other Australian cities exhibit similar patterns (e.g. Morris

Table 5.1 Selected travel data for Sydney statistical division

Travel indicator	Area								
	Inner/ East	North East	South East	Inner/ Central West	North West	South West	Outer West	Central Coast	Total Sydney SD
Private vehicle mode share (all trips) (%)	48.7	67.9	72.3	64.6	80.1	78.7	79.7	77.3	70.0
Private vehicle mode share JTW (%)	49.2	65.2	69.0	64.4	76.8	75.6	77.5	77.3	67.6
Daily VKT per person (km)	10.1	17.9	17.6	14.1	23.2	24.0	33.3	30.1	20.0
Change in VKT per person (%) 1991–2001	–9.9	0.3	9.1	6.0	4.7	23.6	22.8	19.0	11.6

Source: DIPNR (2003: 2)

et al. 2002; Newman et al. 1985), which are also reflected in car ownership rates. Thus, for a given household size Morris et al. (2002) demonstrated that car ownership levels varied by up to 0.79 vehicles depending on whether a household lived in the inner city or at the urban fringe. Dodson and Sipe (2007) calculated that the annual costs of running a small car were more than AU$6000, implying that car ownership is a major cost burden for lower-income households.

Currently available transport or housing datasets are unfortunately not capable of establishing the actual direct costs incurred by households in meeting their transport needs at a scale below the metropolitan level (Dodson & Sipe 2007). However, the Sydney data demonstrate that car dependence retains its historic importance as a dimension of socio-spatial differentiation in Australian cities (Badcock 1984) and is probably accelerating these differences. The social inequities implied by these patterns are compounded by unequal access to public transport (Cheal 2003; Dodson et al. 2006).

The converse of car dependence is the capacity to choose and use other modes of transport for urban travel, especially public transport. Much research demonstrates that households in inner parts of Australian cities (where public transport coverage is dense and frequent) tend to use this mode more than those in middle and outer areas (Mees 2000b; Newman et al. 1985). In inner Sydney motor vehicles are used for only 49.2 per cent of work journeys, while the figures for south and outer west Sydney are 75.6 and 77.5 per cent, respectively (Table 5.1). Other Australian cities show similar patterns (Morris et al. 2002).

Public transport use in dispersed suburban areas is closely related to service availability and quality (Hall 2001; Mees 2000b; Newman 2006). In Australian cities public transport tends to be of highest quality in inner urban locations compared to those on the fringe (Mees 2000b). This spatial difference in the provision of public transport appears to be inversely related to spatial differences in car dependence (Mees 2000a). The gentrification of Australian inner cities in recent decades has enabled higher-income households to spatially 'capture' high quality public transport services, among others (O'Connor & Healy 2002). By comparison, those in the outer areas of Australian cities are typically excluded from access to high quality public transport services (Buchanan et al. 2005; Cheal 2003; Dodson et al. 2004).

Changing urban socio-spatial structures (O'Connor & Healy 2002; Wood et al. 2007) have therefore partly socialised travel costs for wealthier inner-urban residents, while transport costs for car dependent outer suburbanites remain largely private (Dodson & Sipe 2007). This is clearly demonstrated by Table 5.1 where Sydney's wealthier inner-city zones are displaying declining car dependence compared to those in the less prosperous western suburbs. Rising global petroleum insecurity threatens to further increase private motor vehicle transport costs while leaving the social cost for public transport relatively unchanged. Therefore, this differential socio-economic transport geography has substantial implications for households, given recent trends in global oil prices and subsequent interest rate pressures. Those with modest static incomes but rising transport and mortgage expenses are likely to be heavily affected. However, the distribution of housing debt and income status is also strongly socio-spatially patterned. These patterns emphasise the need to better understand the links between transport systems and the social geography of housing debt.

The spatial allocation of home purchase

The 'dream' (Kemeny 1983) of owning a detached suburban house retains strong historic, cultural and economic appeal in Australia. The private dwelling is also the largest asset class held by Australian households (Badcock & Beer 2000), with most households who purchase their dwelling doing so through mortgage finance. Australian mortgages are typically obtained from private providers and cover 90 to 95 per cent of the dwelling value with monthly repayments spread over 30 years and interest rates set at market levels. The homeownership geography of Australian cities is highly differentiated in terms of price, household socio-economic status, mortgage tenure and urban location (Badcock & Beer 2000). Median house prices for Australian suburbs follow consistent 'price decay gradients' (Burnley 1980; Wood et al. 2007) that favour the inner city and which have steepened in recent years (Burke & Hayward 2000; Productivity Commission 2004).

Detached dwellings on individual lots continue to dominate new housing in Australian cities, comprising around two-thirds of new stock (Dowling 2005) compared to multi-unit dwellings, although this dominance is declining, particularly in the case of Sydney, and most Australian metropolitan plans seek to accelerate this spatial shift in development focus (Randolph 2006). New detached dwellings are often concentrated in greenfield residential estates in outer and fringe localities and are often sold as 'house and land' packages. House prices tend to be relatively lower in outer suburban areas when compared to the overall metropolitan market (Burke & Hayward 2001; Burnley et al. 1997; Wood et al. 2007).

Uneven income and housing market structures tend to allocate modest-income home purchasers to outer and fringe localities (Burnley et al. 1997). The popular description of Australia's outer suburbs as 'mortgage belts' reflects this pattern. The proportion of first-home buyers locating in the outer or fringe suburbs of Sydney and Melbourne was 47 per cent and 49 per cent respectively, between 2000 and 2003 (Productivity Commission 2004: 255). In Melbourne the proportion of first-home purchasers locating in fringe areas increased approximately 10 per cent between 1991 and 2003 (Productivity Commission 2004: 255, Figure B.4).

The structure of Australian urban housing markets also differentiates localities in terms of access to public, social and community services. The problem of the relative 'locational disadvantage' that outer suburban households suffer was extensively debated in the early 1990s (Maher et al. 1992; National Housing Strategy 1992). Maher et al. (1992) described locationally disadvantaged areas as lacking the facilities and services necessary to enable a 'satisfactory life' or that required residents to undertake long journeys to access such resources. Maher (1994) argued that locational disadvantage resulted from households making rational housing choices—a view that reflects the 'trade-off' theory (Ball & Kirwan 1977) of residential locational decisions. By comparison, Badcock (1994b) argued that the unfair structure of urban housing markets constrains household choices and that locational disadvantage reflects uneven and inequitable infrastructure and service investment. Burnley et al. (1997) confirmed the Badcock view of Australian suburban disadvantage in their study of households who moved to outer urban areas to achieve homeownership. Transport was a key problem for these households:

> To the extent that people move to outer suburbia to obtain affordable housing, such pricing trends may be socially inequitable unless strong policies to relocate employment and to develop public transport are pursued in tandem.
>
> (Burnley et al. 1997: 1125)

The locational disadvantage debate highlighted the critical role played by transport systems in shaping household socio-economic opportunity in Australian cities, a theme that researchers both in Australia and overseas have only recently begun to address in greater detail (Dodson et al. 2006). The emerging relationship between transport costs and mortgage interest rates could widen urban socio-economic divides. Housing debt will probably be closely implicated in this process, given the fuel price and interest rate link and the role of home purchase location in mediating socio-economic opportunity (Badcock 1994a; Burbidge 2000). Housing debt burdens are also socially and spatially unevenly distributed in Australian cities.

Spatial mortgage debt

Lending for owner-occupied housing in Australia has more than tripled during the past decade (Figure 5.2) in conjunction with a housing market boom (Berry & Dalton 2004; Yu 2005), causing some academic observers to warn of dire economic consequences from any correction in credit markets (Keen 2007). A minority of Australian households are currently purchasing housing, but this proportion has been growing, having increased from 29.6 per cent in 1995 to 35.1 per cent in 2005 (ABS 2006a). These aggregate proportions mask some important socio-economic dimensions, particularly income differences.

Lower-income households are less likely to have a mortgage than those on higher incomes, a phenomenon that is probably due to income-based mortgage lending criteria. Hence only 26.9 per cent of households in the second lowest

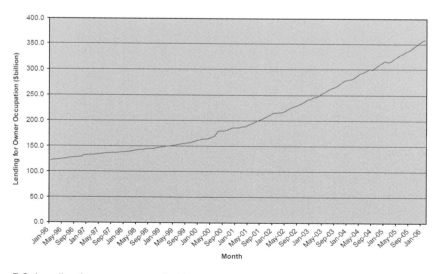

5.2 Lending for owner-occupied house purchase in Australia, 1996–2006.
Source: Reserve Bank of Australia (2006), Table D05

income quintile had mortgages in Australia in 2003–04 compared to 41.2 per cent and 48.5 per cent in the middle and fourth income quintiles respectively (ABS 2005). The distributional burden of mortgage debt varies across Australian households (La Cava & Simon 2005). First and second income quintile households who are purchasing owner-occupied housing tend to have higher levels of debt as a proportion of their income than higher-income households. Of households with housing debt, those in lower income quintiles were also more likely to have financial difficulties than those on higher incomes.

This pattern was also described by Kupke and Marano (2004), who found low income first-home buyers had a greater likelihood of poorer financial wellbeing than those on higher incomes. Given marked house price inflation in the period since the data used in the La Cava and Simon (2005) study were gathered, the trend is probably towards greater levels of household debt and thus greater levels of potential exposure to spatial mortgage and oil risks for some households.

The spatial distribution of household mortgage debt burdens relative to household income levels is poorly documented in Australia—a curious and potentially serious scholarly and policy oversight in a nation that celebrates mortgage funded suburban homeownership. Given what we know about housing markets and urban structure it is probable that while Australian home purchasers constitute a minority of urban households overall, they may nonetheless comprise a higher proportion of households in certain sub-regions, especially the outer suburbs. Thus Randolph and Holloway's (2002) assessment of 'mortgage stress' in Sydney demonstrated that households with high mortgage costs relative to their income were more likely to be found in the suburban outer western areas of Sydney than in the inner and middle areas. The uneven transport geography described above implies that the socio-economic effects of rising fuel costs on home purchasers will also be unevenly distributed.

Previous research has demonstrated that Australian housing markets tend to entrench rather than diminish socio-economic inequality (Badcock 1994a; Badcock & Beer 2000; Burbidge 2000). There is currently little scholarly or policy knowledge about the combination of car dependence, mortgage debt and the socio-economic profile of households at the local scale in Australian cities, and this issue is under-researched internationally. The specific distribution of socio-economic vulnerability to oil price and mortgage interest rises is currently poorly understood beyond the broad spatially aggregated contours. There is a pressing need for empirical assessment that can illuminate the extent of this vulnerability in Australian cities and in other jurisdictions such as the USA. Given that the problems are inevitably also spatially expressed, it is essential that spatial analysis be undertaken to assess the distribution of such risks.

The Reserve Bank of Australia has identified a strengthening relationship between rising fuel prices and price inflation (RBA 2006) and has signalled an intention to raise interest rates to control inflation. This has significant implications for the socio-economic and financial circumstances of households within Australian cities, especially for those on modest incomes with mortgages who are highly dependent on the private motor car and are located in areas with limited alternative transport modes.

An earlier study by the authors (Dodson & Sipe 2007) investigated the areas of Australian cities that would be socio-economically most affected by rising fuel prices and general price inflation through the use of an oil vulnerability index based on Census data. This chapter deploys a similar analytical index that has been adapted to focus on households with mortgages. The index is termed as 'Vulnerability Assessment for Mortgage, Petrol and Inflation Risks and Expenditure' (VAMPIRE). The VAMPIRE assists to understand the spatial distribution of oil and mortgage vulnerability within Australian cities and the prospects for different socio-economic groups, given increasing long-term energy insecurity and increasing medium-term household indebtedness.

The vulnerability assessment for mortgage, petrol and inflation risks and expenditure

The VAMPIRE index is constructed from four variables obtained from the 2001 Australian Bureau of Statistics (ABS) Census that have been combined to provide a composite mortgage vulnerability index that can be mapped at the geographic level of the Census Collection District (CD). Census CDs are area units spatially scaled to contain approximately 200 households. The VAMPIRE assesses the average vulnerability of households within the CD rather than indicating the specific vulnerability of particular households. The variables used are as follows.

- Car dependence. Proportion of those working who undertook a journey to work (JTW) by car (either as a driver or passenger).
- Proportion of households with two or more cars.
- Income. Median weekly household income.
- Mortgages. Proportion of dwelling units that are being purchased (either through a mortgage or a rent/buy scheme).

The use of these variables merits some explanation. The first two variables indicate the extent of car dependence as used in the earlier study (Dodson & Sipe 2007). The JTW figure provides a basic indicator of demand for automobile travel—relative to other modes, particularly public transport—while the number of motor vehicles per household indicates the extent of household investment, and thus dependence, on motor vehicle travel. Together these variables provide an indicator of the extent to which households are exposed to rising costs of travel. A high JTW by car score generally also reflects lower public transport use for a locality because these are proportional responses to the same Census question. The mortgage variable represents the prevalence of mortgage tenure and, accordingly, household exposure to interest rate rises within a locality. The income variable is used to measure the financial capacity of the locality to absorb fuel and general price increases. Together these four variables provide a basic, but comprehensive, spatial representation of household mortgage and oil vulnerability. The authors would have liked to include a variable for the average mortgage cost per CD, however, the Australian Bureau of Statistics Census reports monthly mortgage costs in bands, for example, $300–499, rather than as average figures, thus making inclusion in the VAMPIRE unwieldy.

The VAMPIRE index was constructed by combining the variables according to car dependence, income and mortgage by assigning an index score and weighting according to the percentiles shown in Table 5.2. For a CD to be considered most vulnerable, it would contain high percentages of households that have two or more cars, make their JTW by car, have a mortgage and have low incomes. Using Table 5.2, this CD would have a score of 20 (5 + 5 + 5 + 5).

The four selected variables are not equal in their contribution to VAMPIRE. The variables have been weighted according to their proportional contribution to the overall VAMPIRE score (Table 5.3). Thus of a total possible VAMPIRE score of 30, 5 points are provided by each of the car ownership and JTW variables while 10 points each are provided by the income and mortgage scores. Given the lack of existing research, the index weights car dependence, mortgage tenure and income equally.

The study focuses on Australia's four most populous urban areas: Brisbane, Sydney, Melbourne and Perth. The VAMPIRE results have been mapped by CD within these urban areas based on the Australian Bureau of Statistics definition of the urban area. The ABS defines urban areas as a cluster of contiguous CDs containing a mix of characteristics, including population size and concentration and

Table 5.2 Value assignment relative to Census District percentile for VAMPIRE

Percentile	Value assigned			
	Car own ≥ 2	JTW by car	Income	Mortgage
100	5	5	0	5
90	4	4	1	4
75	3	3	2	3
50	2	2	3	2
25	1	1	4	1
10	0	0	5	0

Table 5.3 Variable weighting for VAMPIRE

Indicator	Proportion of households with ≥ two cars	Proportion of work trips by car	Income level	Proportion of households with a mortgage
Potential points	5	5	10	10
Weighting	combined	33.3%	33.3%	33.3%

mix of land uses (ABS 2006c). The urban areas for the major Australian cities closely match the distribution of population and activity within those cities. VAMPIRE categories for each urban area are consistently shaded as shown in Table 5.4. Maps of the VAMPIRE results are shown in Figures 5.5–5.8. Public transport access has not been assessed directly, although this is partly implicit in the mode of travel variable. In the discussion reference is made to general patterns of public transport service quality, based on the authors' knowledge of the transport systems of each city.

Results of the VAMPIRE

Brisbane

Brisbane has approximately 1.8 million residents and is Australia's fastest growing and third largest city (ABS 2006b: 46). The VAMPIRE results for Brisbane (Figure 5.3) demonstrate a wide variation in mortgage and oil vulnerability levels. The central area of Brisbane immediately surrounding the CBD exhibits the highest concentration of low-vulnerability localities. Other areas with low or moderate VAMPIRE scores were dispersed throughout the middle suburbs, mostly within 15 km of the CBD. The middle suburbs of Brisbane also display some variation in vulnerability with a mix of moderate scores.

The highest VAMPIRE levels were found predominantly in the middle and outer growth corridors to the north, west and east and south-east of the city centre. The overall picture is of broad tracts of outer suburban locations exhibiting high levels of mortgage and oil vulnerability. There are some very small pockets of moderate mortgage and oil vulnerability in outer areas, such as at Ipswich in the

Table 5.4 Assignment of VAMPIRE ratings to map shadings

VAMPIRE value	1 to <10	10 to <15	15 to <17	17 to <19	19 to ≥ 22
Shading:					

Shocking the suburbs 99

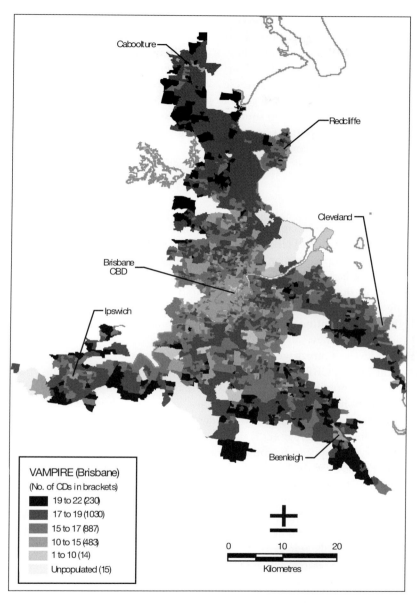

5.3 Oil and mortgage vulnerability in Brisbane.
Source: ABS Census Statistics CD-Rom 2001

outer west. However, these are rare relative to the broad tracts of highly vulnerable densely mortgaged residential areas within the growth corridors.

Inner and middle-suburban localities within 15 km of the CBD demonstrate lower mortgage and oil vulnerability than those further out, which probably reflects higher quality bus and rail services in these zones. Some outer-suburban moderate-vulnerability areas are also located close to good public transport, such as within parts of the western, eastern and southern corridors. Outer suburban and fringe areas where public transport is poor are clearly most vulnerable to the impacts of rising fuel and mortgage costs.

Sydney

Sydney, Australia's largest city, contains 4.2 million residents (ABS 2006b: 46) and is strongly patterned in terms of mortgage and oil vulnerability (Figure 5.4). Two broad areas display low or moderate levels of vulnerability: inner northern Sydney from the harbour mouth in the inner north-east to north of the CBD and to Hornsby in the north; and the area around and east of the Sydney CBD extending through the suburbs approximately 15 km south and west of the CBD. The highest concentration of low vulnerability is immediately around the Sydney CBD and North Sydney.

Higher levels of mortgage and oil vulnerability are found in areas beyond 20 km from the Sydney CBD to the north, south and particularly the west. This effect is particularly pronounced in the greater western Sydney region, especially west of a general line from Baulkham Hills to Liverpool in the south-west. Sydney's most concentrated areas of mortgage and oil vulnerability are located in the outer north and south of the western region.

The rail system appears to confer some oil and mortgage advantage to some Sydney sub-regions. This includes a broad swathe of suburbs along the rail line from lower North Sydney to Hornsby. Despite generally high vulnerability in outer western Sydney some less vulnerable localities are discernable around major rail nodes at Parramatta, Fairfield, Cabramatta, Liverpool and Campbelltown in the south-west and to some extent Blacktown and Penrith in the north-west (Figure 5.4). The role of public transport has not been directly assessed in this study and tenure effects may be implicated in this pattern, given the tendency for renters to locate in these higher-density centres (Randolph 2006).

Shocking the suburbs 101

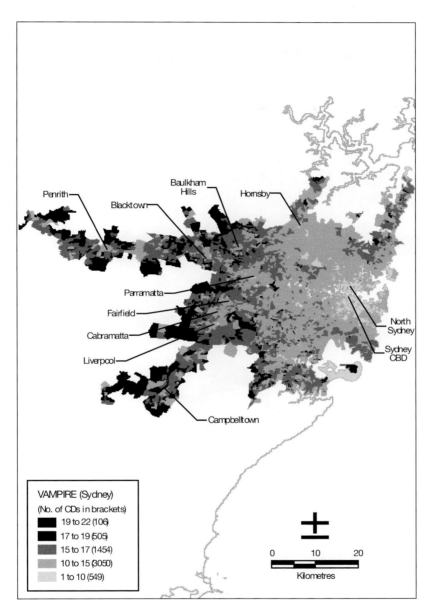

5.4 Oil and mortgage vulnerability in Sydney.
Source: ABS Census Statistics CD-Rom 2001

Melbourne

Melbourne contained 3.6 million residents in 2005 (ABS 2006b: 46) and displays a variegated geography of mortgage and oil vulnerability (Figure 5.5). Localities

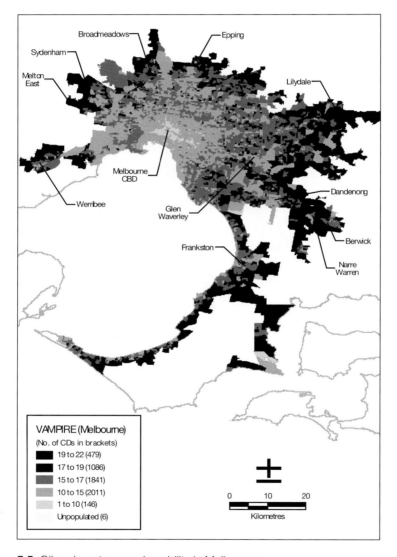

5.5 Oil and mortgage vulnerability in Melbourne.
Source: ABS Statistics CD-ROM 2001

with the lowest VAMPIRE scores are concentrated in close proximity to the Melbourne CBD and extend to the east and north.

Areas of highest mortgage and oil vulnerability in Melbourne predominate in outer suburban areas and are distributed around almost the entire metropolitan fringe. Melbourne's outer urban growth corridors exhibit high VAMPIRE scores. Many new fringe suburban housing estates, such as in the Melton East and Berwick growth corridors, exhibited high VAMPIRE scores.

The middle areas between Melbourne's inner city and the outer suburbs demonstrate considerable VAMPIRE variation, although some patterns can be discerned. Some lower-vulnerability localities strung along the Werribee, Lilydale, Glen Waverly, Dandenong and Frankston rail lines perhaps reflect increasing use of trains for CBD work travel (Moriarty & Mees 2006). Income and tenure effects no doubt also play a role, but it is notable that the public transport system appears to be contributing to higher mortgage and oil resilience in some better-serviced areas.

Perth

Perth contains approximately 1.4 million inhabitants (ABS 2006b: 47) and, like the other Australian cities, it exhibits unevenly distributed mortgage and oil vulnerability (Figure 5.6). Lowest-vulnerability areas are concentrated immediately adjacent to and east of the Perth CBD. A broad arc of middle-ring suburbs displays a mix of low and high mortgage and oil vulnerability. A broad arc of outer and fringe areas including most of the growth corridors to the north, north-east, east and south-east exhibited the highest concentration of high VAMPIRE scores.

The influence of public transport on the levels of mortgage and oil vulnerability in Perth is difficult to discern, but there does appear to be a limited effect arising from the availability of public transport. The best example is the rail corridor between the CBD and Janebrook-Greenmount, where a number of low mortgage and oil vulnerability localities are observed. Such an effect is also apparent, although less marked, in the Armadale and Joondalup corridors. The greater availability of public transport in the middle suburban areas is also likely contributing to relatively lower, albeit uneven, levels of mortgage and oil vulnerability in these localities. This may be due to the greater capacity for circumferential work journeys provided by public transport services in these areas.

Conclusions and policy directions

This chapter has sought to establish whether the interaction of rising fuel prices and socio-economic pressures could generate new forms of socio-spatial differentiation

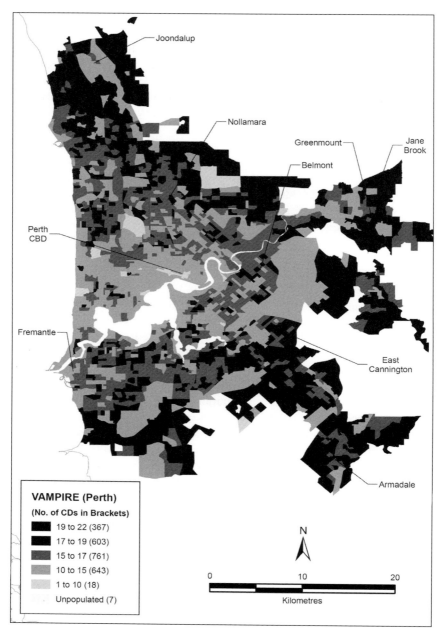

5.6 Oil and mortgage vulnerability in Perth.
Source: ABS Statistics CD-ROM 2001

within cities. The analysis here of patterns in Australian cities has demonstrated that household financial and socio-economic vulnerability to rising fuel prices and mortgage interest rates is socio-spatially differentiated and likely to be more extensively felt across outer and fringe suburban zones. Many outer suburban households are more exposed to these price pressures because of their combined mortgage exposure, modest incomes and high car dependence. The patterns are likely to be more extreme now than when the Census data on which this study is based were collected, given house price inflation and burgeoning household debt since 2001 (Berry & Dalton 2004). Should fuel prices increase beyond their recent near-record high levels, many households in the outer and fringe suburban areas of Australian cities may experience high levels of socio-economic stress. The remainder of this chapter discusses some of the implications and imperatives for urban and housing policies that this analysis suggests, and begins to consider how scholars and policy makers can begin responding to the challenges presented by these imperatives.

The first major challenge is to develop a more comprehensive understanding of the links between urban energy consumption and socio-economic status, not only in Australian cities but also in other jurisdictions, particularly the dispersed metropolitan regions of North America. The VAMPIRE index has been used to demonstrate the distribution of oil and mortgage vulnerability in Australian cities. However, there is a need to increase the range of data included in the VAMPIRE, such as direct mortgage costs and public transport access, and to relate these data in a more comprehensive way by developing and testing appropriate weightings for new variables. It is anticipated that an improved VAMPIRE would reveal similar socio-spatial patterns to those presented in this chapter. However, improved methodological design would provide policy makers with more robust bases upon which to craft responses.

A number of conclusions can be made on the basis of the present study. A major urban policy challenge is to ameliorate the impacts of either transport costs or mortgage costs. Controlling mortgage expenses is the most difficult of these. Australian Government finance and interest rate policy is subject to a wide range of domestic and international pressures of which household sector considerations are only one component. While some influence could be exerted at the aggregate level, there is little likelihood of governments being able to specifically target outer suburban areas through financial mortgage or fuel price relief. The same applies for income or other taxation policies.

Because mortgage and oil vulnerability is spatially expressed, spatial investment strategies are probably required to redress this risk. Australian urban strategies since the late 1990s have, among other objectives, sought to reduce car dependence and increase public transport use by focusing on demand aspects of transport (Forster 2006). The main policy has been to encourage investment in higher-density residential development, particularly around public transport nodes (for example,

DOI 2002) (Forster 2006; Randolph 2006). Unfortunately, urban consolidation policies have limited potential to address suburban mortgage and oil vulnerability. Newman (2006) has shown that residential density is a relatively minor contributor to reduced car dependence, compared to central proximity and availability of public transport. Such policies also depend on private-housing market cycles that are irregularly and unevenly expressed and operate over very long and potentially highly unresponsive investment cycles. Urban intensification is also only indirectly linked to variations in transport costs.

Finally, higher-density housing typically concentrates in inner locations where land values are sufficiently high to justify the expense of multi-storey development (Randolph & Holloway 2002). Densification policies that focus new development around public transport nodes in middle and inner sites will do little for the extensive zones of low-density, dispersed 'mortgage belt' suburbs where public transport is poor. Yates (2001) has already queried the purported capacity of consolidation policies to produce gains in affordability for home purchasers, especially for younger households, with many facing housing choices that constrain their locational options to the fringe. While residents of the new higher-density stock may be less vulnerable to rising fuel costs, the likelihood of densification reducing oil vulnerability among vast tracts of car dependent suburbia is remote in the absence of a suburban transport supply strategy (Mees 2000b). Even prominent consolidation proponents such as Newman and Kenworthy also (1999: 185, Figure 5.3, 188) recognise this problem and have promoted expansion of public transport networks in conventional middle and outer suburban zones. Newman (2006) even reports that public transport access is more significant than urban density in explaining levels of suburban car use. Finally, urban consolidation policy is coming under renewed question in Australia, with critics pointing to the problems caused by blanket policy prescriptions that are insufficiently attuned to socio-spatial differences within cities (Randolph 2006).

Despite the focus on urban consolidation in metropolitan strategies in the past decade there has been relatively little investment in new suburban public transport. The majority of transport investment has been directed to building major roads. There is an urgent need to reassess the supply of public transport services in the outer suburbs relative to service quality prevailing in the wealthier inner areas. New investment could reduce existing transport cost and risk inequalities. There is a long-recognised need to ensure the provision of local services that are closely integrated with broader existing urban public transport networks (Mees 2000b; Newman & Kenworthy 1999). Provision of new outer suburban public transport services could be achieved within existing transport budgets by redirecting funds from major urban road projects to public transport.

Urban scholars should give greater attention to household financial behaviour and housing market conditions. Some critical data gaps need to be overcome.

There is currently no publicly available dataset that includes information on household location, mortgage debt, travel costs and income in Australia. Household travel surveys (HTS) gather information on travel behaviour and household income, but not on transport costs and housing expenses. Household transport costs can be approximated using GIS-based quantification of household journey-to-work origin-destination data (see Newman et al. 1985) but this captures a trip type that is declining as a proportion of travel. We need better understanding of the trade-offs that households make in budgeting for mortgage, fuel, utilities and discretionary expenditure. We need to know what responses households and housing markets will make in the face of higher fuel prices. We also need to investigate how the elasticity of demand for public transport varies geographically within cities relative to the price of fuel and the cost of servicing mortgages.

This research is highly relevant to cities in other jurisdictions such as North America where levels of car dependence are comparable to those in Australia. While it is comprehensive and penetrating, the VAMPIRE assessment is relatively easy to construct from Census data, providing that the relevant variables are available. The authors see no limit on scholars replicating the method presented here and applying the VAMPIRE to cities in Canada and the US, for example, that also have comparable spatial social tenure structures. It is hoped that the investigation will provide the basis for this and much more scholarly inquiry in the future.

A final observation is the need for urbanists to begin identifying the major strategic challenges that cities will face in the coming decades and to begin identifying means to address these through urban research. This chapter has focused on the problem of declining energy security and its socio-economic impacts on cities. There is arguably a comparable challenge to assess how the socio-economic vulnerability to global warming and the effects of government policy responses will be distributed at the urban scale. New processes of global change will probably bring new patterns of urban socio-spatial differentiation which we must, as scholars, begin to attempt to comprehend.

References

ABS (Australian Bureau of Statistics) (2005) Household Income and Income Distribution, Australia, 2003–04 Cat. No. 6532.0 (Canberra: Australian Bureau of Statistics).

ABS (2006a) Australian Social Trends 2006 Cat. No. 4102.0 (Canberra: Australian Bureau of Statistics).

ABS (2006b) Regional Population Growth Cat. No. 3218.0 (Canberra: Australian Government).

ABS (2006c) Statistical Geography Volume 1—Australian Standard Geographical Classification (ASGC) (Canberra: Australian Bureau of Statistics).

AC Nielsen (2006) AC Nielsen Fuel Price Survey Results for Australia—November 2005 (Sydney: AC Nielsen).

Australian Senate (2007) Inquiry into Australia's Future Oil Supply and Alternative Transport Fuels: Final Report (Canberra: Australian Parliament House).

Australian Treasury (2006) International Comparison of Australia's Taxes (Canberra: Australian Government).

Badcock, B. (1984) *Unfairly Structured Cities*, Oxford: Basil Blackwell.

Badcock, B. (1994a) "'Snakes or ladders?' The housing market and wealth distribution in Australia," *International Journal of Urban and Regional Research*, 18: 609–627.

Badcock, B. (1994b) "'Stressed-out' communities: 'out-of-sight, out-of-mind'?," *Urban Policy and Research*, 12: 191–197.

Badcock, B. & Beer, A. (2000) *Home Truths: Property Ownership and Housing Wealth in Australia*, Melbourne: Melbourne University Press.

Ball, M. & Kirwan, R. (1977) "Accessibility and supply constraints in the urban housing market," *Urban Studies*, 14: 11–32.

Bankoff, G., Frerks, G. & Hilhorst, D. (Eds) (2004) *Mapping Vulnerability: Disasters, Development and People*, London, UK & Sterling, VA: Earthscan.

Baum, S., Stimson, R., O'Connor, K., Mullins, D. & Davis, R. (1999) *Community Opportunity and Vulnerability in Australia's Cities and Towns*, Melbourne: Australian Housing and Urban Research Institute.

Berry, M. & Dalton, T. (2004) "Housing prices and policy dilemmas: a peculiarly Australian problem?," *Urban Policy and Research*, 22: 69–91.

Brown, B. (2006) "Petrol price should rule out rate rise," *The Australian*, Sydney, 19 April, p. 29.

Buchanan, N., Evans, R. & Dodson, J. (2005) Transport Disadvantage and Social Status: A Gold Coast Pilot Study, *Urban Research Program Research Monograph 8*, Brisbane: Urban Research Program, Griffith University.

Burbidge, A. (2000) "Capital gains, homeownership and economic inequality," *Housing Studies*, 15: 259–280.

Burke, T. & Hayward, D. (2000) Housing Past, Housing Futures: Melbourne Metropolitan Strategy Technical Report 4 (Melbourne: Department of Infrastructure).

Burke, T. & Hayward, D. (2001) "Melbourne's housing past, housing futures," *Urban Policy and Research*, 19: 291–310.

Burnley, I. (1980) *The Australian Urban System: Growth, Change and Differentiation*, Melbourne: Longman Cheshire.

Burnley, I., Murphy, P. & Jenner, A. (1997) "Selecting suburbia: residential relocation to outer Sydney," *Urban Studies*, 34: 1109–1127.

Burrows, R. (2003) "How the other half lives: an exploratory analysis of the relationship between poverty and home-ownership in Britain," *Urban Studies*, 40: 1223–1242.

Cheal, C. (2003) Transit Rich or Transit Poor: Is Public Transport Policy in Melbourne Exacerbating Social Disadvantage? (Melbourne: Faculty of Architecture, Building and Planning, University of Melbourne).

Commonwealth Bank Research (2006) Petrol Prices: Learning to Live with a Higher Level of Pain. Commonwealth Research Economic Issues (Sydney: Commonwealth Bank of Australia).

Danielsen, K., Lang, R. E. & Fulton, W. (1999) "Retracting suburbia: smart growth and the future of housing," *Housing Policy Debate*, 10: 513–540.

DIPNR (Department of Infrastructure Planning and Natural Resources) (2003) Regional Transport Indicators for Sydney (Sydney: Transport and Population Data Centre, NSW Government).

Dodson, J. & Sipe, N. (2007) "Oil vulnerability in the Australian city: assessing socio-economic risks from higher urban fuel prices," *Urban Studies*, 44: 37–62.

Dodson, J., Gleeson, B. & Sipe, N. (2004) Transport Disadvantage and Social Status: A Review of Literature and Methods, *Research Monograph 5*, Brisbane: Griffith University Urban Research Program.

Dodson, J., Buchanan, N., Gleeson, B. & Sipe, N. (2006) "Investigating the social dimensions of transport disadvantage—I," *Urban Policy and Research*, 24: 433–453.

DOI (Department of Infrastructure) (2000) Challenge Melbourne: Issues in Metropolitan Planning for the 21st Century (Melbourne: Department of Infrastructure).

DOI (2002) Melbourne 2030: Planning for Sustainable Growth (Melbourne: Department of Infrastructure).

Dowling, R. (2005) "Residential building in Australia, 1993–2003," *Urban Policy and Research*, 23: 447–464.

Fainstein, S., Gordon, I. & Harloe, M. (1992) *Divided Cities: New York and London in the Contemporary World*, Oxford, UK and Cambridge, MA: Blackwell.

Fishman, E. (2006) "Petrol's painful formula," *Herald Sun*, Melbourne, 3 May, p. 18.

Forster, C. (2006) "The challenge of change: Australian cities and urban planning in the new millennium," *Geographical Research*, 44: 173–182.

Garnaut, J. & Baker, J. (2006) "Petrol prices hit car sales," *Sydney Morning Herald*, 23 May.

Gittens, R. (2006) "Smile, pain at the pump has pay-offs," *Sydney Morning Herald*, 3 May, p. 13.

Gordon, J. (2006) "Oil rise threatens to hold Australian economy to ransom," *The Age*, Melbourne, 19 April, p. 1.

Government Accountability Office (2007) Crude Oil: Uncertainty about Future Oil Supply makes it Important to Develop a Strategy for Addressing a Peak and Decline in Oil Production (Washington DC: United States Government).

Graham, S. & Marvin, S. (2001) *Splintering Urbanism: Networked Infrastructures, Technological Mobilities and the Urban Condition*, London and New York: Routledge.

Hall, P. (2001) "An old-fashioned solution," *Town and Country Planning*, May: 130–131.

Hamnett, C. (1996) "Social polarisation, economic restructuring and welfare state regimes," *Urban Studies*, 33: 1407–1430.

Heinberg, R. (2004) *Powerdown: Options and Actions for a Post-carbon World*, Gabriola Island: New Society Books.

Keen, S. (2007) "Who's having a housing crisis then?," *Steve Keen's Oz Debtwatch*. Available at www.debtdeflation.com/blogs/ (accessed 11 April 2007).

Kemeny, J. (1983) *The Great Australian Nightmare*, Melbourne: Georgian House.

Klare, M. (2005) *Blood and Oil: The Dangers and Consequences of America's Growing Petroleum Dependency*, London and New York: Penguin Books.

Krizek, K. (2003) "Transit supportive home loans: theory, application and prospects for smart growth," *Housing Policy Debate*, 14: 657–677.

Kunstler, J. H. (2005) *The Long Emergency: Surviving the Converging Catastrophes of the Twenty-First Century*, New York: Grove/Atlantic.

Kupke, V. & Marano, W. (2004) "Job security and first-homebuyers," *Urban Policy and Research*, 22: 393–410.

La Cava, G. & Simon, J. (2005) "Household debt and financial constraints in Australia," *The Australian Economic Review*, 38: 40–60.

Lenzen, M., Dey, C. & Foran, B. (2004) "Energy requirements of Sydney households," *Ecological Economics*, 49: 375–399.

Lucas, K. (Ed.) (2004) *Running on Empty: Transport, Social Exclusion and Environmental Justice*, Bristol: The Policy Press.

Maher, C. (1994) "Residential mobility, locational disadvantage and spatial inequality in Australian Cities," *Urban Policy and Research*, 12: 185–191.

Maher, C., Whitelaw, J., McAllister, A., Francis, R., Palmer, J., Chee, E. & Taylor, P. (1992) *Mobility and Locational Disadvantage within Australian Cities*, Canberra: Department of Prime Minister and Cabinet Social Justice Research Program into Locational Disadvantage.

McMahon, S. (2006) "Shoppers think twice as petrol price and rate rises bite," *The Age*, Melbourne, 9 May, p. 3.

Mees, P. (2000a) Rethinking Public Transport in Sydney, *UFP Issues Paper 5*, Urban Frontiers Program Issues Papers Sydney: Urban Frontiers Program, University of Western Sydney.

Mees, P. (2000b) *A Very Public Solution: Transport in the Dispersed City*, Melbourne: Melbourne University Press.

Moriarty, P. & Mees, P. (2006) "The journey to work in Melbourne." Paper presented at the 29th Australasian Transportation Research Forum, Crowne Plaza Hotel, Gold Coast, 27–29 September.

Morris, J., Wang, F. & Berry, M. (2002) "Planning for public transport in the future: challenges of a changing metropolitan Melbourne." Paper presented at the Australasian Transport Research Forum, Canberra, 2–4 October (Australian Bureau of Transport and Regional Economics).

Motormouth (2006) Average Sydney unleaded petrol prices for past week. Available at http://motormouth.com.au (accessed 19 December 2006).

Musterd, S. & Ostendorf, W. (1998) *Urban Segregation and the Welfare State: Inequality and Exclusion in Western Cities*, London and New York: Routledge.

National Housing Strategy (1992) Housing Location and Access to Services, *Issues Paper*, Canberra: National Housing Strategy.

Newman, P. (2006) "Transport greenhouse gas and Australian suburbs: what planners can do," *Australian Planner*, 43:2, 6–7.

Newman, P. & Kenworthy, J. (1999) *Sustainability and Cities: Overcoming Automobile Dependence*, Washington DC: Island Press.

Newman, P., Kenworthy, J. & Lyons, T. J. (1985) "Transport energy use in the Perth metropolitan region: some urban policy implications," *Urban Policy and Research*, 3: 4–15.

Newman, P., Kenworthy, J. & Lyons, T. J. (1990) *Transport Energy Conservation Strategies for Australian Cities: Strategies for Reducing Automobile Dependence*, Perth: Institute for Science and Technology Policy, Murdoch University, WA.

O'Connor, K. & Healy, E. (2002) The Links Between Labour Markets and Housing Markets in Melbourne (Melbourne: Australian Housing and Urban Research Institute, Swinburne-Monash Research Centre).

Productivity Commission (2004) First Home Ownership: Productivity Commission Inquiry Report (Canberra: Productivity Commission).

Randolph, B. (2006) "Delivering the compact city in Australia: current trends and future implications," *Urban Policy and Research*, 24: 473–490.

Randolph, B. & Holloway, D. (2002) "The anatomy of housing stress in Sydney," *Urban Policy and Research*, 20:4, 329–355.

Randolph, B. & Holloway, D. (2005) "Social disadvantage, tenure and location: an analysis of Sydney and Melbourne," *Urban Policy and Research*, 23: 173–201.

Reserve Bank of Australia (RBA) (2006) Statement on Monetary Policy, 5 May (Canberra: Reserve Bank of Australia).

Sensis (2005) Consumer Report–December (Melbourne: Sensis).

Smith, B. (2006) "RACV expects motorist resistance to petrol at $1.35," *The Age*, Melbourne, 7 April, p. 2.

Somerville, P. (1998) "Explanations of social exclusion: where does housing fit in?," *Housing Studies*, 13: 761–780.

Speak, S. & Graham, S. (1999) "Service not included: private services restructuring, neighbourhoods, and social marginalization," *Environment and Planning: A*, 31: 1985–2001.

Stimson, R., Baum, S., Mullins, P. & O'Connor, K. (2001) "A typology of community opportunity and vulnerability in metropolitan Australia," *Papers in Regional Science*, 80: 45–66.

Taylor, M. (1998) "Combating the social exclusion of housing estates," *Housing Studies*, 13: 819–832.

Troy, P., Holloway, D., Pullen, S. & Bunker, R. (2003) "Embodied and operational energy consumption in the city," *Urban Policy and Research*, 21: 9–44.

Van Kempen, R. & Marcuse, P. (1997) "A new spatial order in cities?," *American Behavioural Scientist*, 41: 285–299.

Wesley, M. (2007) "The geopolitics of energy security in Asia," in: M. Wesley (Ed.) *Energy Security in Asia*, pp. 1–12, Abingdon: Routledge.

Wessel, T. (2000) "Social polarisation and socioeconomic segregation in a welfare state: the case of Oslo," *Urban Studies*, 37: 1947–1967.

Winter, I. & Stone, W. (1998) Social Polarisation and Housing Careers: Exploring the Interrelationship of Labour and Housing Markets in Australia, *AIFS Working Paper No. 13*, Melbourne: Australian Institute of Family Studies.

Wood, G., Berry, M., Nygaard, C. & Taylor, E. (2007) "Community mix, affordable housing and metropolitan planning strategy in Melbourne." Paper presented at the Launch of the International Centre for Housing and Urban Economics (ICHUE), Reading, UK.

Yates, J. (2001) "The rhetoric and reality of housing choice: the role of urban consolidation," *Urban Policy and Research*, 19:4: 491–517.

Yu, X. (2005) "The 'great Australian housing dream' busted on a brick wall: housing issues in Sydney," *Cities*, 22: 436–445.

Chapter 6

The re-enclosure of green space in postmodern urbanism

Michael Hebbert

The chapter compares the roles of green space within modern and postmodern urbanism. Twentieth-century modernism pursued an open space vision inspired by a landscape ideal of unbounded nature. It was implemented in towns and suburbs through urban renewal and highway construction, regulatory standards and open space policies. As the stock of green space increased, so did concern about its environmental and social sustainability, culminating in a paradigm shift around the turn of the century. The new vision of green space sought enclosure instead of openness, and active provision of ecosystem services instead of passive pictorial quality. The chapter describes this rethinking of town greenery, looks for its underlying social and environmental rationale, and assesses its place within a wider planning theory of postmodern urbanism.

Introduction

The tree in Figure 6.1 was one of thousands planted and recently uprooted by Manchester City Council. Its original vision was to create a city of green, open landscapes. Now it wants to rebuild street frontage, and plant a new landscape framed by buildings. The alteration can be described as a process of matrix reversal, a shift from 'out' to 'in' landscape, extensive to intensive form, modernist to postmodern space. However described, it involves a fundamental turn in purpose and method, a paradigm shift.

In this chapter I explore the shift, setting it in historical context and drawing on European and US as well as British planning literature. The story begins with the pulling down of park railings and the opening up of towns to a free-flowing natural landscape. It ends in a process of re-enclosure. Peter Blake once said 'what

---

Submitted by the Association of European Schools of Planning.

Originally published in *Town Planning Review*, 70:1. © 2008. By kind permission of Liverpool University Press.

6.1 Grubbed tree on pile of spoil in Beswick, East Manchester, where a thirty-year-old landscape buffer along the radial Ashton Old Road is being replaced by buildings with active frontage.

Source: Author's photograph

we need more of is less urban open space. We need closed space' (Blake 1981: 11–12). Today that shift is happening. I describe the shift from extensive to intensive provision and explain the paradox of how less has come to be more.

The modernist urban landscape

The twentieth-century landscape vision was to bring nature to town. The US parkway movement experimented from the 1860s onwards with systems of interconnected green space to oxygenate the city and dispel its miasmas (Lubove 1967; Szczygiel and Hewitt 2000). The 'emerald necklaces' of Olmsted and his followers were radical green-space systems that broke down conventional categories of park, urban and rural (Hirsch and O'Hanlon 1995). Very early in the twentieth century the parkway concept clicked into partnership with motor-oriented highway engineering, functional land use zoning and Modern Movement design. Open green space plus speed plus new white architecture became a universal icon of social democracy (Worpole 2000).

A seminal statement of this ideal was Christopher Tunnard's book, *Gardens in the Modern Landscape* (Tunnard 1938; Neckar 1990). The author was the team member responsible for green space within MARS (the Modern Architectural Research Group), the British arm of the international Modern Movement. An active member of the new Institute of Landscape Architects (previously the British Association of Garden Architects), his book was the manifesto for the profession that gave the architectural avant-garde its settings of sunlight, openness and freedom. No more rockeries, streams and contoured lawns, no more regimented lines of trees; nature should not be contained but should flow into the city, connecting playgrounds, recreation areas, home, factory and countryside in a continuous stream of health-giving greenery, a 'garden without limitation' (Tunnard 1938: 166).

This image of the country 'calling in on the town' was a direct homage to eighteenth-century picturesque landscape theory. Tunnard ingeniously transformed a proprietary aesthetic – open vistas from a country landowner's seat to a distant horizon – into a template for landscapes of urban collectivism. He took inspiration from the Claremont estate in Surrey, a great eighteenth-century estate by Sir John Vanbrugh, Charles Bridgeman, William Kent and Lancelot 'Capability' Brown, which was divided by five firms of builders into luxury homes. He caricatured this enclosure process with a sketch of a cow marked up for its carcass joints, entitled the 'Butcher Method'. By contrast (Figure 6.2), he showed how its picturesque setting could have been preserved intact without the loss of a single great tree as an open background for flats and terraced houses, with shared gardens flowing uninterruptedly into a communal landscape (Tunnard 1938: 158).

The idealised pre-industrial pastoral imagery of the picturesque movement fitted well with the technological optimism of the machine age. Nature free of toil was laid out for contemplation and physical enjoyment (Jacques 2000; Worpole 2000).

In mid-twentieth-century planning theory, it was axiomatic that the introduction of vegetation on a large scale to the interior of cities would 'bring life, change and vigour direct to the townsman' (Gutkind 1962: 52). Lewis Mumford saw it as 'giving collective form to the I–Thou relation'. Old-fashioned parks, by contrast, were just 'places of refuge from the urban crowd'. Throughout his influential career, he urged planners to conceive the green public realm on an ambitious scale:

Perhaps the first step towards regaining possession of our souls will be to re-possess and plan the whole landscape . . . In the cities of the future, ribbons of green must run through every quarter, forming a continuous web of garden and mall, widening at the edge of the city into protective green belts, so that landscape and garden will become an integral part of urban no less than rural life.

(Mumford 1964: 173–75)

6.2 Scheme for Claremont.
Source: Tunnard 1938

This vision was international, though its nuances varied from culture to culture. The landscape of US parkways was a setting for the American Dream, giving the suburban commuter a daily taste of the great outdoors (Gandy 2002). The postwar *stadtlandschaft* concept offered a way of repairing Germany's broken historic nexus between folk and homeland (Diefendorf 1993; Mantziaras 2003). Hans Berhard Reichow's *Organische Stadtbaukunst* of 1948, a key text, saw the change from intensive to extensive green space, park to landscape, as a move 'away from the satisfaction of merely aesthetic needs to the fulfilment of general elemental needs, to the establishment of the biologically necessary' (Sohn 2003: 125). In French landscape theory, an appeal to rural vitality and tradition combined with the promise of machine-age mobility for 'horizontal spirits' (Bardet 1949: 121; Rabinow 1989; Cohen 1995). The gigantic public realm of Soviet cities embodied the all-encompassing Communist state (Engel 2006). In their nests of greenery, the neighbourhood units in British reconstruction plans evoked the national nostalgia for the face-to-face world of the village green (see Figure 6.3).

All were variations on a shared theme. It was an idea of progress from enclosure to openness, from the in space of a conventional park to an out landscape of flowing edgeless greenery (Fairbrother 1974: 142); from the confinement and muddle of the street to the geometrical purity of the freeway (Merriman 2004); from the profane confusions of the town to the purity of a Garden of Eden (Tuan 1974: 104).

116 Michael Hebbert

Christopher Tunnard left Britain in 1939 for a career at Harvard and Yale. But his fellow pioneers – Geoffrey Jellicoe, Brenda Colvin, Sylvia Crowe and Nan Fairbrother – continued to take Capability Brown as a reference point for the public landscapes of the welfare state (Crowe 1960; Matless 1998; Jacques 2000). The vision was collectivist and its implementation ran alongside the progress of welfarism. Some progressive British municipalities began to pull down their park railings even before the Second World War made it a patriotic duty (Rasmussen 1982: 416; Fox 1995; Conway 2000). In post-war reconstruction, open-plan was the norm.

In reaction against the brutal mineral environment of the industrial city, green space seemed an unquestionable benefit; the most gentle and universal form of social engineering (Abrams 2003: 119) and the surest basis of public health (Forshaw 1943). The essential building block of the welfare state – the neighbourhood unit – came wrapped in it (see Figure 6.3). National policy statements such as the Dudley Report (CHAC 1944) made generous allowances mandatory within planned estates. Planners and landscape architects took it for granted 'that people inherently like the picturesque, and given the choice would decide to live

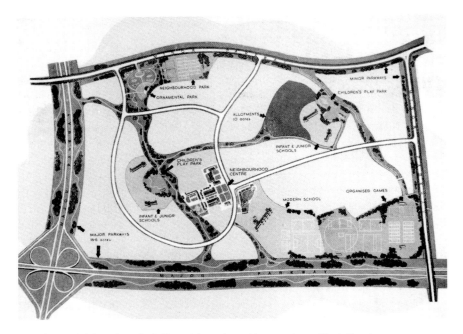

6.3 City neighbourhood, deliberately reduced to quasi-rural isolation by a combination of highway severance and landscape buffer planting.
Source: Nicholas 1945

in a setting not dissimilar to eighteenth-century parkland' (Gilbert 1989: 11). Scaled-down versions of open layout became a standard feature of social housing projects everywhere, even in developing countries (Caminos and Goethert 1978). Expanses of newly laid turf around New York social housing projects were a trigger for Jane Jacobs's *Death and Life of Great American Cities* (1962).

An equally important factor of change was road transport. The central idea of twentieth-century highway design was that roads, like railways, should have their own permanent way for uninterrupted driving. Landscape ideals featured prominently in road construction propaganda. In the US, the American Association of Nurserymen was an active lobbyist for the 1944 Federal-Aid Highway Act and the eventual creation of the $100 billion, 41,000-mile interstate system. It offered landscape planting as a way to win over objectors, reducing 'misunderstandings, oppositions and bad public reactions' (White 1959: 190). Urban roads would become 'elongated parks bringing to the inner city a welcome addition of beauty, grace and green open space' (Snow 1959: xii). The green stuff simultaneously enhanced the driver's pleasure and protected the non-driver's health and safety (Giedion 1941; Crowe 1960; Halprin 1966). From the mid-twentieth century onwards, the principle of a vegetated sleeve or buffer came to be applied by city authorities to roads of every size and grade (Figure 6.4). As the autobahn concept of a roadway without built frontage or side-turns was extended from arterial roads to distributor and local roads, so did the realm of roadside planting. At the most detailed level of site layout, engineering standards specified wide road verges for the convenience of the utilities, and generous visibility splays so that traffic need not slow down on the approach to junctions (Woodford et al. 1976). Road landscapes did more than any other factor to break open the built fabric of cities, reduce densities and increase the area of planted voids. In planned new towns and housing estates, they provided the matrix for complete schemes of landscape design (Higson 1997). Several million tons of spoil were moved to create an invented topography of traffic-screening berms along the highway grid of Milton Keynes, the UK's largest new town (Turner 1998). Twenty-five million trees were planted; the chief planner said his aim was 'to lose the city in a re-created forest' (Walker 2007: 75).

Programmed into the pattern language of the twentieth-century city, the matrix of green space continued to grow with the passage of time. Whereas the size of private gardens was shrinking (Ravetz 1995: 193–94) and the stock of private allotments diminishing (Meller 2005), new development (public and private) steadily added to the stock of public amenity space. Edge planting and amenity strips of grass were often required by planners as a condition or developer contribution (UGSTF 2002: 53). The dominant typology of a building set like a poached egg in its own car-park on an access road multiplied the off-cuts of space left over after (or in) planning (SLOAP and SLOIP) (Casson 1956; AR 1973;

6.4 East Kilbride, near Glasgow: the Westwood neighbourhood, green but still largely treeless 28 years after the designation of the new town.

Source: by kind permission of the Building Research Station

Fairbrother 1974). It was rare for a new building not to be equipped with a vanity patch of evergreen shrubbery to the front or side (Figure 6.5). In Tunnard's original vision, the landscaper (his term) stood equal with architect and engineer as the expert who would open up the city to the flow of nature (Tunnard 1938: 166), but in dispiriting reality the profession seemed little more than a service industry tidying up after property developers and highway engineers (Gandy 2002: 147). The modernist picturesque had declined into a routine application of design standards (Manthorpe 1956). We can hear a thin echo of *Gardens in the Modern Landscape* in the clumps of shrubbery and strips of grass of any suburban business or retail park, and the 'indeterminate land oozes' along roadsides and around housing projects (Jacobs 1962: 102). The entire practice of green-space design and management was on the defensive and under fire from two directions.

6.5 Buffer planting along a flank wall anywhere in the UK. A common requirement of planning permissions, this miserable vestigial vegetation is the direct descendant of the landscape buffers shown in Figure 6.3.

Source: Author's photograph

Social and environmental critiques of urban green space

The green space favoured by modernist urbanism was regarded more for its mastic-like ability to join together the functional zones of the city than for its own functional attributes. As the quantum of urban vegetation grew, so did questions about functionality. Unlike conventional types of street/square/park/garden, its pieces were usually unnamed and weakly identifiable. The earliest criticisms had to do with the amorphousness and lack of urbanity in what Iain Nairn called 'subtopia' (AR 1956: 355). Amenity spaces were often of the wrong scale for human use. They often appeared alongside roads; as Peter Blake put it, where people wouldn't want to use them or couldn't get at them even if they did (Blake 1964: 30). Their very freedom posed new issues of social control (Carr and Lynch 1981). Traditional parks had been enclosed spaces, managed by keepers: the new landscape was intrinsically harder to supervise, whether in its all-too-open patches of grassland or its all-too-closed belts of shrubbery with their undercrofts and detritus of litter. Close planting, for initial impact, often produced dense thickets (see Figure 6.6) that – with the exuberance of nature – would overgrow footpaths and outrun the efforts of maintenance teams (HoC 1999 and 2002–3; Comedia and Demos 2002; DTLR 2002: 10).

As cities' landscape estates expanded, conventional parks and new green spaces were in competition for a diminishing resource base (UGSTF 2002: 53). In a context of continuous downwards pressure on unit costs of maintenance, governments sought economies in batch contracts, replacing site-based groundsmen and specialist keepers with all-purpose maintenance teams who used the same trimmers and gang-mowers in parks as on highway verges. As the quantum of planted space grew, so the ancient art of urban horticulture became coarsened into mechanical crudity (Spirn 1984).

> Too many parks have been reduced to grass, trees and tarmac in the quest for cheaper maintenance. The damage inflicted is quickly and clearly visible, less so the profoundly demoralising effect that a bare park has on its users and the life of the surrounding community.
>
> (HoC 1999)

British parliamentarians examining the state of parks in 1999 were shocked at the levels of decline and abandonment; they found that the keeperless park had become, like the unstaffed railway station, 'one of the ghost zones of modern Britain'. David Nicholson-Lord contrasted the great historical tradition of the picturesque – what Alexander Pope had called 'calling in the country' – with the reality of public open spaces rife with vandalism (Nicholson-Lord 1987: 32). Whatever the actual levels of danger, open-plan landscapes often had the physical attributes of risky places: concealment, exposure, short lines of sight and non-supervision (Coaffee 2003). Maarten Hajer defines the urban public realm as a landscape under continual scrutiny by its occupants for signs of reassurance or threat (Hajer and Reijndorp 2001: 73). Perceptions of threat dominate green-space user surveys (DoE 1996; DTLR 2002; WDS 2004; MKP 2005: 35). Urban landscapes of fear offer an ironic inversion of the historical ideal of the town – a place of sanctuary from the dangerous thickets and lurking predators of the countryside (Tuan 1980: 146).

Meanwhile, a similar process of disenchantment was occurring within the natural environment. The urban picturesque had aimed to bring nature into the city, yet that was not what happened. The Dutch garden designer Mien Ruys observed of Le Corbusier that he worshipped nature but fundamentally misunderstood it (Woudstra 2000: 136). The Modern Movement's image of nature was controlled, improved and gardenesque, and assumed intensive maintenance (Gilbert 1989). It encouraged a perception of green space as an inert construction-industry material, like the green sponge used in architectural models, a static exterior decoration based on nursery-catalogue distinctions: grass, flowers, shrubs, trees (Taylor 1981: 85). As Ian McHarg (1969) showed, this was really an anti-nature attitude. There was a systematic failure to see planted ground as 'a living composition with its own inherent patterns of survival which nature constantly reinstates' (Fairbrother 1974: 25). As those successional processes in a temperate climate would normally tend towards a forest climax, the pre-forest combinations used in the urban picturesque were intrinsically unstable, and so required a continuous input of labour, chemicals and machinery. The conventional mowed, weed-controlled grass of urban amenity space became 'a symbol for everything that is wrong with our relationship to the land' (Hough 1994: 129).

The sterility of public green space was shown up by deindustrialisation. It was extraordinary how rapidly derelict factories, marshalling yards, docks and gasworks could develop a soil structure and a successional vegetation of understorey, overstorey and canopy (Nicholson-Lord 1987). The ecological landscape movement around Ian McHarg, Ian Laurie and Michael Hough argued for parks and other categories of urban green space to be allowed the same ecological freedom, reverting to semi-wild urban commons with vegetation left to seed, grow and decay in natural succession (Laurie 1979). Some Dutch cities successfully applied this successional approach, though it was generally resisted by open-space managers (Spirn 1984).

The most famous case in the UK, and one of the largest in Europe, was the adventitious landscape that colonised an abandoned war munitions factory to the north of Warrington. When the land was developed for housing, much of its semi-wild landscape was retained and incorporated into the system of highway and housing greenery (Figure 6.6). However, the physical layout of the estates was entirely conventional, with clusters of cul-de-sacs surrounded by buffer planting. It was innovative only in the ecological dimension. Unsurprisingly, residents of Warrington's urban wildwoods have proved uneasy about the naturalistic thickets which press too closely around their homes, and fearful of the paths and roadways that run through them (Jorgensen et al. 2007). Other cases reinforce the need to balance biophilia with human liveability. Abrams (2003) found overgrown vegetation projecting 'an image of desolation and despair' in overgrown courtyards of Ralph Erskine's celebrated ecological landscapes at Byker in Newcastle. For the second edition of his book *City Form and Natural Process*, Michael Hough revisited some of the natural landscapes around Dutch housing estates and found that their low-maintenance naturalised vegetation had been radically thinned and simplified in response to mugging and drug-dealing (Hough 1994: 123). Worse still, the arrival of CCTV surveillance often involves complete removal of shrubs and undercroft vegetation.

The re-enclosure movement

The German landscape architect Peter Latz (2000) looked back on the second half of the twentieth century as a period of lost opportunity; an era of buffer greenery

6.6 Uninviting sign on a Warrington distributor road. Pedestrians prefer the hazards of the road verge to the thickets and subways.

Source: Author's photograph

in which open space was defined by anti-urban values. He sensed a change. After half a century of indeterminate land oozes, cities are reinventing and retrofitting morphologies not practised since the early years of the twentieth century: the urban square, the avenue of street trees, the multi-way boulevard, building along frontage lines, the grid-bounded square and park, the corridor (Turner 1996). Urbanism, in the phrase of Trancik's influential study *Finding Lost Space* (1986), has become the task of 'making figurative space out of the lost landscape'. Michael Hough (1990) speaks of a strategy of 'matrix reversal', which turns the green matrix of modernism into a matrix of built form, open into closed space, out to in. Re-enclosure reflects new confidence about cities as places to work, play and live. Its design discourse spans all styles of urbanism from late-modern Rotterdam to retro-romantic Berlin (Hajer and Reijndorp 2001). The variety of approaches can be explored at the Centre de Cultura Contemporàna de Barcelona's web-based European Archive of Contemporary Urban Space, a collection of hundreds of projects entered for the centre's biennial prize (http://urban.cccb.org). For Peter Latz, this archive marks the end of twentieth-century space and the beginning of a twenty-first-century urban landscape (Latz 2000: 5).

The re-enclosure phenomenon goes beyond local circumstances and cultures; like earlier phases of green-space conversion, its reach is global (Clark 2006: 6). Tjallingii (2003) speaks of a fundamental shift from 'green' to 'red' dominance in the struggle for urban land. In the former Communist bloc, the process of infilling and partitioning open space is driven by a sometimes chaotic process of privatisation (Beer et al. 2003; Engel 2006). Re-enclosure in market economies is also often linked to neoliberal shifts of ownership and management of a previously undifferentiated public realm (Webster 2007). In France the process is called 'residentialisation', and it involves a mix of demolition, low-rise infill, internal fencing and privatisation. Phillipe Panerai's reworking of the Teisseire district of Grenoble (Figure 6.7) is a characteristic example (Panerai 1999; Desfontaines 2004). In Holland, the once-iconic modernist landscape of Amsterdam's Bijlmermeer has become an equally iconic example of an architecture of addition (Figure 6.8) to form streets in the green void (Docter 2000: 203–13; Bruijne

6.7 New owner-occupied housing with detached garages inserted around the edge of the Teisseire estate in Grenoble with the aim of désenclavement (breaking social and physical isolation). The prospects for the tree seem poor.

6.8 New terraces built over the arcadian landscape of the Bijlmermeer estate in south-east Amsterdam, to form a connective street fabric.

Source: Author's photograph

et al. 2003). In Germany, Berlin's city authorities have pursued an ambitious and controversial strategy to replace free-flowing urban landscape (*stadtlandschaft*) with bounded urban space (*städtischer Raum*) (Burg 1997: 75; Hajer and Reijndorp 2001). In Ireland, Dublin's high-rise satellite estate of Ballymun has been transformed by a process of architectural enclosure, which reclaimed 170 acres of housing land from bleak amenity grassland to create a townscape of gardens, parks, pocket-parks and tree-lined boulevards (Figure 6.9) (BRL 1998). In England, the open amenity spaces of the Hulme estate in Manchester were recolonised after only twenty years to make a grid reminiscent of the nineteenth-century street lines (Rudlin and Falk 1999). Campbell Park in Milton Keynes,

6.9 Street trees receive attention in the new central high street of Ballymun.
Source: Author's photograph

designed to blend seamlessly into a continuous planted landscape, is acquiring a hard frame of high-density flats (MKP 2005). Everywhere, the once 'sturdy, versatile, recognisable and timeless concept' of modernism's urban picturesque is beginning to seem like an endangered habitat (Docter 2000; Beer et al. 2003).

Enclosure and functionality

We opened with a disconcerting image of nature uprooted. The site of that tree in East Manchester was part of an aerial photograph included in Richard Rogers's Urban Task Force report over the caption 'land going to waste'. Some of Manchester's wasteland was derelict industrial property (brownfields), but most was public open space.

Some urban areas have too much public space, much of which is poorly designed, managed and maintained. Many twentieth-century residential developments have a public realm that is simply SLOAP (space left over after planning) – soulless, undefined places, poorly landscaped, with no relation to surrounding buildings (UTF 1999: 57).

126 Michael Hebbert

Excess open space exacerbated the problems of abandonment in neighbour-hoods where almost half the shops and a quarter of dwellings were empty. The remedy is compaction (Echenique and Saint 2001). East Manchester's amorphous landscapes are being reduced and re-enclosed into public spaces framed and overlooked by building, as the Urban Task Force prescribed: 'from the front door to the street, to the square, the park and on out to the countryside . . . a hierarchy of public spaces that relate to buildings and their entrances' (UTF 1999: 71). If the defining feature of the twentieth-century urban landscape was its freedom of definition – everything flowed into everything else – the emergent paradigm is based on spaces with definition and multiple purposes, serving both human use and as an infrastructure for ecosystem services.

This concept applies at every scale. Rooftops, balconies and gardens take on roles in water management and temperature regulation. The grassed amenity spaces of housing estates are 'residentialised' into enclosed gardens (Desfontaines 2004). Gardened interiors of urban blocks – one of the most ancient types of green space in European cities (Sitte 1900) – are revived, Barcelona's Villa Olimpica and Amsterdam's Eastern Harbour District leading the way with innovative mixes of communal space and private gardens, car-parking, playspace and natural planting.

Beyond the dwelling curtilage, streets are being prised from the grasp of the highway engineer and redefined as multi-functional space (DfT 2007). As the art of urban arboriculture revives, more is being discovered about the benefits that street landscapes can bring (Randrup et al. 2002; Nowak et al. 2004; Konijnendijk et al. 2005). Trees reduce vehicle speeds, dampen noise, form a psychological safety barrier between lanes of movement, moderate the microclimate and reduce the traffic-induced pollution of ozone, sulphur dioxide and particulates (Beckett et al. 1998; Jacobs et al. 2002). They have equally important functions within the larger web of a town's green infrastructure: they provide a significant share of the urban forest; their continuous canopy acts as an ecological corridor; and each tree pit (as Zurich demonstrates, Figure 6.10) has the potential to be a mini-wildflower patch (Bonamoni 1990: 60). Sustainable drainage requirements are also beginning to transform that old cliché, the tree-planted amenity grass strip, into a new type of linear green space, a vegetated open swale whose wetland planting adds fresh strands of ecological interest to the street environment (Konijnendijk et al. 2005).

Moving upwards in scale, the next category of postmodern green space is the local park, now seen not as a diffused background but as a precious enclosure within a predominately mineral urban matrix. Again, the postmodern approach combines human and ecosystem functions. For humans, the emphasis is on safe accessible spaces for relaxation, physical exercise and play. As Jane Jacobs realised, fifty years ago, people shun spaces that are just landscape oozes, they like the enclosure of parks (Jacobs 1962). A generation used to the lifestyle of an urban flat needs outdoor spaces that function as public or shared realms. Hajer and Reijndorp

6.10 The city of Zurich allows the free growth of adventitious weeds around the base of street trees and along parking verges.

Source: Author's photograph

(2001: 9) argue that unless such spaces are fenced, defined and identifiable they cannot give structure to social encounters amid the cosmopolitanism and social diversity of the modern city.

Neighbourhood green space has important ecosystem functions too in temperature regulation and water management. The use of green space to manage surface run-off is revolutionary for urban design. At the outset it implies much closer attention to topography and the lie of the land in the layout of new development (Gordon and Tamminga 2002). Green spaces engineered for water retention will tend to be located centrally in relation to the buildings and spaces they drain. Much of the initial work on sustainable drainage was suburban in spirit, with wide swales and filter strips that pushed buildings apart and reduced site density (CIRIA 2000). But surface-water management can enhance streetscapes and reinforce the relation between local parks and their neighbourhoods, as illustrated in the designs for recent extensions to the British new town of Milton Keynes (Figure 6.11).

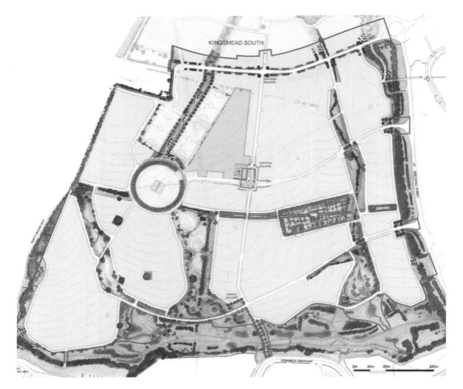

6.11 Sustainable drainage integrated with compact urban design in the landscape plan for Tattenhoe Park, Milton Keynes.
Source: By kind permission of English Partnership

Next on the scale of urban green spaces are the bigger parks, commons, valleys and landscape corridors that shape a town's overall structure. Again, they are being asked to serve many purposes at once. Greenways open up the city for cyclists and encourage sustainable commuting. Structural green space mitigates heat-island effects through air movement. Large open spaces combine with the filigree of gardens, parks and streets in a mosaic of ecological patches and corridors (Dramstad et al. 1996). The connectivity and heterogeneity of an urban ecosystem can to some extent be purposively planned and managed (Forman 1995: 448). Its biodiversity depends on an effective spatial zoning of patches in which vegetation is allowed to grow freely: annually as a summer meadow or through the years as a successional plant community (CABE 2006). Successful urban wildernesses may not have railings and keepers in uniforms but need to be as artfully designed so that undisturbed areas are zoned with margins, and clear routeways are offered that

The re-enclosure of green space **129**

connect up, have good visibility and look cared for (Gilbert 1989; Konijnendijk et al. 2005; Kowarik and Körner 2005). As CABE puts it (2006: 17), 'litter picking is as important in a wildlife area as in a formal rose bed'.

The theme running through all these typologies is that less can be more. The quality and functionality of planted space count for more than its extent in hectares. One of Europe's most environmentally aware municipalities, the Dutch town of Breda, has downsized its public open space, selling land for development and giving the proceeds to its parks department (Tjallingii 2003: 112–13). England's Commission for Architecture and the Built Environment (CABE) has encouraged local councils to do the same, releasing the parts of their estate that fail to provide 'local identity, character and delight' (CABE 2004: 87; Bennett 2004). Unfortunately some have built over gardens, allotments, playing fields and amenity landscapes without a strategy to enhance what remains – less simply is less. The policy gap, extending through all levels of English government, was severely criticised in the Royal Commission on Environmental Pollution's report on the urban environment (RCEP 2007: 82). In Scotland local councils are already required to frame a positive policy for their unbuilt spaces (Campbell 2001; SG 2007): this is the way forward.

Conclusion

This chapter has told a before-and-after story. We started with the twentieth-century vision of an unbounded natural realm within towns and cities, and ended with an alternative vision of enclosure – in the words of the post-unification plan for inner Berlin (Burg 1997), *städtischer Raum nicht fließendoffene Freiräume* (urban-type space, not open-flowing free space).

Of course, neither type has been universal. The rise of the modern landscape concept hardly affected the cities of Mediterranean Europe, where the late-twentieth-century experience was of a prolonged assault on the landscape by speculative development and self-built housing, much of it unauthorised; see, for example, Gabriella Corona's terrifying account of the disappearance of nature in modern Naples (Corona 2005). The postmodern decline of the picturesque landscape should not be overstated. Despite the privatisation of their assets, most of the UK's new towns maintain their extensive landscape areas to a high standard, thanks to the setting up of special trusts with significant revenue-generating endowments of offices, shops and pubs to cover upkeep costs (Higson 1997).

In the best of Scandinavian post-war settlements, such as Vallingby outside Stockholm, Christopher Tunnard's vision of a democratic landscape of open layout and interweaving greenery can still be seen and enjoyed. But even these famous settings are showing the strain (Figure 6.12), while the best new Scandinavian

6.12 Today, the strains of maintaining an 'out' landscape show even in the iconic settings of the British and Swedish new towns, as seen here in Vallingby.

Source: Author's photograph

development – such as Stockholm's Hammarby Sjöstad – follows a radically different strategy of enclosure.

Ken Worpole's book *Here Comes the Sun* (2000) celebrated the passing of the modernist landscape in an optimistic spirit. He wrote hopefully about the new green-space agenda:

> Increasingly, the issue of the provision of high-quality public amenities and public spaces, together with a much greater attention to issues of environmental quality, ranging from air quality to the reduction of traffic and greater walkability of urban settings and amenities, is bringing back together for the first time since the 'heroic' period of early modernism, issues of urban planning, social equity and public health policy in the name of 'ecological modernisation'.
>
> (Worpole 2000: 131)

Today's enclosure movement, like early modernism, crystallises a conception of social and environmental progress into physical form. It is another heroic moment for urban landscape design.

References

Abrams, R. (2003) "Byker revisited," *Built Environment*, 29: 117–31.

AR (Architectural Review) (1956) "Subtopia" (December), Special Issue.

AR (Architectural Review) (1973) "SLOAP – space left over after planning," *Architectural Review*, 920: 201–66.

Bardet, G. (1949) *Mission de l'Urbanisme*, Paris: Les Edition Ouvrières.

Beckett, K. P., Freer-Smith, P. H. and Taylor, G. (1998) "Urban woodlands: their role in reducing the effects of particulate pollution," *Environmental Pollution*, 99: 347–60.

Beer, A., Delshammar, T. and Schildwacht, P. (2003) "A changing understanding of the role of green-space in high density housing," *Built Environment*, 29: 132–43.

Bennett, A. (2004) "Is there too much space?," *Green Places*, 13: April.

Blake, P. (1964) *God's Own Junkyard: The Planned Deterioration of America's Landscape*, New York: Holt Rinehart & Winston.

Blake, P. (1981) "The Surgeon General has determined that urban open space is dangerous to your health," in Taylor, E. (ed.) *Urban Open Spaces*, London: Academy Editions, pp. 11–12.

Bonamoni, L. (1990) *Le Temps des Rues: vers un nouvel aménagement de l'éspace rue*, Lausanne: IREC.

BRL (Ballymun Regeneration Ltd) (1998) Master Plan for the New Ballymun, Dublin: Ballymun Regeneration Ltd.

Bruijne, D., von Hoogstraten, D., Kwekkeboom, W. and Luijten, A. (2003) "Amsterdam Southeast: Central Area Southeast and Urban Renewal in the Bijlmermeer 1992–2010," Bossom: Thoth.

Burg, A. (ed.) (1997) "Planwerk Innenstadt Berlin: ein erster Entwurf," Berlin: Kulturbuch-Verlag.

CABE (Commission for Architecture and the Built Environment) (2004) Creating Successful Masterplans: A Guide for Clients, London: Commission for Architecture and the Built Environment.

CABE (Commission for Architecture and the Built Environment) (2006) Making Contracts Work for Wildlife: How To Encourage Biodiversity in Urban Parks, London: Commission for Architecture and the Built Environment.

CHAC (Central Housing Advisory Committee) (1944) The Design of Dwellings: Report of the Central Housing Advisory Committee, London: HMSO.

CIRIA (Construction Industry Research and Information Association) (2000) Sustainable Urban Drainage Systems: Design Manual for England and Wales (CIRIA C522), London: Construction Industry Research and Information Association.

Caminos, H. and Goethert, R. (1978) *Urbanization Primer*, Cambridge, MA: MIT Press.

Campbell, K. (2001) "Rethinking Open Space: Open Space Provision and Management, A Way Forward," Edinburgh: Kit Campbell Associates for Scottish Executive Central Research Unit.

Carr, S. and Lynch, K. (1981) "Open space: freedom and control," in Taylor, E. (ed.) *Urban Open Spaces*, London: Academy Editions, pp. 17–18.

132 Michael Hebbert

Casson, H. (1956) "Outrageous postscript," *Journal of the Town Planning Institute*, 42: 187–91.

Clark, P. (ed.) (2006) *European City and Green-space*, Aldershot: Ashgate.

Coaffee, J. (2003) *Terrorism, Risk and the City: The Making of a Contemporary Urban Landscape*, Aldershot: Ashgate.

Cohen, J.-L. (1995) *Scenes of the World to Come: European Architecture and the American Challenge 1893–1960*, Paris: Flammarion.

Comedia and Demos (2002) *Park Life: Urban Parks and Social Renewal*, Stroud: Comedia.

Conway, H. (2000) "Everyday landscapes: public parks from 1930 to 2000," *Garden History*, 28: 117–34.

Corona, G. (2005) "Sustainable Naples: the disappearance of nature as resource," in Schott, D., Luckin, B. and Massard-Guilbard, G. (eds) (2005) *Resources of the City: Contributions to an Environmental History of Modern Europe*, Aldershot: Ashgate, pp. 97–112.

Crowe, S. (1960) *The Landscape of Roads*, London: Architectural Press.

Desfontaines, M. (2004) "Un nouveau concept, la résidentialisation des HLM," *La Gazette des Communes*, 1735: 40.

DfT (Department for Transport) (2007) *Manual for Streets*, London: Department for Transport.

Diefendorf, J. (1993) *In the Wake of War: The Reconstruction of German Cities after World War II*, Oxford: Oxford University Press.

Docter, R. (2000) "Postwar town planning in its midlife crisis", in T. Dekker (ed.), *The Modern City Revisited*, London: E & FN Spon, pp. 197–213.

DoE (Department of the Environment) (1996) *People Parks and Cities*, London: Department of the Environment.

Dramstad, W. R., Olson, J. and Forman, R. (1996) *Landscape Ecology Principles in Landscape Architecture and Land Use Planning*, Washington, DC: Island Press.

DTLR (Department for Transport, Local Government and the Regions) (2002) *Improving Urban Parks, Play Areas and Open Spaces*, London: Department for Transport, Local Government and the Regions.

Echenique, M. and Saint, A. (eds) (2001), *Cities for the New Millenium*: London: E & FN Spon.

Engel, B. (2006) "Public spaces in the blue cities of Russia," *Progress in Planning*, 66: 3.

Fairbrother, N. (1974) *The Nature of Landscape Design*, London: Architectural Press.

Ford, L. (2000) *The Spaces Between Buildings*, Baltimore, MD: Johns Hopkins University Press.

Forman, R. (1995) *Land Mosaics: The Ecology of Landscapes and Regions*, New York: Cambridge University Press.

Forshaw, J. H. (1943) *Town Planning and Public Health* (Chadwick Public Lecture), London: P. S. King and Staples.

Fox, C. (1995) "The battle of the railings," *Architectural Association Files*, 29: 50–60.

Gandy, M. (2002) *Concrete and Clay: Reworking Nature in New York*, Cambridge, MA: MIT Press.

Giedion, S. (1941) *Space Time and Architecture: The Growth of a New Tradition*, Cambridge, MA: Harvard University Press.

Gilbert, O. L. (1989) *The Ecology of Urban Habitats*, London: Chapman & Hall.

Gordon, D. and Tamminga, K. (2002) "Large-scale traditional neighbourhood development and pre-emptive ecosystem planning: the Markham experience 1989–2001," *Journal of Urban Design*, 7: 321–40.

Gutkind, E. A. (1962) *The Twilight of Cities*, New York: Free Press of Glencoe.

Hajer, M. and Reijndorp, A. (2001) *In Search of New Public Domain*, Amsterdam: NAI Publishers.

Halprin, L. (1966) *Freeways*, New York: Reinhold Publishing.

Higson, N. (1997) "Landscape," in *New Towns Record 1966–1996* (CD-ROM), Glasgow: The Planning Exchange.

Hirsch, E. and O'Hanlon, M. (1995) *The Anthropology of Landscape: Perspectives on Place and Space*, Oxford: Clarendon Press.

HoC (House of Commons) (1999) *House of Commons Select Committee on the Department of Environment, Transport and the Regions: Town and Country Parks*, London: HMSO.

HoC (House of Commons) (2002–03) *House of Commons Select Committee on the Office of the Deputy Prime Minister, Eleventh Report: Town and Country Parks*, London: HMSO.

Hough, M. (1990) *Out of Place: Restoring Identity to the Regional Landscape*, New Haven, CT: Yale University Press.

Hough, M. (1994) *Cities and Natural Process*, London: Routledge.

Jacobs, A., MacDonald, E. and Rofé, Y. (2002) *The Boulevard Book: History, Evolution, Design of Multiway Boulevards*, Cambridge, MA: MIT Press.

Jacobs, J. (1962) *The Death and Life of Great American Cities*, London: Jonathan Cape.

Jacques, D. (2000) "Modern needs, art and instincts: modernist landscape theory," *Garden History*, 28: 88–101.

Jorgensen, A., Hitchmough, J. and Dunnett, N. (2007) "Woodland as a setting for housing . . . in Warrington New Town UK," *Landscape and Urban Planning*, 79: 273–87.

Keeble, L. (1956) *Principles and Practice of Town and Country Planning*, London: Estates Gazette.

Konijnendijk, C. C., Nilsson, K., Randrup, T. B. and Schipperijn, J. (eds) (2005) *Urban Forests and Trees*, Berlin: Springer.

Kowarik, I. and Körner, S. (eds) (2005) *Wild Urban Woodlands, New Perspectives for Urban Forestry*, Heidelberg: Springer.

Latz, P. (2000) "Structuring the avant-garde landscape," *Garden History*, 28: 4–5.

Laurie, I. (1979) *Nature in Cities: The Natural Environment in the Design and Development of Urban Green-space*, Chichester: Wiley.

Lubove, R. (1967) *The Urban Community: Housing and Planning in the Progressive Era*, Englewood Cliffs, NJ: Prentice-Hall.

McHarg, I. (1969) *Design with Nature*, New York: Natural History Press.

Manthorpe, W. (1956) "The machinery of sprawl," *Architectural Review*, 120: 409–19.

Mantziaras, P. (2003) "Rudolf Schwarz and the concept of Stadtlandschaft," *Planning Perspectives*, 18: 147–76.

Matless, D. (1998) *Landscape and Englishness*, London: Reaktion Books.

Meller, H. (2005) "Citizens in pursuit of nature: gardens, allotments and private space in European cities 1850–2000," in Schott, D., Luckin, B. and Massard-Guilbard,

134 Michael Hebbert

G. (eds) *Resources of the City: Contributions to an Environmental History of Modern Europe*, Aldershot: Ashgate, pp. 80–96.

Merriman, P. (2004) "Freeways", in Harrison, S., Pilr, S. and Thrif, N. (eds) *Patterned Ground*, London: Reaktion Books, pp. 86–88.

MKP (Milton Keynes Partnership) (2005) *The New Plan for Milton Keynes: Environment and City Design*, Milton Keynes: Milton Keynes Partnership.

Mumford, L. (1964) *The Highway and the City*, London: Secker and Warburg.

Neckar, L. M. (1990) "Christopher Tunnard's gardens in the modern landscape," *Journal of Garden History*, 10: 237–46.

Nicholson-Lord, D. (1987) *The Greening of Cities*, London: Routledge and Kegan Paul.

Nowak, D. J. (2002) "The effects of urban forests on the physical environment," in Randrup, T. B. et al. (eds) *Urban Forests and Trees*, Luxembourg: Office for Official Publications of the European Communities, pp. 22–42 .

Nowak, D. J., Kuroda, M. and Crane, D. E. (2004) "Tree mortality rates and tree population projections in Baltimore, Maryland, USA," *Urban Forestry and Urban Greening*, 2: 139–47.

Panerai, P. (1999) "Convictions et références," *Projet Urbain*, 18: 19–21.

Rabinow, P. (1989) *French Modern: Norms and Forms of the Social Environment*, Cambridge, MA: MIT Press.

Randrup, T. B. et al. (eds) (2002) *Urban Forests and Trees*, Luxembourg: Office for Official Publications of the European Communities.

Rasmussen, S. E. (1982) *London: The Unique City*, Cambridge, MA: MIT Press.

Ravetz, A. (1995) *The Place of Home: English Domestic Environments 1914–2000*, London: E & FN Spon.

RCEP (Royal Commission on Environmental Pollution) (2007) *The Urban Environment*, Norwich: HMSO.

Rudlin, D. and Falk, N. (1999) *Building the 21st Century Home: The Sustainable Urban Neighbourhood*, London: Architectural Press.

SG (Scottish Government) (2007) *Open Space and Physical Activity*, Edinburgh: Scottish Government.

Sitte, C. (1900) "Greenery in the city," in Collins, G. R. and Collins, C. C. (translated and introduced by) *City Planning According to Artisitic Principles*, London: Phaidon Press, pp. 303–21.

Snow, W. B. (ed.) (1959) *The Highway and the Landscape*, New Brunswick, NJ: Rutgers University Press.

Sohn, E. (2003) "Hans Bernhard Reichow and the concept of Stadtlandschaft in German planning," *Planning Perspectives*, 18: 119–46.

Spirn, A. W. (1984) *The Granite Garden: Urban Nature and Human Design*, New York: Basic Books.

Szczygiel, B. and Hewitt, R. (2000) "Nineteenth century medical landscapes: John H. Rauch, Frederick Law Olmsted and the search for salubrity," *Bulletin of the History of Medicine*, 74: 708–34.

Taylor, E. (ed.) (1981) *Urban Open Spaces*, London: Academy Editions.

Tjallingii, S. (2003) "Green and red, enemies or allies? The Utrecht experience with green structure planning," *Built Environment*, 29: 107–16.

Trancik, R. (1986) *Finding Lost Space*, New York: Van Nostrand Reinhold.

Tuan, Y.-F. (1974) *Topophilia: A Study of Environmental Perception, Attitudes and Values*, Englewood Cliffs, NJ: Prentice-Hall.

Tuan, Y.-F. (1980), *Landscape of Fear*, Oxford: Blackwell.

Tunnard, C. (1938) *Gardens in the Modern Landscape*, London: Architectural Press.

Turner, T. (1996) *City as Landscape*, London: E & FN Spon.

Turner, T. (1998) *Landscape Planning and Environmental Impact Design*, London: UCL Press.

UGSTF (Urban Green Spaces Task Force) (2002) *Green Spaces Better Places*, London: DTLR.

UTF (Urban Task Force) (1999) *Towards an Urban Renaissance: Final Report of the Urban Task Force Chaired by Lord Rogers of Riverside*, London: Department of Environment, Transport and the Regions.

Walker, D. (2007) "Milton Keynes at 44," *Urban Design*, 104: 75

WDS (Women's Design Service) (2004) Making Safer Places: Interim Reports, available at www.wds.org.uk.

Webster, C. (2007) "Property rights, public space and urban design," *Town Planning Review*, 78: 81–102.

White, R. P. (1959) "The functional uses of plants on the complete highway," in Snow, W. B. (ed.) *The Highway and the Landscape*, New Brunswick, NJ: Rutgers University Press, pp. 182–92.

Woodford, G., Williams, K. and Hill, N. (1976) *The Value of Standards for the External Residential Environment*, London: Department of the Environment.

Worpole, K. (2000) *Here Comes the Sun: Architecture and Public Space in Twentieth Century European Culture*, London: Reaktion Books.

Woudstra, J. (2000) "The Corbusian landscape: arcadia or no man's land?," *Garden History*, 28: 135–51.

Chapter 7

Safe urban spaces

Security issues for city design

Maria Julieta Nunes de Souza and Rose Compans

This chapter reports on a recent trend in Brazilian urban design: the incorporation of spatial strategies aiming at public security. Based on a methodology developed in Europe and North America and disseminated by multilateral agencies, these strategies have been adopted in Brazil by the National Public Security Department, via the National Public Security and Citizenship Program. We outline the main features of this program, its methodological basis, and its underlying theoretical approaches. As an urban design methodology, this tool has been applied in many Brazilian cities through the implementation of the 'Safe Urban Spaces' project – which consists of physical interventions in low-income urban environments. In Rio de Janeiro – a large city in Brazil – this methodology was used in 30 poor areas designated as 'slums', during the preparations for the Pan-American Games (2007). This implementation is the focus of our study.

Introduction

This chapter discusses the recent emergence of 'public security' as a major theme in the practice and theory of national and international urban planning. This specific issue was incorporated by Brazilian public departments in 2003 and was specially adopted in 2007, when the National Public Security Department (SENASP – related to the Ministry of Justice) released the National Public Security and Citizenship Program (PRONASCI), responsible for implementing the methodology called 'Safe Urban Spaces'. Working together with interventions in the areas

---

Submitted by the National Association of Urban and Regional Postgraduate and Research Programmes (ANPUR), Brazil.

This paper was presented at the XIII National Meeting of ANPUR, held in Florianópolis, Brazil, 25–29 May 2009.

Originally published in Portuguese in *Revista Brasileira de Estudos Urbanos e Regionais* 11:1 (2009).

and sectors traditionally related to security – i.e. prisons and police groups – the idea of PRONASCI is to add a socio-environmental dimension that leads 'space' to be seen as a place of intervention in the fight against crime and for security, rather than a place where crime occurs.

Given the rapid growth of public concern about security and its enforcement in this country, it appears that the pursuit of 'public security' is to be a mandatory focus in urban planning from now on, particularly with reference to spaces where low-income families live. Some well-known procedures put forward by 'Safe Urban Spaces' form the basis for interventions within the ambit of the Brazilian Federal 'Growth Acceleration Plan' (PAC), the largest technical and economic investment in poor areas in Brazil to date.

As Martins noted (2003), many official documents and urban plan updates clearly highlight the high level of international concern about public security. The institution of Law n. 95–73 in France (in 1995) – which inaugurated the inclusion of public security criteria into the proceedings for urban development approvals – shows the extent of this worldwide preoccupation in the very choice of its name: *Loi d'Orientations et de Programmation relative à la Securité*. England (in 1994) instituted a similar urban measure titled 'Planning Out Crime'. This measure signaled the incorporation of 'security' into urban development regulation as something outside the bounds of police intervention, as a kind of prevention in the design of the city. The recent prominence of this topic has generated a wide range of research concerning the relationship between public security and planning.

This research takes two different approaches in Latin America. On the one hand is INVI – a pioneer in researching the role of the physical and architectural environment on human behavior, particularly the deterrence of acts of delinquency. On the other hand, Juntas Comunales has contributed to the development of methodologies for popular participation in interventions in low-income areas.

For urban planning, security has become a theme for physical interventions at different 'urban design' scales. As it is believed that human behavior – individual or collective – plays a crucial role in space and is conditioned by physical configurations of space, different mechanisms and tools have been adopted by researchers as means to improve the level of security in city housing spaces.

We approach the topic by asking two questions: (1) Is there a strict relationship between environmental physical configuration and security? (2) If there is a relationship, what are the mechanisms and ways to promote security? We decided to trace a double path in our investigation: examining the origins of the emergence of this practice in Brazil – and worldwide – and following the course of its introduction in Brazil, outlining its characteristics and applications.

We begin by presenting the main characteristics of PRONASCI as the Brazilian version of the international model. Then we examine the concepts underlying the model, focusing on three levels of theoretical debate that have influenced it. Finally, we present the experience of Rio de Janeiro with the 'Safe Urban Spaces' project, as a way to elucidate in concrete form an example of the model's application in our country (Brazil) during the lead-up to the Pan-American Games in 2007.[1]

PRONASCI – national public security and citizenship program

In October 2007, Law n. 11530 was instituted in order to implement the National Public Security and Citizenship Programme (PRONASCI) in Brazil, in response to the perceived necessity for a new instrument that could respond immediately to the catastrophic security situation in the country.

This event was preceded by a partnership between SENASP (Ministry of Justice) and the United Nations Development Programme (UNDP) in 2003, in order to implement the 'Segurança Cidadã' Project, which aimed to initiate "a group of actions directed towards diminishing violence in Brazil". Based on the principle that security should not be limited to police efforts, it is clear that the aim is "being able to walk safely in urban areas, solving conflicts in a peaceful way and integrating communities so as to avoid rumors and disturbances".[2]

'Segurança Cidadã' is an idea that has been adopted by many international agencies and signals the creation of "an adequate conceptual landmark" in dealing with security. The document prepared by the Inter-American Development Bank (IDB) says:

> It is no exaggeration to remind us that displacement of the concept of national security (associated with the concept in a non-democracy) and the concept of state security (associated with the concept centered on the state) has taken place in the wake of the entrenchment of the concept of citizenship security (which places the problem under the umbrella of a concept centered on the citizen and the community).
>
> (Alda and Beliz 2007: xix)

This document assumes that police efforts to repress crime should be expanded in order to incorporate inclusive actions towards citizens. It advances the idea that a reformulation of the idea of public security is necessary, and that responsibility for police and prison systems should be shared with community groups.

In the same document, IDB declares its intention to sponsor initiatives that take 'Segurança Cidadã' as a motto in Latin America and the Caribbean Islands, committing over 200 million dollars for the implementation of this project in many different countries on the continent, as the proceedings at the Interamerican Forum on Security and Citizenship Interaction in Medellion (Foro Interamericano de Seguridad y Convivencia Ciudadana (September 2005)) affirm.

It is worth mentioning that a team composed of specialists was formed a few years ago – including Luiz Eduardo Soares and Antonio Carlos Biscaia. One of the results of their collective effort was a 'bitter' document titled "Public Security Project for Brazil", in which the researchers recognized the impotence of governmental institutions to respond effectively to the existing challenges. In their diagnosis, some data on crime was linked, on the one hand, to its socio-economic meaning and on the other hand, to recurrent governmental problems (i.e. corruption, financial extortion). These problems were presented as symptoms of a deeper societal malaise. In addition, they pointed to qualitative and quantitative evidence of inadequacies in the operation of the prison system in Brazil.[3] Even though this document was ultimately ignored, it did influence subsequent debates on this topic.

Our perception is that PRONASCI was shaped by two factors: the anxiety about, and the societal demand for, a political revision in public security affairs; and the call to balance the policies of international agencies (e.g. UNDP and IDB) towards Latin-American countries (policies derived from Canada and European countries).[4] The Brazilian example reflects both factors.

The official PRONASCI documents demonstrate a dedication to integrating security policies with social actions; a preoccupation with 'prevention'; and the search for causes of violence in society, with a focus on strategies for maintaining public security and social order. These documents focus on evaluating the performance of public security professionals; restructuring of the penitentiary system; the fight against corruption; and the involvement of the community in the prevention of violence.

Since the diagnosis based on a "Public Security Project for Brazil" was released (2004), 94 measures have been presented, addressing two priorities: (1) structural actions, consisting of modernizing public security institutions and prison systems; (2) local programs, which are actions of a social and normative nature, developed in regions selected by PRONASCI.

With regard to public security, PRONASCI has promoted strategies to benefit and give value to the police profession – ranging from creating specific housing programs, to the acquisition of housing, to providing scholarships to stimulate professional development.

With regard to local social programs, the targets of preventive actions are young men and women between 15 and 30 years old and black people, identified as the greatest victims of violence and discrimination. Therefore, the programs were aimed at protecting these groups and avoiding their seduction by a 'life of crime'.

With regard to the prison system, some measures provide options for rehabilitating prisoners at a young age, and to avoid their having relationships with tough, older prison inmates. In order to accomplish this, it has been necessary to segregate prisoners according to their age and the seriousness of their crimes.

What is most significant about these issues, is that every local program is designed to analyze and solve problems related to spaces characterized by high levels of delinquency – identified as 'social de-cohesion territories'. The program developed by PRONASCI foresees the implementation of various measures aimed at the enhancement of educational projects within local communities, with an emphasis on cultural and leisure facilities.

The 'Segurança Cidadã' project started in 2003, but PRONASCI joined it in 2007. At that time, the project itself was running some local programs that contributed to the creation of 'Peaceful Territories' and to the integration of young people's families into the project 'Security and Living' – an improved version of a very successful implementation of 'Segurança Cidadã' in other Latin-American cities, especially in Colombia. The successful Colombian experience is known for its strong sense of community and its emphasis on the implementation of cultural facilities, mainly libraries that serve as meeting places and as focal points for many other activities in the surrounding area.

Seeking to adapt this model to Brazil, SENASP signed an agreement with the Ministry of Culture in 2007, focusing on the propagation of Cultural Landmarks, Museums, Modern Libraries and Skilled Centers for Community Digital Inclusion. Rehabilitation of sports and leisure spaces completed the repertoire.

Even though the 'carioca' (born in Rio de Janeiro) style of 'Segurança Cidadã' techniques is limited to the establishment of these facilities in poor neighborhoods, the range of socio-spatial interventions supported by international agencies is much larger. There has been a consensus among many countries since the 1970s that calls for recognition of the relationship between spatial configuration and its technologies (urban equipment, vegetation, etc.) and the role of 'security' (measured by the incidence of crime). This agreement reflects the perception of violence by the local population. This knowledge integrates such disciplinary fields as sociology, integrationist psychology, architecture and urbanism into both academic and political practices, as we discuss below.

PRONASCI has made its metropolitan actions, and those relating to national places with a high level of criminality, a priority.[5] Considering the few years PRONASCI has been in existence, its activity and strength on the national level are impressive. In September 2006 an agreement was reached between the National Mayors Front and PRONASCI, and in November 2006 some municipal governments in the metropolitan region of Rio de Janeiro joined the agreement.

Once again, the priority of this effort was to include the social legacy of benefits promoted by the Pan-American Games in Rio de Janeiro (which we will speak of later) and, more recently, the programs, proposals and actions provided to help the Growth Acceleration Plan in run-down areas of the city.

In the last month of December 2008, the 'Safe Urban Spaces' seminar, held in Brasilia, Brazil, outlined the main results of the implementation of this program in Brazil. Even though most of the projects presented did not conform to the proposed model, they represented progress towards introducing the procedures of 'Segurança Cidadã' into urban planning.[6]

Safe urban spaces: 'security' in city design

At first sight PRONASCI may look like an isolated initiative – either an attempt to simply imitate a very successful experience in Bogota, or an unreflective implementation of 'good practices' recommended by international agencies. However, a deeper analysis of the underlying foundations of the program leads to a different understanding.

Some theoreticians have already highlighted the link between space and human behavior. Foucault described the implications of 'panopticism' in his 1975 work *Discipline and Punish: the birth of the prison*, asserting that urban interventions hide security strategies. As many other authors have already shown, the disciplinary systems of regulation in the spaces of the city have also been concealed in urban interventions since Haussmann's times.

At the end of 1970s Foucault identified the replacement of the 'disciplinary society'[7] by the 'society of control', in which sophisticated techniques are based on the extension of knowledge and on the employment of new information technologies – empowered by the world of the electronic eye (cameras, satellite images, GPS, etc.). The precision of these new gadgets offers a wide range of control, as they make possible digital statistics and new tools for evaluating and quantifying. In Foucault's opinion, the society of control is firmly built on political dimensions that incorporate public conditioning – or *biopolitics* – in which the term 'security' changes its scale and is applied on a microphysical level. As security assumes a strategic role in the structure of power and behavior, surveillance and control rise in importance. In recent decades there has been an explosion of research that explains these mechanisms, e.g. the works of Mike Davis and Michael Sorkin. From this perspective, we become aware of global policies tending to control poverty levels in metropolitan cities that are understood as risk factors in security issues. Loïc Wacquant, a famous researcher on peripheral contexts in North American big cities, had previously situated the smooth transfer of surveillance from police corporations to society and other institutions – something he has labeled as social panopticon (Wacquant 2008: 14). For Wacquant, the broader context of increasing informality in the economy in big cities raises the specter of the HYPER-GHETTO – the contemporary form of the post-war ghetto – which no longer provides neighborhood districts with extensive reserve labor, like a "mere repository of surplus categories", as in past times, but now has "lost its positive role of collective damping, becoming a killing machine of pure social banishment" (Wacquant 2008: 56).

The transformation of a community ghetto into a hyper-ghetto is dramatically described by Wacquant as a combination of "daily processes of 'non-pacification'" – revealed by the infiltration of violence into the social fabric – with neoliberal penalties, "of which the idea of 'zero-tolerance' spread all over the world by the action of politicians, government officers and academic people is a badge" (Wacquant 2008: 13).

Meanwhile, on the other hand, punishment is on the increase, as a result of contemporary malaise and the 'construction' of a generalized fear by news media. As Cavalcanti et al. (2005) and Machado and Leite (2007) argue, social danger is usually blamed on the poor (and is linked to this condition through analysis of their place of living, usually the 'slums') who embody the diffused fear that embraces the upper and middle classes.

Taking Rio de Janeiro as our spatial evidence for these considerations, it is clear how much of Carioca's common sense is related to this kind of generalized information and to the microphysical action of the news media – which turn the 'slum' into a global space of problems and social deviation and blames the poor for high levels of criminality in the city. In reality, these problems are a result of more complex processes – a view endorsed by most criminologists.

It was no surprise that at the end of the 1970s there was a convergence of knowledge from many disciplines, and that at the beginning of the twenty-first century this knowledge began to focus on large low-income areas in big cities, having the goals of transparency, constancy and collective surveillance that constrain the coming and going of people in space and assure the repertoire of the 'Safe Urban Spaces' methodology.

Theories of safe urban spaces

Originating in Europe, this approach arrived on the international scene with the 1st International Conference on Crime Prevention Through Environmental Design (CPTED), in Calgary, Alberta, 1996.[8] Building on this background, theoretical-conceptual and practical application in Latin-American countries led to two centers of excellence in architecture and urbanism – PUC-Chile and Institute La Vivienda/ INVI, both in Santiago do Chile.[9]

The theories consolidated around the theme of Space and Security have two things in common: first of all, the belief that there is a direct relationship between physical configuration and the incidence of crime which makes "the location of delinquency [. . .] an architectonic fact that shall be analyzed and thought by architects" (Rau 2004: 13). Perez Aravena (cited in Rau 2004: 13) suggests that the "understanding of the relation between form and life is crucial in order to articulate the reality of the project and to inhibit the location of some types of crime, as well as to enhance the sense of security". According to these authors, we should understand delinquency as an act that involves a specific space.

Second, these theories articulate the same variables: (1) the urban physical configuration, including plans, uses/activities, accessible routes, typology and the general aspect of public space; (2) the existence of behavior profiles based on specific spatial configurations that are linked to some symptoms of delinquency and to more common crimes; (3) the location of crimes in the city. As a matter of fact, poor areas are usually the places in which crimes most frequently occur; thus, the procedures developed in Europe and Latin America by 'Safe Urban Spaces' are not deployed in all the spaces in the city, but concentrated in low-income neighborhoods.

In spite of differences of opinion, the works that pursue the theoretical and practical construction of this theme agree that the Chicago School was crucial to this debate. This school originated thinking regarding the impact of physical and psychosocial issues on collective spaces in order to "let architectonic knowledge – specially related to security dwelling issues – set itself around the re-organization of spatial forms so as to analyze changes in behavior and social structures" (Sepúlveda et al. 1999: 21).

Jane Jacobs sets down a fundamental reference for the construction of this idea when she evokes residents' positive attitudes and the generation of values such as 'confidence', 'respect' and some others of crucial importance for the promotion of security, although her ideas went beyond the physical environment, towards developing the sense of community required for security.

The fundamental ideas of Jacobs are based on collective surveillance (the 'eyes' of the street), the idea of diversity and the constant movement of people and activities necessary to deter delinquency, as well as the effective appropriation of places so as to awake 'natural' control actions.

In developed methodologies subsequently, these items are converted into proposals for the reformulation of space so as to amplify the axes of visibility and the exercise of a collective consciousness – which results in a 'natural surveillance mechanism', in contrast to the official 'public surveillance' provided by governmental agents (police).

The first author to formulate a consistent theory around this subject was the American criminologist Schlomo Angel, whose thesis "Discouraging Crime through City Plan" (defended in 1968) utilized the ideas of Christopher Alexander's urban patterns. His study, applied to certain spaces in Oakland, California, focused on the process of inhibiting so-called 'street crimes'.

The literature review presented in Macarena Rau's thesis highlighted the following three theories related to this concept. These theories were subsequently transformed into methodologies and techniques for redesigning space:

- the theory of Defensible Urban Spaces, which presented the first hypothesis and developed methodological instruments for design application and new urban spatial configurations;
- the Situational theory, which built on Crime Prevention through Environmental Design (CPTED) techniques. This theory has spawned an international movement of the same name;
- the urban theory of Space Syntax, which examines the role of space in the generalized perception of fear.

Defensible spaces

The greatest contribution to this area of urban studies came from the architect Oscar Newman, when *Defensible Spaces* (1972) was released. His starting point was the idea that environment plays a fundamental role in determining people's behavior and may stimulate delinquency.

Newman's vision was based on the simplest of propositions: that acts of delinquency result from occasions in which three basic elements are gathered in time and space: (1) a potential delinquent, (2) an 'appropriate target' and (3) the absence of sufficient dissuasion. This last element would be melded with urban design.

From the very beginning of his research on Residential Peripheral Groupings with a high incidence of crimes, Newman postulated three causes for the increase in 'anti-socialized conducts': (1) the anonymity of residents (because of the large size of groupings); (2) the lack of surveillance (that once occurred naturally in building premises); and (3) the lack of alternative routes in the 'maze' formalized by urban design. The techniques proposed by Newman focused on the definition of public and private domains, on the idea that surveillance is conditioned by visibility and clarity of definition, and on the norm that the territory is everyone's responsibility.

Newman defines 'defensible space' as "the physical space taken by an individual and defended against others" (Sepúlveda et al. 1999: 21), calling for relations of similarity versus difference, identity and ways of spatial appropriation. This 'formula' makes evident that 'space' is differentiated by its design, and that residents play an active role in defense of their territory. His interventions are practical: the separation of spaces, the creation of visible borderlines, and the promotion of a typological scale: public, semi-public, private and semi-private spaces.

Situational (crime prevention) theory

Macarena Rau distinguishes two 'generations' of propositions that outline the so-called CPTED 'school'. This school emerges from the work of Ray Jeffrey (1971) – a criminologist at Florida State University – one year before Newman. Through the publication of his work, CPTED (Crime Prevention through Environmental Design) became a primary source of inspiration for this subject. Jeffrey's work originated in research that aimed to improve the level of security in schools – specially focusing on the elimination of crime in young populations.

His 'opportunity theory' emphasizes the importance of three pillars for security: (1) the figure of the delinquent, (2) the victim and (3) the situation. In addition, there is the idea that some crimes are linked to certain kinds of spaces.

According to Rau, these were the four concepts of 'first generation' CPTED: (1) natural control of accesses; (2) natural surveillance; (3) maintenance of urban spaces (referring to cleaning, gardening and general care of public spaces); and (4) territorial reinforcement (referring to the affection each resident develops towards the surrounding space). Regarding this last concept, some studies showed that a positive effect would be achieved by "finding safe activities to be developed in unsafe areas".

One of the differences between Newman's and Jeffery's works is the emphasis placed on 'social' matters. Newman constrains his studies to the development of collective surveillance by residents. In reality, the debate in the following decades centered on this idea. The 'social' component, albeit restricted to the size of the 'community', plays a central role in the discussion, which confirms the crucial role of resident involvement in 'defending' the place. The engagement of residents in this kind of discussion is a key part of the process, one that results in better proposals for intervention.

As a matter of fact, one of the structural differences between *Defensible Spaces* and the 'Situational Theory' is that the second emphasizes aspects related to the social structure of the population, so that necessary interventions include the development of sensitive bonds, good relationships with neighbors, historical occupation, demographic data, income levels, levels of poverty and scholarly targets, sanitary conditions, etc.

Concerns about this type of question are more evident in 'second generation' CPTED, developed with four new categories of analysis: (1) community-scale development: a physical variable associated with the perception of fear which is, consequently, associated with control over space; (2) assessment of community meeting spaces; (3) involvement of existing community organizations; (4) and the active participation of residents.

Space Syntax

Space Syntax was developed by Bill Hillier in the 1970s and emerged from some personal questions about the social relationships promoted by different configurations of built spaces. The idea of security (and safeness) in Space Syntax was developed some time ago (Hillier and Hanson 1984).[10] The primary element in this study is INTEGRATION, understood as the level of accessibility reached by pedestrians and vehicles around the city. This integration stimulates people's movement and activities within a region.

This emphasis strengths the importance of meeting spaces, reflecting the idea that "bodies, movement and face-to-face communication are embodied in space". Bill Hillier names the potential contact among people in urban spaces as VIRTUAL COMMUNITY – a concept akin to the concept of natural surveillance, as it encourages the presence of many people in the same place.

Rau emphasizes three reasons (taken from Hillier) that justify Space Syntax as a good instrument for studying spatial patterns in urban crime: (1) it allows identification of the possibilities of movement at a singular urban configuration – which would work as a substitute tool for the management of natural surveillance; (2) investigating the patterns of crimes in distinct areas or even inside of them; (3) quantifying spatial variables, as well as social and economic ones.

Synthesis

The above-mentioned theories have four fundamental characteristics in common: (1) 'natural surveillance' (the idea that every potential delinquent is under observation); (2) increasing the presence and movement of people (improvement of transportation systems and accessibility; variety of uses; connection of spaces; ease of crossing areas); (3) territorial reinforcement (responding to the natural instinct to dominate space); (4) collective control over accesses and territories (control of entrances and presence of people; transparency of boundaries). The objects of intervention, amongst others, are: public lighting; design of streets; reduction of block sizes; care for the quality of environment; diversification of uses; gardening; paving of sidewalks; enlargement of windows and traffic ways. It is clear that the necessity of residents' engagement is fundamental – their surveillance guarantees the maintenance and proper functioning of facilities. Our study emphasizes the similarity between the previous approaches and the measures proposed by the 'Safe Urban Spaces' project in Brazil.

The 'Carioca' experience

The 'Safe Urban Spaces' project (SUSP) was conceived by the National Public Security Department (SENASP – related to the Ministry of Justice) during preparations for the 2007 Pan-American Games in Rio de Janeiro. The project was first applied to the construction of Olympic villages in three 'vulnerable' zones of Rio de Janeiro, next to the stadiums: Complexo da Maré, Complexo do Alemão and Cidade de Deus.

The strategy was based on taking advantage of the 2007 Pan-American Games – as an instrument of mobilization – in order to fully apply the concept of 'Segurança Cidadã', which is the basis of the National Public Security and Citizenship Plan. As the plan would involve both residents and the state in the task of constructing socially inclusionary public policies, the participation of the community was fundamental to guaranteeing the peaceful appropriation of collective public spaces.[11]

The primary goal of constructing Olympic villages ended up being displaced as a result of the redefinition of what some community leaders called "vulnerable areas" (due to their proximity to places used for criminal purposes). The planned investment was to be in the region of US$1.5 million, using resources from the United Nations Development Programme (UNDP). This money would be offered to almost 200,000 people; it was initially controlled by the City Hall – who were in charge of construction. Later on, the State Government replaced City Hall and assumed responsibility for the administration of the program, with the aid of the Social Assistance and Human Rights Office.

Some training workshops for the implementation of SUSP (dealing with 'administration' and 'urban design') were offered in July 2007 to 100 community leaders. The partnership between SUSP and the Low Income Families Syndicate ('Central Única das Favelas' (CUFA)) resulted in the elaboration of rehabilitation projects for urban physical spaces. Another training workshop – in partnership with the Municipal Federation of Industries (FIRJAN) – was offered to a thousand teenagers and adults, aiming at training them in civil construction.

At the end of 2007, 30 projects (out of 100 originally considered) were selected in different regions of the city. While the call for proposals did not include any criteria regarding proximity to the arenas and stadiums that would accommodate the Pan-American Games, half of the interventions (15 projects) were concentrated in neighborhoods of the northern zones, while 10 were applied to communities in the western zone, 2 to southern zones and 2 in the central areas of the city.

Due to some financial adjustments, the basic projects were modified and some were not realized, e.g. in Morro da Providencia and Gamboa neighborhoods. Because of this, only 29 communities were beneficiaries of urban interventions:

constructing sports courts, placing community and leisure equipment and rebuilding plazas.

Some of the interventions were based on remodeling existing spaces: plazas with a playground for kids (13) or a sports court (4). However, in 10 other areas the interventions were on a larger scale: new plazas (7) and new sports courts (3) in previously unutilized spaces or soccer grounds. Besides those similar interventions, there were cases in which a part of a sidewalk was repaired and a community garden was planted.

Less than half of the interventions involved the construction or remodeling of classrooms for children's academic support and teenagers' skill training (3), barbecue areas (2), changing rooms (2), a covered space for events, a recycling center and a track for playing 'bocha'.

The construction was characterized by low complexity and low investment. The greatest difficulty was in preparing the ground, because the sites were usually located near muddy and uneven places – generally subject to slippage in rainy weather. In other cases, the difficulty was in accessing the areas.

Community reviews

Interviews with leaders and community residents enabled us to assess the level of satisfaction and community involvement with the interventions. Although the informants had previously identified necessary equipment and public services – related to basic infrastructure, health care centers, kindergartens – in the end, the different interventions proposed by governmental efforts were widely accepted by them.

In general, when reviewing projects, the community always focused on the quality or details of implementation, rather than on the concept itself. A frequent complaint was related to the fact that there was no protection against heat and direct sun in recreation areas – because of either the lack of vegetation or the lack of covered areas. The sports courts were also criticized by the community for the lack of surrounding protective netting, as well as the absence of changing rooms and complementary equipment for the courts.

However, some misunderstandings concerning the aims of the interventions were apparent. In Jardim do Itá community an unusual situation occurred when a track for playing 'bocha' and a space for elderly people were constructed. The problem was that there were no spaces for teenagers, nor recreational areas for children – who were the major populations in that community.[12]

In Nova Sepetiba community, the problem was related to lack of accessibility – no ramps to access higher levels – even though the intervention was a housing community for people with mobility difficulties.

In Vila Comari, the selection of a community garden rather than a sports court was perplexing to the community. Neither the garden nor the sport courts had been the highest rated projects. The community had selected recreation areas for teenagers and the development of professional skills for teenagers as most important.

The residents of Quitungo community stated that the previous playground area was better than the new one. They had complained about various problems during construction: the lack of security for children (the playground was on a different level than the street); the absence of protective devices next to the playground area (which were later put in place by the contractors); the bare earth surface that replaced the former grass surface, and may have been responsible for the flooding of nearby areas on rainy days. The community also disagreed about the location of the playground; they had once considered building it next to the poorest zone of the community in order to improve the quality of the environment and ease future interventions in that area. Their idea was completely ignored.

The interventions in Alagados de Sepetiba and Rocinha were also criticized as 'inadequate' by the community. As the first name suggests (*alagados* = wetlands), the basic intervention should have been focused on sanitary and better sewer installations – but this did not happen. The second was located in an area considered at risk by GEO-Rio Foundation, outside of the Eco-limits demarcated by the city to control the expansion of the slum into the remaining existing Atlantic Forest, i.e., in an area of environmental preservation.[13]

In Cachoeirinha and Morro do Borel, two sports courts were built near houses that had earlier been identified as being on risky slopes. Likewise, in Borel, a steep site would put the courts at permanent risk from landslides in rainy weather. There were no governmental actions to take care of these problems. A formal report was sent to GEO-Rio to recommend containing and retaining the slopes, but it did not result in any action.

Reaching the goals

Although post-occupancy evaluations were not carried out, the level of satisfaction and anxiety with the opening of those spaces indicates a high degree of utilization by the local population – at least in the mornings and afternoons. At night the use of plazas and open spaces was still difficult because of the absence of lighting, e.g. in Parque Itambé and Parque Maré.

In Parque Itambé a positive aspect of the environmental renovation was that there were no thefts in the mornings. In addition, scrap metal, litter and illegal parking lots were eliminated. In Fazenda Viegas, the improvement of security by proper lighting enabled residents to start using a bus stop that they had previously considered to be a 'dangerous and dark place'.

However, the satisfaction of Viegas' and Vila Esperança's residents did not convert them to an awareness of cleanness and protection. The new areas have clearly been adopted as a huge open 'garbage can'.

What do these projects have in common? The peripheral location of the community that was the principal beneficiary. This situation was also found in Anchieta, Cachoeirinha, CHP-2, Inhaúma, Ramos (Parque Itambé), Turiaçu and Vigário Geral.[14]

This is not necessarily a negative aspect; interventions on the borders of a formal urban fabric are sometimes recommended as a strategy for integrating segregated realities, be they informal/formal, poor/rich or city/slums. Nevertheless, these strategies should always focus on the perceptions of those affected, otherwise they might decline to become involved.

The choice of areas for intervention seems to have followed some random criteria, and did not help to generate potential ways of addressing the problem of 'urban barriers', as would be expected of a project that aims to reduce the physical domain of crime. The lack of a detailed study of the barriers virtually nullifies the meaning of the project actions. In this case intervention in the central areas would have been a preferable way to better meet project goals. This is because the internal areas of communities are more accurately identified by their residents and therefore more easily controlled.

However, it is worth mentioning that most of the interventions occurred in places that had been in constant use by the community, as the only leisure facility in the area. This factor should give rise to better expectations of maintenance by the community, although this alone is not sufficient to deter criminal activity. It should be easier to maintain the functioning of these spaces, once they are seen as a common good.

This hypothesis does not exclude the necessity of shared administration among NGOs (non-governmental organizations), community associations and the public. Such a partnership is necessary to assure proper maintenance of the equipment, the promotion of cultural and sports events to bring together children, teenagers, adults and elders, the crucial work of environmental education and an agenda for social advancement, including job training programs leading to income generation. This administrative aspect was not included in the experimental phase of SUSP, but it is essential to guarantee the achievement of goals.

Even though possible partnerships were not explicit in the formalization of projects, the residents' associations began to take care of them – and they relied upon themselves exclusively. The efforts of local actors to preserve what has been achieved can be noted in Cidade de Deus – the rides placed in the playground for the children are removed and locked up at night, so as to prevent deterioration and vandalism. In Muzema and Vila Comari, the residents have also been concerned about the future generation of new spaces and have started encouraging positive actions for the preservation of equipment – fearing a return to the previous situation.

Another fundamental aspect of the efficacy of social appropriative strategies in public spaces is environmental quality. In the 'carioca' experience of SUSP, this quality has been compromised by solutions focused on a spatial-functional repertoire. The indiscriminate establishment of sports courts reflects this limited perspective, as in Parque Maré and Praia de Ramos.[15]

Besides the lack of spatial and architecturally effective contributions, the scale of interventions ended up by generating uninteresting urban spaces. In many cases, the adaptations consisted of a mere repair of paving or damaged equipment, such as retouching paint; sometimes it was the construction of changing rooms for the courts, something we do not consider to be a significant urban benefit.

Final considerations

The language that characterizes all the 'Segurança Cidadã' documents is clearly addressed to the 'general population', rather than to the residents of low-income neighborhoods and 'slums' (its immediate beneficiaries). It seems that they are aimed at the urban middle classes – whose thinking is usually influenced by the news media.

The way 'favelas' appear in the news media (Cavalcanti et al. 2005; Machado and Leite 2007) gives the impression that drug dealers are the major cause of crime in the city, and that 'favelas' are the places for these people. Because of that, some authors conclude that "rather than fearing somebody, we ought to fear somewhere. However, for a prudent middle class [. . .] 'favela' is not a dangerous place but, mainly, a place where crimes come from" (Cavalcanti et al. 2005: 8).

There is a transfer of responsibility for the problem of security or danger. The metropolitan city was once seen as the 'cradle of risks'; now the concept is that criminality looks not for the city, but for certain places in the city. These places do not give birth to 'dangerous people', but they themselves are dangerous. This danger may flood as a lethal virus into other spaces of the city. Police forces have to operate constantly (violently, if necessary), in order to attain middle- and upper-class ideas of peace.

In this sense, the 'Safe Urban Spaces' project consists of a group of interventions directed at unsafe neighborhoods and housing settlements, which means working on peripheral places where the immigrant and poor populations live. In Brazil, those places are called 'favelas'. The intentions of these interventions are to improve residents' living conditions and to pacify them (this appeal relates 'living well' to the lack of violence – as the 'Peace Territory' project illustrates).

The idea of 'natural surveillance' incorporated by this methodology shows a curious leap from an ordinary person to a wise, unrecognizable and multiple 'public agent' – it can be done by 'anyone'. Foucault once said that the secret of the efficacy of this system (connected to the idea of panopticism) is due to the internalization of surveillance, which efficiently precedes punishment and inhibits undesirable actions. In Bentham's *Panopticon*, this goal is realized by the construction of a single central tower that represents the constant presence of an observer, even when he is not actually there. The 'natural surveillance', as well as the use of video cameras, aims at the same objective: creating a doubt (is the observer there?). When performed by members of the community, this fact may interact with other kinds of embarrassment that are related to public retaliation. In the community's opinion, this is even more frightening than being approached by a corrupt policeman.

As for the concrete forms of intervention, one wonders how far the spatial discipline imposed by applying the techniques of Safe Urban Spaces – for the removal of 'noise', making them somewhat simpler and more visible – is appropriate for these settlements. To the extent that they were built spontaneously by their own residents, will these areas now be able to maintain the same identity and emotional relationships of the past? This is a question that can only be answered through post-occupation research that will be undertaken in the future.

Notes

1 The data used in this chapter is part of the Final Report prepared for a research project on the implementation of 'Safe Urban Spaces'. This project was made possible thanks to the partnership between the State Social Action and Human Rights Department and the Bennett Methodist Institute.

2 Extracted from "Projeto Segurança Cidadã" 4 September 2007, accessed at Portal Segurança com cidadania, <www.segurancacidada.org.br> 24 July 2008.

3 This document shows the great complexity of the problem facing us, as it demands effective and contextualized actions.

4 Canada was a pioneer in exporting this kind of knowledge into the Americas. From that point, Chile took advantage of it, and the idea of 'segurança cidadã' was founded.

5 PRONASCI priority interventional States: Acre, Alagoas, Bahia, Ceará, DF e entornos, Espírito Santo; Minas Gerais; Pará; Paraná; Pernambuco; Rio de Janeiro; Rio Grande do Sul; São Paulo. PRONASCI settled agreements: Rio Grande do Norte; Maranhão; Piauí; Sergipe; Tocantins.

6 Except for 'Camaragibe Saudável' Project, which covers the Rivers Passarinho, Jacarezinho, Vazadouro and Capilé located in affluent Camaragibe, in Pernambuco. This Project is known for its concern with the environment and procedures related to 'segurança cidadã'.

7 According to Foucault, the 'disciplinary society' is that society in which social command is constructed through a diffuse network of dispositives or apparatuses

that produce and regulate customs, habits, and productive practices. Putting this society to work and ensuring obedience to its rules and its mechanisms of inclusion and/or exclusion are accomplished through disciplinary institutions (the prison, the factory, the asylum, the hospital, the university, the school, and so forth).

8 We count 75 professors, architects, criminologists, security professionals and police officers attending this conference. They were responsible for founding the ICA (International CPTED Association), host of the data collected with the theoretical debate over the practical application of this knowledge. The ICA now has more than 700 members.

9 These centers seem to move forward against the traditional speed of others in the continent – due to the partnership between the city of Toronto and the government of Santiago. Since 2000, they have implemented the 'Paz Cidadã' Project and started including crime-prevention measures through a different form of environmental design – based on the revision of security and criminal policies. Since 2004, Macarena Rau, a Professor at PUC-Chile and author of a relevant thesis in this matter, has been running the International CPTED Association in Latin America.

10 The concept of space syntax encompasses a set of theories and techniques for the analysis of spatial configurations. The general idea is that spaces can be broken down into components, analyzed as networks of choices, and then represented as maps and graphs that describe the relative connectivity and integration of those spaces.

11 According to the architect Cláudia Muniz, coordinator of the 'Safe Urban Spaces' Project (Projeto Espaços Urbanos Seguros), this was inspired by an experiment in Bogota in 2003, with excellent results for the reduction of crime. For more information, visit <www.agenciabrasil.gov.br/10jul2007>.

12 The objective of this project was also point of discussion, as nobody in the community would play 'bocha'.

13 GEO-Rio Foundation is a local government agency responsible for risk analysis of geotechnical slopes. ECO-limits is a program for local government implementation of physical barriers that prevent the expansion of slums.

14 The most representative cases of peripheral locations in Rio de Janeiro are: Inhaúma – a neighborhood partially located along an expressway (Linha Amarela); Ramos – situated along another expressway (Ave. Brasil), outside the limits of the closest community (Parque Itambé); and CHP-2 – a community constructed along a principal avenue in the neighborhood (Ave. D. Helder Câmara), a region known for the high incidence of conflicts between rival communities.

15 Within the limits of 'Favela da Maré' neighborhood, this area is located between two CIEPs (schools designed by Oscar Niemeyer); these schools offer covered play courts and are close to the Olympic Stadium of Maré. On 'Praia de Ramos' there is a complex of open swimming pools specially built for low-income families, and also sports courts.

References

Alda, E., and Beliz, G. (eds) (2007) "Banco Interamericano de Desarollo. Introducción," in: *Cuál es la salida? La agenda inconclusa de la seguridad ciudadana*. Washington DC: Inter-American Development Bank.

Cavalcanti, M., Oliveira, L., and Sá-Carvalho, C., Vaz, P. (2005) *Pobreza e Risco: a imagem da favela no noticiário de crime*. Anais da COMPÓS.

Hillier, B. and Hanson, J. (1984) *The Social Logic of Space*. Cambridge: Cambridge University Press.

Jeffrey, C.R. (1971) *Crime Prevention Through Environmental Design*. Beverly Hills, CA: Sage.

Machado, L.A., and Leite, M.P. (2007) Violência, crime e polícia: o que os favelados dizem quando falam desses temas? *Sociedade e Estado, Brasília*, 22: 3, 545–591, set./dez.

Martins, A.M.M. (2003) Segurança e espaço: novas idéias francesas acerca desta relação. In: ENCONTRO NACIONAL DA ANPUR 10, 2003, Belo Horizonte (MG). Encruzilhadas do Planejamento: repensando teorias e práticas, Anais. Belo Horizonte: ANPUR.

Newman, O. (1972) *Defensible Space*. New York, NY: Macmillan.

Rau, M. (2004) Vigilancia Natural en límites de apropriación comunitária. Tesis de Magister PUC, Santiago, Chile.

Sepúlveda, R., de la Puente, P., Torres, E., and Tapia, R. (1999) Seguridad residencial y comunidad Universidad de Chile/Facultad de arquitetctura y Urbanismo/INVI/ Facultad de ciências sociales – Departamento de Sociologia; Chile.

Wacquant, L. (2008) *As duas faces do gueto*. São Paulo: Boitempo.

Chapter 8

Public space and conservation of a historic living city

Melaka, Malaysia

Samira Ramezani, Zul Azri Bin Abd Aziz, and Syed Z.A. Idid

In the historic living city of Melaka, recent attempts to revitalize as well as to conserve the historical entities of this world heritage town seem to lack an orientation towards socio-cultural historical assets. Since Melaka is known for its historical past and rich heritage, it has become a popular tourism destination among international and domestic tourists. The state government has implemented a number of tourism infrastructure and enhancement projects in its efforts to attract more tourists and thus maximize revenue from tourism. Unfortunately, there is a lack of research on residents' lifestyle and use of the outdoor environment of their city. The local residents' culture needs to be taken into consideration more sensitively if the goal is to conserve the character of Melaka as a living city. Therefore this chapter aims to describe how public space – as a place of social interactions and cultural activities – is significant in achieving the goal of conserving Melaka as a historic living city. In tracing the current patterns of outdoor space usage, this study points to the need for conserving various forms of public spaces, including taken-for-granted mundane public spaces. These spaces include streets, linear space, semi-nodal space and nodal space. The main objective is to contribute to formulating future approaches to planning, including design guidelines that will be more effective in terms of enhancing the liveliness of the inanimate areas of the city, and the quality of life in the outdoor environment in a manner that enhances and suits the residents' culture.

Introduction

In the living heritage city of Melaka, recent conservation attempts – aimed at conserving the urban heritage and enhancing tourism – have paid little attention to the socio-cultural issues of local residents. The day-to-day activities of the local

---

Submitted by the Asian Planning Schools Association (APSA).

Paper delivered at 2009 annual conference of APSA.

residents are an important part of urban heritage and should be taken into consideration. Since Melaka is known for its historical past and rich heritage, it has become a popular tourism destination among international and domestic tourists. The state government has implemented a number of tourism infrastructure and enhancement projects in its efforts to attract more tourists and thus maximize revenue from tourism. A plethora of research has been conducted on the tourism issues. However, there is a paucity of research aimed at understanding the locals' day-to-day use of the outdoor environment. If the goal is to conserve the character of Melaka as a living city, then there should be more sensitivity to local residents' cultural needs. Principles for the development of effective conservation guidelines are lacking, especially with regard to enhancing the liveliness of one semi run-down urban area that was once the hub of a particular town (Idid 2008). There is a lack of emphasis on "*area conservation*"; the focus of conservation is centred on the building as "*a single piece of monument*". This emphasis fails to deal with matters of communal interest and with the key "*place making*" precept of conserving elements (tangible and intangible forms) that constitute significant "*places*". It is important to understand local use of the public spaces, because the extent to which the residents use the outdoor environment affects the liveliness of the city.

Published literature on open space use in Melaka is either devoted to preservation of land with cultural and historical significance (e.g. Cartier 1993; Cartier 1997), or to restoration of the urban green environment based on theories of vegetation ecology (e.g. Miyawaki 1998). In contrast, this chapter aims to develop an understanding of how the local residents use the open spaces in the heritage zone. The main objective is to contribute to formulating future approaches to planning, including design guidelines that will be more effective in terms of enhancing the liveliness of the inanimate areas of the city, and the quality of life in the outdoor environment, in a manner which enhances the residents' culture.

The chapter starts with a brief review of the role of public space in increasing the social interaction of residents, which revitalizes the city. It then argues for the importance of public space in historic living cities, outlining the relationship between conservation and the socio-cultural function of public space. The next section introduces the *Melaka Conservation Zone*, and demonstrates that different patterns of spaces have the potential to be used as active communal space, thus revitalizing the liveliness of the Conservation Zone. A number of policy implications are discussed in the last section.

The significance of public space in the revitalization of cities

Public spaces are a fundamental feature of cities – they increase liveability by encouraging everyday activities (as well as special events such as festivals), which results in increasing social interaction. They represent sites of sociability and face-to-face interaction; at the same time their quality is commonly perceived to be a measure of the quality of urban life. Ideally they are places that are accessible to everybody, where difference is encountered and negotiated (Young 1990). Public spaces can be viewed as '*locales*', settings in which social relations and a sense of place are constituted (Eyles and Litva 1998).

Georg Simmel first emphasized the sociological importance of taken-for-granted social routines and practices (Wolffe 1950). According to Simmel, social relationships and forms of association were best understood by referring to their spatial context. He was particularly concerned with minor, less obvious forms of social interaction that take place in everyday social settings (Lechner 1991). Moreover, earlier studies explored the effects of more mundane sources of interaction and engagement on social relations. They pointed to the importance of opportunities for casual interaction afforded – through such local features as street markets, residential squares, sitting-out areas and canal-side walks, or journeys on foot to a school or workplace – to perceptions of inclusion and a sense of community (Cattell and Herring 2002). Public space is the focus of different needs, demands and desires. To explain how this fundamental feature of public space relates to its potential for social integration, Jacobs observed that it was casual contact with people different from oneself which was essential for integration and "*exuberant diversity*" (Jacobs 1961: 70). Often quite mundane places attain symbolic significance for people through social relations that take place there. Public spaces are more than just simply containers of human activity, they possess subjective meanings that accumulate over time. Spaces can contribute to meeting needs for security, identity and a sense of place (Cattell et al. 2008). Ehrenreich (2007) has argued that traditional collective activities such as communal celebrations and festivals have been highly pleasurable for people over the centuries.

The aforementioned literature emphasizes the importance of public space in revitalizing the city in that it generates a sense of place, as well as a sense of community, by encouraging local activities and special events that are an integral part of urban heritage.

The importance of public space and communal activities in historic living cities

According to Litvin (2005), an historic living city is a combination of places, which have been blessed by its people's heritage values. Its environment can be described as a relic of past events, occupations and ownerships. In the context of an historic living city, the communal activities that have been inherited over hundreds of years are the source of liveliness. This can be seen as a reflection of culture, values and worldview resulting in lifestyle. Lifestyle is the main factor in understanding the operative system of the city, and the way people behave in it (Rapoport 1977). These influences create a rich blend of cultures, sub-cultures, customs, traditions, artworks. This manifestation of folk-culture and sub-cultures creates a juxtaposition of a very unique society and its built environment (Idid 2008). The interaction of humans with the past and the present surroundings produces an urban dynamism, and creates the spirit of the place.

Daily lifestyles have a strong impact on the usage of space. The little plaza, the wall along the street, the space under a family tree, the streets, the spaces under the roof of the house entrance doors, and all of the spatial surroundings create the possibility of social life. The existence of the place contributes to the development of social life in the community (Dumreicher and Kolb 2008). In other words, public spaces, in accommodating daily as well as communal activities, are an important source of liveliness in living historic cities.

Urban conservation and the socio-cultural function of public space

Urban conservation provides a set of values and methods by which to preserve and renew the important elements of the city (from which the city's unique sense of place is derived) that often represent the world's multicultural heritage. Planning for urban conservation assures that the cities of the world – with their unique structures, places and districts – will be sustained in their irreplaceable role as the realm of vibrant life, culture and civil society.

According to Cohen (1998), the five issues to address in assessing the urban conservation potential of a particular site are:

- Define the *character* of the urban setting and clarity of the borders of the site. This is a measure of the extent to which boundary and structure are recognizable as urban elements, such as city squares, parks, side streets and elements of nature.
- Locality and *sense of place*.
  This is a measure of the site's regional and local character, atmosphere and its urban spaces, with links to the context of the city.
- *Internal space*, proportions and relations.
  This is assessed by the connections of urban spaces that are created by the volumes of the built environment, but also by the internal continuity of function and uses.

- Style and *design*.
  The overall design approach, character and style prevalent at the site, comprised of buildings, land and uses.
- *Construction* methods and materials.
  The level of performance achieved by authentic building technology.

The ensuing environmental assessment emphasizes the fact that planning for urban conservation should consider the socio-cultural assets of the historical cities in addition to focussing on conserving the heritage buildings and materials. Human activities basically contribute to local identity enhancement, which is based on the patterns of the local community's culture and lifestyle. According to Relph (1976), the identity of a place is determined not only by its appearance, but also by observable activities and their meanings. Similarly, Idid (1998) argues that experiencing activities in social spaces is most significant in determining the meaning of a place. Moreover, in the context of a living city, the lifestyle of people utilizing their environment is much more meaningful than the buildings; it shows that the city is alive and vibrant.

Culture is manifested through daily individual and communal activities (people's behaviour) in cities, which mostly take place in public spaces. Thus, planning for conservation goals must focus on the generation of vibrant public spaces which facilitate conservation of the cultural assets of the community. If the cultural needs of residents are met, their strong sense of place and attachment will result in an interest from local dwellers in the perpetuation of the valued qualities of the place. This interest contributes to the goals of conserving the urban heritage.

The study area

According to Idid (2005), what makes Melaka significant in the context of World Heritage is a crucial question; it will form the basis of the rationale for Melaka's urban conservation approach. Nowhere else in the world can be found a complex system of urban cultures such as that which can be seen in many Malaysian cities. History and the spirit of tolerance among various ethnic groups (natives or immigrants) has enabled various cultures to co-exist. Each has retained its own identity and customs, without merging or blending completely into one another (although some local habits became intermingled and assimilated). Some root cultures have remained resistant to changes even though places where they originated may have undergone significant changes. This has been an essential starting point for the complexity of cultural entities which we see in Malaysia today. Melaka is an icon for the emergence of this type of urban culture.

Melaka can be characterized as a multicultural society in a living city. It is a historic city inhabited by many different societies that have brought about a unique blend of culture. Starting with external influences, it developed its own unique style that started from as early as the 1400s, right through to the present day. This style formed the actual makeup of the current Malaysian urban society; it exhibits the gradual growth of cultural diversity found in all of Malaysian towns. In comparison with other similar settlements around the region, Melaka is undoubtedly unique. Although various urban settlements in the region of Southeast Asia may resemble each other in terms of their physical features, none contains the complex structure of urban societies that is exhibited in Melaka.

Since Melaka is a historic living city that is full of life, public space that generates activity has become part of the town's pattern and character. At times, it can be seen that efforts of urban conservation to promote the cultural entity have been too hasty. More often than not, this has resulted in superficial physical enhancement, aimed only at improving amenities for visitors to the heritage sites. Less emphasis has been given to enhancing communal spaces and to promoting a sense of belonging amongst the local residents. Such schemes were also ineffective in augmenting the value of heritage as an educational source to younger generations. In aspiring to improve the prestige and the quality of life for visitors who come to witness this unique place, overzealous authorities made drastic changes to the tangible and intangible context of the city, destroying the very essence of its character.

Introducing a new scheme or improvement sometimes brings negative effects to a place that is already unique and special. Changes are inevitable and will occur in any place, but these changes should be managed so that the physical and the non-physical entities that have shaped and characterized the particular place are maintained and preserved. The cultural context of the study area does not merely depend on the characteristics of its buildings, but also relates to how humans interact with their surroundings. Efforts to reinstate the ambience of the local surroundings through urban conservation sometimes fail to recognize the significance of social space in preservation of its cultural value.

Within the heritage zone in Melaka, there are many unique localities that can be distinguished by the different types of activities which occur there. Every road in the conservation area has its own unique character, often reflected by unique activities as well as its architectural features and historical context. Because these features are most apparent, they generally become the focus of attention for the visitors and for the authority in its efforts to improve and enhance the physical context. The local communities have received less attention, especially in improving the communal spaces, which had become an integral part of their local lifestyles and culture.

Public space and conservation of a historic living city 161

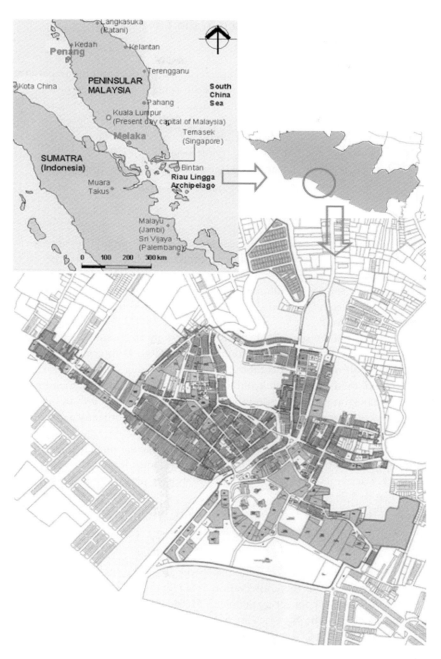

8.1 Central Melaka map.

Usage patterns of outdoor space in the conservation core zone in Melaka

The physical fabric of the Melaka heritage zone exhibits several unique patterns of spaces. Each pattern can be interpreted in the context of its uniqueness by the composition of its physical character and usage. The purpose of discussing these patterns of outdoor space here is to emphasize the need for conserving and enhancing such spaces, not only with regards to conserving the physical entities of this World Heritage City, but also conserving local lifestyles. These public spaces can be described as streets, linear spaces, semi-nodal spaces and nodal spaces.

Street

The streets in the study area are unique; each has its own character. They are narrow – allowing a single passage of movement by motorized vehicles – because they were designed to cater only to light volumes of traffic, especially animal-driven carriages. In the present context, heavy volumes of traffic pass through these streets, causing much discomfort to the residents, with vibrations which threaten the stability of building construction.

Streets were public spaces, which served not only as connectors to various other localities within the heritage zone, but also acted as intermediate spaces connecting the fronts of one neighbour house to the other. In such a situation, children played on the street, while old folks would sit by the front of their houses chatting with neighbours. Even now most of these streets cater not only to vehicular movements, but are generally shared between pedestrians and other forms of street-centred activities.

One particular street, *Jonker Street*, is closed to vehicular movements entirely during the evenings of Friday, Saturday and Sunday, in order to host an array of night market activities. Others are closed for shorter periods, e.g. Jalan Hang Kasturi (Figure 8.2).

Linear space

In the study area, *linear spaces* include the spaces along the riverfront of Melaka River that stretch around the conservation area (Figure 8.3a). This space is the essential component in connecting people to Melaka River, which itself is another significant character of the Historic City of Melaka. *Linear spaces* also include those spaces behind the shops/townhouses – the back lanes between the blocks of buildings. Although these spaces can be seen as simple interconnected passages, they also function as meeting places for individuals and communities (Figure 8.3b).

Public space and conservation of a historic living city 163

8.2 Jalan Hang Kasturi is turned into a pedestrian mall every Sunday morning.

8.3a River walkway that stretches along Melaka River.

With the advent of motorised transport, the "*verandahs*" of the old shops/townhouses became less conducive for the occupants to enjoy the evening sitting and watching the world go by. As more vehicles passed by the fronts of the shops/townhouses, noise and smoke emitted by the vehicles became more acute, and eventually intolerable. The occupants resolved this by using the back part of their shops/townhouses and the back lanes as alternative places to enjoy their private moments away from the busy streets in front.

In some instances, the occupants prefer to stay indoors, especially during the weekdays. This gives an impression that the heritage zone is a dead city. The spirit of liveliness in the heritage zone could only be realized by proper alleviation of the intolerable noise and smoke in the communal or social spaces. This would further strengthen the identity of the historic living city that is the main characteristic and uniqueness of Melaka. The primary goal of conservation efforts in this city should be to acknowledge this key factor.

8.3b Back lanes between shophouses.

Semi-nodal space

Semi-nodal spaces are vacant spaces along the route or path or river within the heritage zone (Figure 8.4). Most of these spaces are pocket spaces and are generally neglected or converted into parking spaces. They have the potential to be used as communal spaces within the heritage zones. The *linear spaces*, especially the back lanes, can be interlinked with these semi-nodal spaces to facilitate better utilization and connectivity.

Nodal space

These free-standing types of spaces offer opportunities to create organized public spaces within the heritage zone. They also have the potential to become important elements that provide robust and attractive spaces and activity venues to the people of the Historic City of Melaka. One example of such spaces is located beside *Tan Kim Seng Bridge* by the riverbank adjacent to the *Dutch Square* (currently an open car-park), which has strong potential to be used as a meeting point and as a river taxi terminal.

8.4 Semi-nodal space adjacent to Kampung Pantai Street.

Another example of this kind of space is located at *Kampung Pantai Street* where a peculiar but interesting "*wedge-shaped*" space was formed. The space was formed incidentally from the way the layout of the area responds to two conditions. One portion of the layout conforms to the existing grid of the old town, whilst the other portion conforms to the meandering shape of the Melaka River. As a result, a unique space was created where the broader area of the wedge gives a strong visual impression – it accentuates the gable-end of a row of shophouses that form its backdrop. Although this space is unfortunately utilized at present for parking space, there is a strong potential that this could become proper public and communal space in the future.

Conservation policy implications

Enhancing and regenerating a variety of spaces

Debates about public space have been shaped by the need to reverse the "*decline*" of public space by providing high-quality urban design solutions. They have tended to disregard the role that "*unexceptional*" spaces play in people's everyday lives. The economic and commercial focus of regeneration programmes should not overshadow the social and cultural value of "*mundane*" spaces. Nor should they ignore people's needs for a degree of stability in their physical and social environment. A policy challenge is to ensure a balance between the need to sustain the cultural entity, which is pertinent to the local residents, and the need to generate new opportunities for tourist attraction. In order for communities to be both sustainable and inclusive, and to realize the beneficial effects of diversity, facilities are needed that encourage the use of public open space and contact between different members of the community. Ordinary public spaces like streets and markets can provide the main place for daily outdoor recreation. "People need a variety of spaces within an area to meet a range of everyday needs: spaces to linger as well as spaces for transit; spaces that bring people together; and spaces for solitude" (Cattell et al. 2008).

Towards residents' participation in planning processes

- Planning decisions have consistently failed to consider the locals' perceptual, affective and cognitive responses to the outdoor environment. In fact, decisions that are related to physical-ecological, functional-behavioural, and aesthetic-visual factors are often left to the designer. In dealing with such considerations, designers of outdoor spaces tend to impose their own

8.5 Social space pattern in the study area.

judgement instead of considering the preferences of the expected users. In other words, they base their decisions on the a priori expert-judgement of experts (e.g., Appleyard and Lintel 1972; Carson 1972; Peterson 1967) and sometimes make naïve assumptions (with questionable relevance) about user preferences (Abu-Ghazzeh 1996). As Canter (1977) asserts, if we are to understand people's responses to places and their actions within them, it is necessary to understand what (and how) they think.

- In the context of Melaka, people participating in the planning process for conservation would likely feel more responsible to protect the entities of their world heritage city as possessing a strong sense of place and as belonging to their locality. It would help to create guidelines which are more responsive to residents' lifestyle and culture, and therefore more practical in enhancing liveliness in areas of the conservation zone which currently seem inanimate.

Conclusion

Public spaces are a fundamental feature of a city. Apart from hosting everyday activities of the local community, these spaces also accommodate special public events such as festivals. Social interactions that take place in these spaces will inevitably promote liveliness in the city. To the Historic City of Melaka, these kinds of activities are in fact cultural assets, which must not be neglected, because they are part of the "*outstanding universal value*" that has put Melaka on the World Heritage list. In managing this cultural property, the local authority must make attempts to ensure that emphasis is given to enhancing the quality of these cultural assets and to regaining the activities and functions of these communal spaces. This not only will restore the vivacity of the place but also enhance the residents' quality of life.

The spaces that are discussed in this chapter have different forms. They are *streets*, *linear spaces* as well as *semi-nodal* and *nodal* spaces. Therefore, in the planning process for revitalization of the study area, the role of various public spaces as places of day-to-day activities of local communities should not be ignored. The local community must be encouraged to participate in the planning process, even at the initial stage, so that they can assist the planners in understanding their patterns of outdoor space use. The participation of the local community in the planning and management process will help to promote their awareness of the importance of conserving their locality.

References

Abu-Ghazzeh, T. (1996) "Reclaiming Public Spaces: The Ecology of Neighborhood Open Spaces in the Town of Abu-Nuseir, Jordan," *Landscape and Urban Planning*, 36: 197–188.

Appleyard, D. and Lintel, M. (1972) "The Environmental Quality of City Streets: The Residents' Viewpoint," *Journal of the American Institute of Planners*, 38: 84–101.

Canter, D. (1977) *The Psychology of Place*, London: Architectural Press.

Carson, D. H. (1972) "Residential Descriptors and Urban Threats," in: J. F. Wohlwill and D. H. Carson (Eds.), *Environmental and the Social Sciences: Perspectives and Applications* (pp. 154–168) Washington, DC: American Psychology Association.

Cartier, C. L. (1993) "Creating Historic Open Space in Melaka," *Geographical Review*, 83:4, 359–373.

Cartier, C. L. (1997) "The Dead, Place/Space, and Social Activism: Constructing the Nationscape in Historic Melaka," *Environment and Planning D: Society and Space*, 15:5, 555–586

Cattell, V. and Herring, R. (2002) "Social Capital and Well Being: Generations in an East London Neighbourhood," *Journal of Mental Health Promotion* 1:3, 8–19.

Cattell, V., Dinesb, N., Geslerc, W., and Curtisd, S. (2008) "Mingling, Observing, and Lingering: Everyday Public Spaces and their Implications for Well-Being and Social Relations," *Health & Place*, 14: 544–561.

Cohen, V. F. (1998) *The Encyclopaedia of Malaysia: Architecture*, Archipelago Press.

Dumreicher, H. and Kolb, B. (2008) "Place as a Social Space: Fields of Encounter Relating to the Local Sustainability Process," *Journal of Environmental Management*, 87: 317.

Ehrenreich, B. (2007) *Dancing in the Streets: a History of Collective Joy*, London: Granta Books.

Eyles, J. and Litva, A. (1998) "Place, Participation and Policy: People in and for Health Care Policy," in: Kearns, R. A. & Gesler, W. M. (Eds.), *Putting Health into Place: Landscape, Identity and Well-being*, Syracuse, NY: Syracuse University Press.

Idid, S. Z. (1998) "Aspek Aktiviti Manusia di dalam Konteks Imej Bandar" (Human Aspects in the Context of Urban Imageability), Non-periodical paper, Faculty of Built Environment, U.T.M.

Idid, S. Z. (2005) "Urban Conservation Approach for a Multi Cultural Historic City – the Urban Planning and Design Perspective – case study on the Conservation Guidelines for the Historic City of Melaka, Malaysia." PhD thesis, Department of Urban Engineering, the University of Tokyo, Japan.

Idid, S. Z. (2008), "Melaka as a World Heritage City," Malaysia: The Melaka State Government and Melaka Historic City Council.

Jacobs, J. (1961) *The Death and Life of Great American Cities*, London: Penguin.

Lechner, F. J. (1991) "Simmel on Social Space," *Theory Culture and Society*, 8: 195–201.

Litvin, S. W. (2005) "Streetscape Improvements in an Historic Tourist City: a Second Visit to King Street, Charleston, South Carolina," *Journal of Tourism Management*, 26: 421–429.

Miyawaki, A. (1998) "Restoration of Urban Green Environments Based on the Theories of Vegetation Ecology," *Ecological Engineering*, 11:1–4, 157–165.

Peterson, G. L. (1967) "A Model of Reference: Quantitative Analysis of the Perception of the Visual Appearance of Residential Neighbourhood," *Journal of Regional Science*, 7: 19–31.

Rapoport, A. (1977) *Human Aspects of Urban Form: Toward a Man–Environment Approach to Urban Form and Design*, Pergamon Press.

Relph, E. (1976) *Place and Placelessness*, London: Pion Limited.

Wolffe, K. (1950) *The Sociology of Georg Simmel*, New York: Free Press.

Young, I. M. (1990) *Justice and the Politics of Difference*, Princeton, NJ: Princeton University Press.

Chapter 9

New urbanism, social equity, and the challenge of post-Katrina rebuilding in Mississippi

Emily Talen

In October 2005, the Congress for the New Urbanism (CNU) was enlisted to produce rebuilding plans for eleven towns along the Mississippi Gulf Coast that had been devastated by Hurricane Katrina. The plans they produced are microcosms of New Urbanist social doctrine: an accessible public realm, neighborhoods that are socially diverse, and walkable access to life's daily needs—design principles that are essentially aimed at promoting social equity. This chapter examines the rhetoric and reality of the social equity goals of the New Urbanist plans for the Mississippi Gulf Coast region. While social equity goals are both implicitly and explicitly stated and visualized throughout the plans, the realization of social equity goals will require more than physical designs. Without the policy, institutional, programmatic, and process requirements that go along with the New Urbanists' physical design proposals, the designs may lose their connection to social equity goals. Given the intensity of development pressure in the region, realization of social equity goals may require an unprecedented level of effort and commitment.

Introduction

When Mississippi Governor Haley Barbour asked the Congress for the New Urbanism (CNU) to develop plans for rebuilding the Mississippi Gulf Coast following Hurricane Katrina, he may not have understood the full implications of his request. In social terms, the eleven city plans formulated during a week-long charrette in October 2005 are microcosms of New Urbanist social doctrine: an accessible public realm, neighborhoods that are socially diverse, and walkable access to life's daily needs. The vision is to be accomplished by implementing design principles aimed at promoting social equity: the mixing of housing types and land uses, the provision of a wide variety of public facilities, and an emphasis on walkable neighborhoods.

Originally published in (2008) *Journal of Planning Education and Research*, 27: 277–294.

By kind permission of Sage Publishers.

172 Emily Talen

Critics of New Urbanism view these social equity goals as rhetoric. They argue that, despite the stated commitment to social equity in the Charter of the New Urbanism (Congress for the New Urbanism 2000), New Urbanist developments have mostly failed to achieve their social vision. According to their detractors, the New Urbanists have accomplished only a facade of social improvement, promoting instead quaint architecture and a "yuppie infantalist fantasy" for the upper middle class (Risen 2005). Some critics are horrified not only that the New Urbanists were put in charge of formulating rebuilding plans for Mississippi but that they will also play a strong role in the redesign of major parts of Louisiana (National Public Radio 2005). A recent article in the *New York Times Magazine*, "Battle for Biloxi," dismissed the New Urbanist plans in Mississippi as quaint and irrelevant (Lewis 2006).

Throughout planning history, there have been tensions between those whose focus is urban aesthetics and those whose focus is social welfare (Talen 2005). In the current manifestation of an old debate, New Urbanism is accused not only of failing to address that discord but of leaning too heavily on the side of design. New Urbanists counter that political realities are behind the failure to translate their social equity goals into reality. Such politics are rapidly playing out in Mississippi, where local officials are now calling for widened highways, larger casinos, and high-rise condominium development (Rivlin 2005). New Urbanist plans calling for development that is neighborhood centered, transit oriented, and supportive of social diversity are marked departures from development patterns predicated on global capitalism, the home building industry, and the highway lobby, all of which are powerful forces in the Mississippi Gulf Coast region.

This chapter examines the rhetoric and reality of the social equity goals of the New Urbanist plans for the rebuilding of the Mississippi Gulf Coast region. For each of the eleven Mississippi communities damaged by Hurricane Katrina, plans formulated by the CNU call for mixed-income, mixed-use, pedestrian-oriented, transit-linked traditional neighborhood development. The social equity goals are both implicitly and explicitly stated and visualized throughout these plans, loyally reflecting the social orientation of the CNU's 2000 Charter of the New Urbanism.

With this social vision in mind, my purpose is to lay out what it will take to achieve social equity via New Urbanist-style planning in the Gulf Coast region. I start by articulating how the plans envision the achievement of social equity through physical design. I then discuss the resistance these plans are likely to encounter and then outline in broad terms what it will take to realize the social equity goals they contain. While social equity is the ultimate aim, realization of this vision will require more than physical plans. Given the level of development pressure in the region, it may require an unprecedented level of effort and commitment. As one critic remarked, "New Urbanism, like all utopian ideas, is great at depicting a better life, but rather poor at showing how to get there" (Risen 2005). To a large extent, this inability to "make it happen" confronts New Urbanism wherever it has tried to

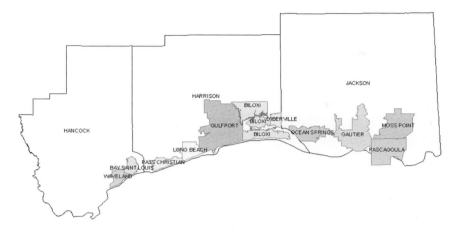

9.1 The three counties and eleven municipalities of coastal Mississippi that were the subject of the Mississippi Renewal Forum charrette held October 11–18, 2005, in Biloxi, about six weeks after Hurricane Katrina struck.

implement its vision. The Mississippi context presents a particularly clear challenge. To date, New Urbanism has been a movement more focused on new green-field projects or unencumbered redevelopment sites in inner city locations. The Mississippi plans are about the revitalization of existing places. Clear understanding of what it will take to realize social equity goals is crucial for the success of rebuilding Mississippi in particular and the success of New Urbanism more generally.

Background

Hurricane Katrina struck the 120-mile Mississippi Gulf Coast on August 29, 2005. The entire coastline was devastated, and property was damaged up to 200 miles inland. Most of the destruction was sustained by eleven coastal cities in three counties (see Figure 9.1). Figures 9.2 and 9.3 show examples of the extent of the devastation. Three cities to the west—Waveland, Pass Christian, and Bay St. Louis—were almost completely destroyed. Although no exact numbers have been collected, thousands of homes, businesses, and essential public facilities were leveled. One estimate is that Hancock County, which was closest to the eye of the storm, lost 50% of its housing stock. For individual cities directly on the coast, the numbers were higher. In Pascagoula, for example, 85 to 90% of the housing stock was damaged (Thompson 2005). Bridges, roads, communication systems, and water and sewer services were wiped out. Flooding rendered standing homes uninhabitable, and tornadoes and hurricane-force winds tore off roofs. Casinos

9.2 Pass Christian, Mississippi, looking toward the site of the marina, now destroyed.

Source: Photograph by Sandy Sorlien

which had been permitted on the coast on floating barges were thrust onshore and created significant damage to other structures.

Days after the hurricane, initial meetings between the governor's representatives and leaders of the CNU began. With funding from the Knight Foundation, CNU was enlisted to conduct a weeklong planning charrette at the damaged Isle of Capri Hotel in Biloxi from October 11 to October 18. About 100 CNU-affiliated architects and planners participated, partnering with a similar number of local architects, planners, and officials from the eleven cities. Several additional teams were formed to address particular issues—transportation, retail, regional, architecture, and social issues. The charge of the eleven city teams was to produce a rebuilding plan for each city. In addition to showing the locations of future roads, parks, and civic spaces, the plans delineated neighborhood structure (in the form of "pedestrian sheds" or areas encompassing a five-minute walk), open space systems, and architectural renderings of revitalized downtowns. All plans conformed to the New Urbanist ideal of mixed-use, mixed-income, walkable neighborhoods. In short, an emulation of "the once marvelous, walkable, villages, towns, and urban neighborhoods of our country— places that organically included the richer and the poorer, the younger and the older"

New urbanism, social equity, and post-Katrina rebuilding 175

9.3 Biloxi Bridge.
Source: Photograph by Sandy Sorlien

(Duany 2005: 4). An editorial by a member of the Governor's Commission on Recovery, Rebuilding and Renewal further emphasized the vision:

> We must preserve the character of damaged neighborhoods, and focus our attention on those who need our help most—citizens in low-income areas, senior citizens, and workers who do not have their own cars to drive to work.
> (Winter 2005)

It would be impossible not to notice that most New Urbanist developments are located in relatively well-off parts of the country, and Mississippi therefore represented a very different type of venue. Mississippi is often referred to as the poorest state in the nation, ranking lowest on many economic, educational, and health indicators. The three coastal counties that were hardest hit by the hurricane are considered to be somewhat wealthier than the rest of Mississippi, but on wealth and other indicators, they still rank below most national averages. For example, median household income and median housing value for the three coastal counties are higher than the state but lower than the national average. About 14% of the coastal county population is below the poverty line, which is lower than the state but higher than the national average. Like many parts of the South, there was strong population growth in the 1990s, but growth slowed significantly after 2000. Tables 9.1 and 9.2 provide basic demographic information for the eleven cities and the region as a whole.

The Mississippi Gulf Coast is socially diverse. About one-fourth of the population is minority—black, Asian, or Hispanic. Furthermore, the minority population percentage is growing in all three counties according to 2004 estimates. Although blacks are by far the largest minority (approximately 20% of the total population), in some cities there is a significant Vietnamese population, many of whom were employed in the shrimping industry. There is also significant income diversity. The population of the Gulfport-Biloxi area, for example, was categorized in 1999 as being 26% low income, 63% middle income, and 11% high income. These numbers had changed slightly since 1969, when the area was approximately 30% low income, 60% middle income, and 10% high income (see also Mississippi Development Authority 2005).

In terms of the social geography of the region, the area is not dissimilar from many other parts of the country—low-income and minority populations toward the older urban centers and wealthier, white populations in the peripheral suburbs. Figure 9.4 shows an example of the typical distribution for housing value in the coastal cities. Figure 9.5 shows an example of the distribution of housing tenure, which tends to be clustered, sometimes with rental units grouped along the coast. Note that spatial segregation is somewhat less relevant in the smaller towns, where relatively wide income disparities are contained within a small geographic area (Pass Christian, for example, is very racially and economically mixed within an eight-square-mile area).

Table 9.1 Population, income, and race statistics for the eleven cities along the Mississippi Gulf Coast

	Population, 2000	Median family income	% White	% Black	% Below poverty line
Gulfport	71,127	$39,213	62	34	17.7
Biloxi	50,644	$40,685	71	19	14.6
Pascagoula	26,200	$39,044	67	29	20.7
Ocean Springs	17,224	$56,237	88	7	5.3
Moss Point	15,851	$37,712	28	70	17.8
Gautier	11,681	$46,835	68	28	17.3
Bay St. Louis	8,209	$41,957	80	17	13.2
D'Iberville	7,608	$40,347	78	11	11.7
Waveland	6,674	$38,438	85	11	13.7
Pass Christian	6,579	$46,232	66	28	10.8
Long Beach	17,320	$50,014	87	7	9.0

Source: U.S. Census, 1990, 2000, and recent updates

Table 9.2 Basic demographics for the three-county Gulf Coast region, Mississippi, and the nation

	Hancock	Harrison	Jackson	All three counties	Mississippi	Nation
Population, 2004 estimate	45,933	192,393	135,436	373,762	2,902,966	290 million
Population, net change, April 1, 2000 to July 1, 2004	2,966	2,792	4,016	9,774	58,310	9.5 million
Population, percentage change, April 1, 2000 to July 1, 2004	6.90%	1.50%	3.10%	2.69%	2.00%	3.30%
Population, 2000	42,967	189,601	131,420	363,988	2,844,658	281 million
Population, 1990	31,760	165,365	115,243	312,368	2,573,216	248.7 million
Population, net change, 1990 to 2000	11,207	24,236	16,177	51,620	271,442	32.3
Population, percentage change, 1990 to 2000	35.30%	14.70%	14.00%	16.53%	10.50%	7.70%
Population 65 years old and over, 2000	6,009	21,002	13,547	40,558	343,523	35 million
Persons 65 years old and over, percentage, 2000	14.00%	11.10%	10.30%	11.14%	12.10%	12.40%
White persons, percentage, 2000	90.19%	73.10%	75.40%	75.96%	61.40%	75.10%
Black or African American persons, percentage, 2000	6.83%	21.10%	20.90%	19.33%	36.30%	12.30%
American Indian and Alaska Native persons, percentage, 2000	0.60%	0.50%	0.30%	0.43%	0.40%	0.90%
Asian persons, percentage, 2000	0.88%	2.60%	1.60%	2.02%	0.70%	3.60%
Persons of Hispanic or Latino origin, percentage, 2000	1.80%	2.60%	2.10%	2.33%	1.40%	12.50%
Black or African American persons, percentage, 2004 estimate	7.08%	22.95%	22.19%	20.73%		12.80%
Asian persons, percentage, 2004 estimate	1.06%	3.43%	2.21%	2.70%		4.20%
Persons of Hispanic or Latino origin, percentage, 2004 estimate	4.67%	5.49%	5.13%	5.26%		14.10%
Households, 2000	16,897	71,538	47,676	136,111	1,046,434	105.5 million
Persons per household, 2000	2.52	2.55	2.72		2.63	2.59
Median household income, 1999	$35,202	$35,624	$39,118		$31,330	$41,994
Per capita money income, 1999	$17,748	$18,024	$17,768		$15,853	$21,587
Persons below poverty, percentage, 1999	14.40%	14.60%	12.70%	13.53%	19.90%	12.40%
Housing units, 2002	22,363	83,631	54,035	160,029	1,195,133	119.3 million
Housing units, net change, April 1, 2000, to July 1, 2002	1,291	3,995	2,357	7,643	33,180	3.4 million
Housing units, percentage change, April 1, 2000 to July 1, 2002	6.10%	5.00%	4.60%	4.78%	2.90%	2.90%
Homeownership rate, 2000	79.60%	62.70%	74.60%		72.30%	66.20%
Median value of owner-occupied housing units, 2000	$92,500	$87,200	$80,300		$71,400	$119,600

Source: U.S. Census, 1990, 2000, and recent updates

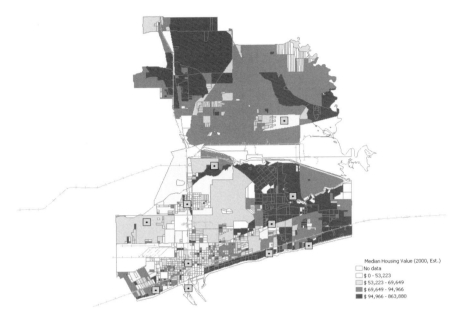

9.4 Gulfport, Mississippi, showing low to high housing value; squares with a dot in indicate proposed neighborhood centers.

Source: www.mississippirenewal.com/documents/Rep_ SocialIssues.pdf

Still, black neighborhoods along the coast tend to be dispersed (more so than in the North) but segregated, as is typical in southern cities (see Goldfield 1982, 1997).

Also of interest are the pre-Katrina projected growth patterns computed by the coast's Gulf Regional Planning Commission (www.grpc.com). In these projections, the core areas of most towns were associated with population and economic decline, while the outlying suburban areas were associated with growth and higher income housing. In the larger towns such as Biloxi and Gulfport, growth since the 1970s has been in the form of sprawling, automobile-based commercial and residential development. The growth projections extrapolate these trends.

The social equity of the Mississippi rebuilding plans

Social equity can be variously defined. Robert Putnam (2000), in his "Social Capital Community Benchmark Survey," defines social equity on the basis of civic involvement across class lines, that is, the equality of civic engagement across a community. But social equity can also be defined as distributive equality— the issue of balancing who gets what—in a planning sense, the spatial distribution of

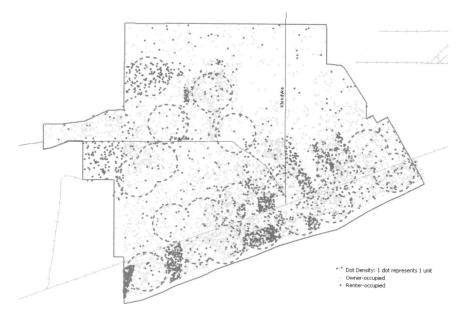

9.5 Long Beach, Mississippi, showing mix of owner- and renter-occupied housing. Proposed neighborhood pedestrian sheds are shown as dotted red circles.

Source: www.mississippirenewal.com/documents/Rep_ SocialIssues.pdf

people and resources. Under this definition, the goal of social equity is the equalization of access to resources. Conceptually, this definition is not dissimilar from other strategies for rectifying the inequity of class divisions. In city planning, this has been extended primarily through the writings of equity planners such as Norman Krumholz (1982; see also Krumholz and Forester 1990).

The provision and distribution of resources as the basis of social equity in planning has been approached in different ways. Lucy (1981) identified five categories of equity of relevance in planning for local services. Subsequent taxonomies relevant to planning have been offered by Truelove (1993) and Marsh and Schilling (1994). The categories of equity that have evolved essentially amount to differences in the methodology that should be used for distribution. Relatedly, social equity in a New Urbanist context uses physical design strategies to put people of all income levels in proximity to the goods and services they require on a daily basis. The design implications of this basic goal are far-reaching. Incorporating services and facilities in residential communities requires design principles that minimize the potentially deleterious effects of combining uses; putting people of different income levels and house size requirements in one location means employing appropriate design to ease the transitions; providing equitable access means that public transit must be appropriately designed to integrate within the community, and so on.

Sometimes, design principles for social equity conflict with one another. Density is the most obvious case. High-density residential form may be good for the social equity goals of access to public transportation but less effective at retaining affordability. One study of the ability of the compact city to address the goals of social equity compared higher density and lower density cities in terms of their ability to improve public transport and access to facilities, reduce social segregation, and increase the supply of affordable housing. From a sample size of twenty-five, the study concluded that "higher urban densities may be positive for some aspects of social equity and negative for others" (Burton 2000, 1969).

Despite potential conflicts, social equity goals in New Urbanist design are clearly defined. The overall goal of social equity can be broken down into three interrelated subcategories: community, diversity, and access. Each of these dimensions is evident in the Mississippi rebuilding plans.

Community

While there are a number of meanings of *community* in planning (Talen 2000), the concept is used in the Mississippi plans to leverage strong investment in public space (Figure 9.6). Equity is promoted because investment in the public realm benefits all inhabitants equally regardless of wealth. This is an idea with long-

9.6 Public space in the form of a village green proposed for Moss Point, Mississippi.
Source: Prepared by the HOK Planning Group with Judson & Partners. Illustration by David Carrico. www.mississippirenewal.com/documents/Rep_MossPoint.pdf

standing roots in the city planning profession (for example, this was the position advocated by City Beautiful era planners a century ago; see Burnham and Bennett (1909/1970)). Assuming public space is kept truly public and access to it is equitable, an enhanced public realm is seen as a progressive asset.

There are a number of ways in which a well designed, located, and activated public realm is to be promoted, and the New Urbanist plans call for a variety of strategies. For example, the insistence on minimal street setbacks is based on the idea that the public realm is greatly improved when "outdoor rooms" are created. This motivation is seen in the plan for Pass Christian, for example, which calls for a redesigned WalMart intended to enhance the public space in front of it (Figure 9.7).

Another way of connecting community and equity is to argue that collective space is essential for providing a venue for community engagement. The idea is that the social and economic diversity of urban places requires some basis for commonality. For New Urbanists, that commonality is the public realm, which fulfills the function of providing a collective identity—if it is designed and located properly. This has been deemed especially important in devastated communities, where restoration of the civic realm is seen as a way of supporting and uniting a fractured community.

This casting of community as essential for social equity is directly related to Putnam's (2000) notion of equity. The New Urbanists, however, put more

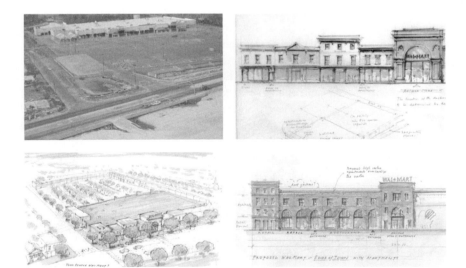

9.7 Architectural rendering of a redesigned WalMart in Pass Christian, Mississippi.
Source: Illustrations by Ben Pentreath. www.mississippirenewal.com/documents/ Rep_Pass Christian.pdf

Diversity

One of the most important tenets of New Urbanism is the notion that human settlement must be patterned in such a way that social and economic diversity is allowed to flourish at the neighborhood level. Particularly undesirable is the creation of communities that, in social terms, are homogeneous (Kunstler 1993). Interestingly, most of the principles New Urbanism espouses are in some way linked to this diversity principle. The call for mixing housing types is obvious, but the call for an adequate provision of public space, public transportation, and neighborhood-based facilities and services within walking distance are all elements that, on a theoretical level, are geared toward the creation and maintenance of social diversity.

That socioeconomic diversity within one small geographic area or neighborhood is a worthwhile goal is driven by two theoretical notions. First is the idea that social mixing in one place is more equitable because it promotes fairer access to resources for all social groups—it nurtures what is known as the "geography of opportunity" (Briggs 2005). In the second sense, diversity is viewed as the basis of a better, more creative, more tolerant, and more peaceful and stable world. Under the first objective, diversity in one location is a matter of better distribution and improved access to resources. Under the second, even those in higher income brackets can take advantage of the creativity, social capital, and cross-fertilization that occurs when people of different backgrounds, income levels, and racial and ethnic groups are mixed.

Sarkissian (1976) and Cole and Goodchild (2001) have reviewed the history of policies that attempt to mix social groups in place. They identified the various goals of social mixing, such as to raise the standards of the lower classes, to encourage aesthetic diversity and cultural cross-fertilization, to increase equality of opportunity, to promote social harmony, to improve the physical functioning of the city (better access to jobs and services), and to maintain stable neighborhoods, whereby one can change housing type and remain in the same neighborhood.

Emphasis on the effects of social mixing in place is still prevalent. A recent report called "Katrina's Window" by the Brookings Institution discussed the benefits of social diversity and the costs of concentrated poverty (Berube and Katz 2005). The authors argued that Hurricane Katrina exposed the flaws of American development patterns by revealing to the world the extent and consequences of concentrated poverty. Social progressives called for the immediate expansion of voucher programs and other methods for deconcentrating poverty and creating geographically based opportunities. Following this principle, planners involved in New Orleans reconstruction advocated the reintegration of the poor (Rosen 2005).

New Urbanists are committed to this same diversity principle, but their emphasis is on using design to achieve it. Specifically, diversity is to be achieved by mixing a range of housing sizes and price levels within the same neighborhood. The approach is evident throughout the Mississippi rebuilding plans. The plans employ such mechanisms as backyard cottages and apartments above shops to allow proximity of dwelling types. In addition, the mixing of housing types and sizes creates the added requirement that units be similarly designed; for example, housing designed for low-income residents is supposed to look as good as housing designed for middle-income residents.

Access

Finally, social equity is to be achieved by providing equal access to neighborhood-based facilities and services. Access, which is defined largely in terms of distance, involves equity because distance has a significant effect on the time, effort, and resources that are required to obtain a given service, good, or facility. This is sometimes referred to as spatial equity—that everyone should have to travel a similar distance to derive benefit from a given resource. In New Urbanism, the term *traditional neighborhood* is based on the idea of access—that diverse peoples and the goods and services they require should be contained within a system of walkable neighborhoods.

Equitable access implies a mix of uses and is in turn connected to social diversity. Diversity of use and social diversity go together since a mixed-use environment is best supported by a mix of different types of people and vice versa. Obviously, if access to services and facilities is only obtained by one segment of the population, the goal of social equity will have been undermined.

New Urbanism promotes equity through equal access by organizing residential environments within a neighborhood structure of pedestrian sheds, representing a five-minute walk to goods, services, or other neighborhood-based assets (Figure 9.8). This promotes equitable access because such goods are likely to be within reach of all residents regardless of their ability to drive. This is especially beneficial for children, the elderly, and the poor who are likely to be more impacted by distance than those with access to a car. Compactness is part of the equation too, because it makes public transit more feasible, and public transit is an important part of maintaining access for all income groups.

Accessibility in the Mississippi plans is essentially promoted through the principle of mixed use—creating a balanced mix of residential and nonresidential

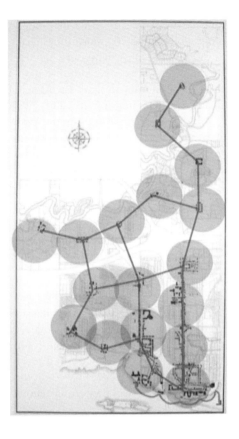

9.8 Neighborhood structure in D'Iberville, Mississippi. Within each one-fourth mile pedestrian shed, there is to be a mix of unit types and walkable access to goods and services.

Source: www.mississippirenewal.com/documents/Rep_ D-Iberville.pdf

uses within one area. In addition, access is enhanced by promoting better street connectivity and shorter blocks, which minimize distance. Correspondingly, streets are engineered to provide not only for automobiles but equally for pedestrians and bicycles.

Successes and failures

In all of these ways, the New Urbanist plans for the Mississippi coast are aimed at promoting social equity. Yet the plans are highly idealized formulations, and many factors will likely work against the implementation of social equity in these terms. This section identifies the main sources of this resistance and includes an assessment of what has transpired along the coast so far. Whether plans can be deemed successful in terms of their implementation is a complex subject (Talen 1996), and some elements of the New Urbanist plans, such as housing type diversity accomplished through long-term application of codes, cannot be assessed in the short term. However, one year after the Mississippi Renewal Forum event, it is possible to make some preliminary assessments. Overall, implementation of the plans has so far been mixed; there are some successes and some failures.

To begin with, the social equity-related goal of community—achieved by enhancing public space—requires significant investment in the public realm in a state not accustomed to high levels of public expenditure. It is generally acknowledged in New Urbanist circles that "the private sector alone cannot be trusted to make places of lasting value" and that the making of community via enhancement of the public realm must be accomplished by a forthright commitment to public sector investment (Duany et al. 2000: 220). The requirement for parks, civic spaces, carefully designed boulevards, and design controls on private buildings are all aspects of community that will be met with the very practical challenge that the political climate in Mississippi is focused on achieving goals largely through the private sector. So far, publicly funded improvements along the coast have not been in keeping with the New Urbanist vision. The biggest disappointment has been the inability to change the course of major transportation projects in the region. The construction of the Biloxi Bridge (sometimes referred to as "bridgezilla") and the widening of the scenic coastal highway (U.S. Route 90) go directly against the New Urbanist plans for a pedestrian-oriented public realm. Now under way, these projects represent an endorsement not only of conventional road building but of the conventional, automobile-based development that accompanies it—most important, high-rise condominiums and casinos. Conventional land use development patterns require the construction of an eight-lane bridge, which is in direct conflict with the New Urbanist plans for a pedestrian-oriented boulevard and transit system. Other transportation planning in the region is continuing this focus on road widening, with little attention given to pedestrian space or alternative transportation modes (see, for example, Gulf Regional Planning Commission 2006). So far, there is little indication that the New Urbanists will be able to promote alternative transportation modes and traffic calming measures that were intended to enhance the public realm.

Indeed, commercial development of casinos and condos along the coast is proceeding conventionally in many places, especially Biloxi. Because the casinos are an important source of revenue, cash-strapped cities along the coast have found it politically infeasible to force casino operators to take a revised, more integrative and less traffic-generating approach to development. One local activist related that the opposition to the New Urbanist plans in favor of casinos and condos has come from local politicians with a vested interest in zoning that favors certain individuals and particular parcels. Gayla Schmitt of Pass Christian commented, "At this point, the citizens of Pass Christian, as well as other towns, feel betrayed. We helped develop a plan which could take us forward in a great way and we have been so far sold out to condo developers."[1]

The ability to use design to enhance community is also impacted upon by an inability to elevate neighborhood planning as a key concern. Other than short-term planning aimed at large public works projects, the embrace of neighborhood planning has been slow. The editor of the largest newspaper along the coast, Ricky Mathews of the Biloxi *SunHerald*, lamented in a recent editorial that "Biloxi seems on a course to manage its growth by granting time-consuming variances every step of the way" (Mathews 2006). For the New Urbanists, resistance by elected officials to socially oriented planning limits the community-building potential of their designs. While academics worry that community building efforts through design are linked to efforts to promote social homogeneity and exclusion (Silver 1985) and planners have admonished the attempt to socially engineer particular types of "balanced" communities (Banerjee and Baer 1984), opposition in the coastal region seems more a matter of opposition to planning in general. From the New Urbanist perspective, overcoming resistance will not be about getting people to think of community as a matter of equity rather than coercion but getting them to think about the importance of planning to begin with.

Elected municipal leadership has failed to endorse the neighborhood organizational structure of the Mississippi plans, but there is evidence of grassroots interest in neighborhood planning. While sociologists question the ability to link individuals and place in fixed ways, arguing in particular against the importance of geographical proximity in fostering notions of community (Janowitz 1952; Suttles 1972), the idea of neighborhood structure presents both a clear vision of what everyday life for Mississippians could be like and simultaneously, hope for localized control. Through a vision of place-based, villagelike neighborhood units, the "new citizen planners" of Mississippi are emerging "more aware now than ever of their options," which is "remarkable for a place that had very little planning outside of conventional zoning administration," according to a planner with the Mississippi Development Authority.[2] It is likely that the plans—their pattern, design, and resource allocation—resonate because they constitute the "things that really count"—schools, security, jobs, property values, and amenities (Pattillo-McCoy 1999: 30). The idea of neighborhood is believed to be more effective if rooted in notions of self-governance as opposed to fixed plans (see, for example, Jacobs 1961), but the Mississippi experience may be evidence that the two conceptual-izations can be valued simultaneously.

It would seem that the social equity goal of diversity would have even less chance of success in a place with little experience in socially oriented planning. There is a wide variety of challenges to planning for diversity in any location, especially given the dynamics of consumer choice, discrimination in institutions and governance, and macro-based phenomena involving political economy and social change (van Kempen 2002; Grigsby et al. 1987). Economists have argued that "even with elimination of all institutional practices that hinder spatial integration, market-based factors would still drive some forms of spatial segregation in a metropolitan area" (Wassmer 2001: 2). The reality in Mississippi as elsewhere is that wealthier residents may be unlikely to consider living near people unlike themselves (poorer and of a different color), particularly if there are issues regarding school quality or perceptions about public safety.

Thus, it is not surprising that the proposal to mix housing unit types within neighborhoods challenges the conventional mode of development along the Mississippi coast, which is characterized by market segmentation of unit types. While Mississippi towns along the Gulf Coast had some degree of residential mixing in their older sections, the mix was at a different scale than that proposed by the New Urbanist plans and tended not to be integrated at the block level. A recent effort by Habitat for Humanity to construct seventy affordable housing units in Waveland is an example of the kind of resistance that occurs. Habitat was seeking a variance to construct houses on 60-foot-wide lots instead of the 100-foot zoning currently in place, a proposal coincident with the New Urbanist plans for more compact development on smaller lots. Habitat was denied the variance because, as one alderman bluntly stated, "large-scale development on small lots could jeopardize property values for existing homeowners" (Habitat Permits Denied 2006). This attitude reflects the historical pattern of settlement in Mississippi, which, not unlike many other parts of the United States, has not been particularly geared toward the maintenance and support of social diversity (Goldfield 1997).

Despite these setbacks, the two arenas in which the New Urbanist plans have met with some success—designing for small affordable housing units and introducing new types of zoning—both address the diversity goal of the plans, specifically affordability and housing type mix. First is the "Katrina Cottage," a small, easily constructed housing unit that was designed during the charrette and meant to be a more permanent replacement to Federal Emergency Management Agency (FEMA) trailers. The cottage has received widespread national publicity (featured in *USA Today*, the *New York Times*, and the *Washington Post*) and was recently awarded the Cooper-Hewitt National Design Museum's People's Design Award. Recently, the U.S. Congress appropriated $280 million under an Alternative Housing Pilot Program in support of Mississippi's Katrina Cottages initiative. The 308-square-foot cottages are now being sold as a kit by Lowe's home building centers. One Gulf Coast community, Ocean Springs, is constructing a cluster of the cottages in its downtown centered around a two-acre parcel dubbed "Ocean Springs Cottage Square."

The second notable success is the widespread consideration of form-based codes, specifically the "SmartCode," a New Urbanist version of a form-based code first developed by Duany, Plater-Zyberk and Co. According to the Mississippi

Renewal Forum's (2006) Web site, eight of the eleven communities along the coast are considering customized SmartCodes, and every community has had at least one SmartCode planning charrette or workshop in the last year. The Web site argues that "no region in America has been so actively engaged in discussions about appropriate zoning" (Miller et al. 2006). The SmartCode goes substantially further in its support of social diversity than conventional zoning by allowing a range of housing types and price levels within each zone. There is also proaction: percentages of housing types are required within each zone. In the General Urban Zone, for example, a minimum residential housing mix of three types is required (SmartCode 2006).

The final dimension of social equity included in the plans—access to neighborhood-based services—may be the most difficult to assess in terms of success only one year after the plans. While most people would agree that walkable access to neighborhood-based goods, services, and facilities would be ideal, there is skepticism that such access can occur in neighborhoods that do not have very high residential densities (a characteristic that applies to most places in the United States). Researchers have found links between urban form and travel behavior (Handy 1996; Boarnet and Crane 2001), but as Levine (2005) points out, clear policy directives are elusive. Some argue that there are specific reasons the traditional walkable neighborhood model was replaced by something completely different, most important because of transportation and technological changes (Marshall 2000).

Still, success on the question of increasing pedestrian access to neighborhood-based services could be measured on the basis of whether the plans are stimulating more mixed-use environments. This is again being accomplished primarily through the application of the SmartCode, which is specifically geared to promoting mixed use. The SmartCode encourages neighborhood facilities and services by permitting multiple functions—a variety of lodging, office, retail, and civic uses—in each of four zones. While many zones within the small towns of the Gulf Coast are composed of the less intensive of these zones, adoption of the SmartCode would permit substantially more land use heterogeneity than current zoning allows.

If the more permissive SmartCodes are adopted for the communities along the coast, the hope is that future development will gradually start to look and feel like the kinds of places envisioned by the New Urbanist plans—walkable, diverse, community oriented. In some respects, it will mean that the form and pattern of New Urbanism can in fact be generated irrespective of the processes that created these forms in the past. Some will see it as a matter of saving appearances, "Disney-style parodies of the places they once were" (Hall 1989: 280). Others will see it as a quest to retain walkable urban forms and patterns that will necessarily have to be validated through development control rather than technology or underlying economic processes.

Future realization

Opposition to community, diversity, and access goals on both practical and theoretical grounds means that any hope of realizing the social equity vision of the rebuilding plans will require strong commitment on the part of local officials, residents, and socially attuned business leaders. Residents will be required to think as citizens and not solely as taxpayers, and in addition, they will need to be convinced that living in mixed-income neighborhoods can be a positive experience. Bureaucrats will need to think as generalists who value social goals, not as specialists with narrow technical interests; and the private sector will need to think in terms of new ways of realizing profits. All of these constituencies will need to recognize that social equity goals do not undermine their interests. For their part, the New Urbanists will need to keep up their visibility and their unequivocal, public commitment to social equity principles.

The commitment begins with a basic recognition that social equity is not achieved by relying solely on the private sector. The two early successes of the plans—the Katrina Cottage and the SmartCode—are important but not sufficient. A variety of approaches will be required for full realization of social equity principles—from the public sector, private sector, and the nongovernmental, nonprofit sector. Individual property owners as well as developers will need to play an important role by agreeing to accept new land development codes that, among other things, create the physical forms important for community, diversity, and access. Local residents will be called on to give increased support to the need for investment in the public realm and active participation in civic institutions. There must be political backing to enforce codes, require a mix of housing types, and motivate the private sector to conform to various development standards.

These facts contradict the rhetoric coming out of the Mississippi governor's office, at least in regard to the rebuilding efforts. The governor has repeatedly stated that the implementation of the rebuilding plans is to be substantially market driven. On the surface, his statements envisioning the rebuilding of the Gulf Coast as "bigger and better" are more aligned with the economic values of casino owners than with the social equity values contained in the plans. The chairman of the Governor's Commission on Recovery, Rebuilding and Renewal, Jim Barksdale (2005), wrote in the opening pages of the Renewal Forum's Summary Report that "after all, it's private investment dollars that will make all this happen" (p. 2). Almost everyone recognizes that both the public and private sector will be needed, but the rhetoric tends to have a tone that downplays the requirement for strong public sector investment.

In reality, the implementation of social equity goals is going to require a lot more than simply turning things over to the private sector. While the complexities involved in realizing equity goals are great, I discuss three requirements that are paramount: maintaining affordability, strengthening public facilities and civic institutions, and committing to full, inclusive public engagement. These will be essential for promoting social equity: the need to link design ideas with the policies, institutions, and programs required for equity and the need to ensure the full engagement of the public in a truly participatory rebuilding process.

Maintaining affordability

In social equity terms, New Urbanism has often been the victim of its own success. In the case of rebuilding Mississippi, replacing devastated middle-income and low-income communities with expensive resort towns such as Seaside, Florida, seems unfathomable. New Urbanism is to be a solution to the displacement caused by the hurricane not another form of it. If their plans are put in place in Mississippi— the boulevards of Biloxi, the transit villages of Gautier, the urbanized Wal-Mart in Pass Christian—New Urbanists run the risk that such efforts will price everyone but the wealthiest out of the market. A recent study by the Rand Corporation sounded the alarm that affordable housing along the Mississippi Gulf Coast was already a critical challenge (Bernstein et al. 2006). If beautiful places are to be created, there will necessarily need to be a political commitment to maintaining the social equity goals that have been so central to the rebuilding vision. For this reason, the realization of social equity will require more than a physical plan; it will require a concomitant set of policies and programs originating from the public and nonprofit sectors.

The fact that a better marriage between policy and design is needed has been increasingly recognized (Deitrick and Ellis 2001). Both the New Urbanists and policy/process oriented neighborhood planners recognize the need to integrate "people-focused and place-focused strategies" (Blackwell and Bell 2005: 291). Those working to provide affordable housing, like many community development corporations (CDCs), believe they are tackling both types of strategies, producing both a large number of affordable housing units as well as focusing on strategies devoted to building community, delivering social services, and generally acting as agents of social renewal (Briggs et al. 1997). Organizations such as PolicyLink and the Institute for Community Economics provide assistance specifically with this merger in mind. Much of the work is focused on delineating actionable tools to promote social equity in urban revitalization strategies (see especially Harmon 2003, 2004; Karlinsky 2000).

Despite calls from New Urbanists to take a policy approach to housing affordability—explicitly called for in both the Mississippi plans and the Louisiana regional plan recently completed by a team of New Urbanists—the integration of design and policy to maintain affordability is not easily accomplished. In conservative states such as Mississippi, the political requirements may be especially difficult. A focus on architecture and design may be viewed as much more palatable to a conservative constituency, while an attempt to implement policies that either regulate the private market or call for the expenditure of public funds may not. Furthermore, while many New Urbanists recognize the need for progressive strategies such as housing vouchers, the earned income tax credit, and the targeting of affordable housing to low-income neighborhoods, the active dialogue and subsequent implementation devoted to these policies are not typically a major part of the New Urbanist agenda.

Because many of the policies and programs needed to achieve social equity goals will require the active participation of community development and affordable housing advocacy groups, New Urbanists and policy planners need to be working in tandem. Unfortunately, the planning charrette that took place in October 2005 did not include the active participation of those who would be in a position to advocate for the kinds of policies needed: linkage fees, profit sharing, and community land trusts, for example. For a variety of logistical and political reasons, this kind of up-front engagement with policy advocates did not occur, and New Urbanist planners and architects worked exclusively on physical designs. In effect, rebuilding plans were limited to supply-side rather than demand-side schemes, with little attention given to the many policies, programs, and neighborhood strategies that are needed to support the vision. As critic Michael Sorkin observed, the rebuilding plans offered "vague recommendations about 'social issues,'" but they were "ghettoized in their own separate—and easily ignored—report" (Sorkin 2006: 47).

As investments are made, construction costs rise (as a result of increased FEMA building requirements) and property values increase, it will be particularly important to implement strategies that ensure affordability for low- and middle-income residents. Many groups involved with coastal recovery planning have asserted this need, most recently the "Reviving the Renaissance" initiative led by the mayor of Biloxi, A. J. Holloway. Housing recommendations in the "Coming Back" and "Moving Forward" reports are in line with the New Urbanist planning goals: utilizing the Biloxi Housing Authority to build units "using the HOPE VI model," stepping up "scattered site" tax credit projects, and calling for an assessment of "workforce housing" needs (Reviving the Renaissance Steering Committee 2006; Living Cities and Goody Clancy 2006).

However, the reports stop short of recommending decisive action on the part of local governments, especially the need to adopt inclusionary zoning regulations. They instead rely on voluntary private and nonprofit efforts. Rising land and construction costs will weaken the ability of these sectors to compete. Local governments along the coast may ultimately find it necessary to follow the model advocated by the 1000 Friends of Florida: establish partnerships in which local governments adopt inclusionary zoning regulations and use community land

192 Emily Talen

trusts—essentially operating as property management companies—to provide a way for developers to meet the requirement. Developers need only to donate some portion of their land to fulfill the requirement (1000 Friends of Florida 2006).

The ability to connect policy and design in this way for the purpose of increasing diversity is or should be the New Urbanist *raison d'être*. The New Urbanist plans and their accompanying codes provide the necessary framework for a partnered approach to occur—a way to successfully integrate market-rate and subsidized housing successfully within the same development. Yet there is often a disconnect with those whose focus is on policy. The recent *Geography of Opportunity: Race and Housing Choice in Metropolitan America* (Briggs 2005) does not mention New Urbanism, nor the design aspects involved in the attempt to plan for diverse human settlement. While conceptually the remedies of those concerned with design solutions and those concerned with policy solutions are the same—both are moved "to build a meaningful response" to the problem of "uncontrolled, sprawling inequity" (Blackwell and Bell 2005: 295)—there is a need to operationalize a more effective integration. Housing policy activists have acknowledged that one reason their organizations have not had more success is "a lack of strong, organized constituencies" (Blackwell and Bell 2005: 306). A stronger connection to design and its regulation might be one way to get it.

There is more the New Urbanists—specifically their national organization, CNU—could do to promote these linkages and ensure that New Urbanist recovery plans do not simply foster exclusive resort development. First, they could step up their efforts to give the issue more visibility by sponsoring publications, Web site resources, conferences, and other approaches to making the connection between diversity, design, and policy explicit and realizable. Efforts to link policy and design have occurred in the past, but there has been little follow-through. In 2000, for example, a conference at Seaside, Florida, was held that brought together a variety of policy and design experts from the Department of Housing and Urban Development, the Urban Land Institute, and CNU to discuss the implementation of mixed-income housing. Robert Davis, the developer of Seaside, conceded that, while he once thought the mix of housing types alone would sustain affordability, he now recognized that policies would also be needed. Recommendations coming out of the conference included the need for "training, education, health care, and other services that lift lower-income people into the ranks of the middle class," and the need for government to "require local communities to accommodate inclusionary housing" (Langdon and Bodzin 2000: 2). According to conference organizers, no further activity took place, beyond the publication of a booklet by the Department of Housing and Urban Development.[3]

Another role for CNU would be to act as a resource for developers who may be interested in partnering with an affordable housing (tax credit) developer or a community land trust or who may be interested in taking advantage of whatever tools and incentives are available to them. Of the 450-plus New Urbanist projects in the United States in various phases of development, it is estimated that approximately 15–20% of them make use of some sort of government or quasi-government program to provide affordable housing as part of their development (Steffel 2002). This includes most notably HOPE VI projects but also Low Income Housing Tax Credits, Block Grants, state affordable housing funds, TIF monies, property tax abatements, and housing trust funds. Although many New Urbanist developers are likely to be open to partnering with nonprofits and/or taking advantage of various government subsidies, they may lack information and experience. Financing for mixed-income communities is often extremely complex, with developers having to piece together funding from a long list of sources. CNU could facilitate the coming together of New Urbanist-oriented developers with lenders and affordable housing financial planners. They could start by researching and making public information on what New Urbanist projects have been most successful in combining programs to produce stable, mixed-income developments. They could publish a "toolkit" specifically geared to the New Urbanist development community. They could showcase mixed-income projects that seem to be particularly replicable, bestow a special award, or host a series of workshops on mixed-income finance.

Public facilities and civic institutions

In addition to the need for affordable housing options to attract and sustain a diverse range of residents, neighborhoods must also be well serviced and safe. There must be good parks and schools, and the spaces and streets of diverse communities must be kept secure. In Mississippi as elsewhere, this usually requires two things: adequate funding for public facilities and a healthy and active array of civic institutions. Civic institutions are generally defined as nongovernmental, faith-based, and/or community-based organizations of all kinds that act as essential players in the democratic process (Kretzmann and McKnight 1997). To help sustain a diverse neighborhood, such institutions must operate at the neighborhood level, even if their interests reach a much broader geographic constituency.

Well functioning public facilities and neighborhood-sustaining civic institutions go hand in hand. Good facilities are not only a matter of sufficient public expenditure— although that is essential—but there must be some active level of collective, grassroots citizenship that can be leveraged to sustain those public facilities. While this is critically important in disadvantaged neighborhoods, it may be just as important in the socially diverse neighborhood, where bonds among residents are likely to be stressed for reasons having to do with cultural, social, and economic differences.

Schools are one category of public facility that is seen as especially crucial for attracting and retaining a diverse range of residents (Varady et al. 2005). One of the most common critiques of the attempt to engender social diversity in a neighborhood is that wealthier residents are unlikely to want to move into an area that contains affordable housing or even housing occupied by people with lower incomes than themselves based on fears about the quality of local services, especially schools. The repeated dismantling of this worry in terms of the effect of lower-income housing on property values (Galster et al. 2003) notwithstanding, the perception that parents will have to send their children to underperforming schools and expose them to higher levels of crime and other dysfunctional behaviors is a very significant issue.

What specific policies are needed to ensure good quality schools in diverse neighborhoods is unclear. Some have argued for completely new school financing methods, advocating for "school choice," such as vouchers and charter schools, as a way of overhauling the current system. In fact, it has been documented that education finance policies are having the effect of spatially segregating households by income (Nechyba 2001). The alternative view is that school choice in the form of charter schools and vouchers has only increased the problem of school segregation, and exacerbated the problem of school inequality.

The need for the "right" public facilities extends beyond schools. There is concern that low-income groups should not be forced into a suburban package of services and amenities that is car dependent and irrelevant to the needs of low-income residents. Special attention needs to be paid to the match of services to population. As Goetz (1996) put it, "the poor relate to [neighborhood] amenities in ways fundamentally different from more affluent families" (p. 3). For example, public transportation and affordable day care are likely to be much more important to poor families. This is not confined to low-income residents. Bayer (2000) argued that African Americans have different tastes in neighborhood attributes, and since these attributes are not always available in high-quality school districts, African Americans may be forced to trade off school quality for neighborhood amenities. Thus, it is necessary to try to ensure that the type and spatial location of amenities in diverse areas is appropriate to the needs of residents with different sets of requirements.

Sustaining diverse communities in the Gulf will also require finding ways to emphasize the positive aspects of diversity that are immediately recognizable. It will require concerted focus on the ways in which improvements in public facilities for some groups translate to improvements in public facilities for all groups. Civic institutions will be needed to help get the message out that diversity as a generator of economic health and improved access to all kinds of services ties directly to diversity intended to foster social equity objectives. As it stands, residents may view increases in housing as an unwelcome addition. The connection that more people may mean better public services and facilities for all needs to be made much more explicit.

Public safety is also a particular concern in diverse neighborhoods, and civic institutions will be needed to help engage residents in ways that overcome their fear of others, help them take control of whatever public safety issues arise, and work to build social connectedness rather than isolation and exclusion. Historically, the Gulf Coast area has had relatively high rates of property crime.[4] In Biloxi, there has recently been some organizing in response to a perceived increase in crime following Hurricane Katrina. Although a rise in crime has not been confirmed by police, residents are now promoting neighborhood watch groups as a way of combating what they say is a post-Katrina spike in crime (Phillips 2006). This encouragement of resident control may be especially important for fostering more positive attitudes about safety in socially mixed neighborhoods. Residents need to be assured that the streets and public spaces of areas mixed in use and mixed in population are safe places and, as one resident of East Biloxi put it, "The cops can't be here all the time . . . people need to look out for one another" (Phillips 2006). The formation of neighborhood watch groups ties into the idea that citizens need to take responsibility for crime prevention and that nongovernmental institutions play a significant role (Peterson et al. 2000).

Community-based institutions—schools, neighborhood associations, faith-based organizations, and the like—are essential for promoting the level of participation and responsibility necessary for this kind of involvement. While policing partnerships are usually aimed at reform of failed methods, they are likely to be essential in places where residents are particularly anxious and fearful of others. In a diverse neighborhood, locally driven efforts are not about transforming the neighborhood but about connecting diverse interests in a way that is sustaining.

Participatory planning

The Mississippi Renewal Forum charrette was an unusual event in terms of public process. The typical planning charrette is widely advertised, and active participation from the public is considered critically important to the success of the event. In Mississippi, this approach was not feasible for several reasons. First, there were few residents available, since many people had been displaced and were living elsewhere. Second, the residents who were still in the Gulf Coast area were consumed with basic questions of survival—fixing their homes, finding jobs, and dealing with insurance claims and FEMA officials. Many were not prepared to engage in a long-range planning process. Third, many of the civic, faith-based, or neighborhood organizations that would normally be a critical part of the planning process were consumed with meeting the basic needs of their constituents. Fourth, the site of the charrette event, which was on the upper floors of a casino hotel that had been heavily damaged, was not a venue capable of accommodating open public access. Finally, communications infrastructure in many of the towns was severely disrupted, and thus, it would have been difficult to properly advertise the charrette. The charrette, therefore, relied solely on the participation of the elected and appointed officials of each town—the mayors, councils, planners, and various other local officials. The idea was that the public would become involved after the plans were formulated and then have a chance to accept, reject, or modify the plans.

Thus it was curious when at the opening of the Mississippi Renewal Forum event, Governor Haley Barbour stated, "I hope that people of the coast will feel that it is their plan because they were so thoroughly involved in crafting it" (Barbour 2005). It signaled a failed understanding, at least on the part of the governor, of what is involved in a truly participatory plan-making process. My point, however, is not to critique the process that occurred but rather to argue that for the achievement of social equity goals, widespread involvement of the public will be critically important. This is because local residents, as opposed to large corporations, casino and condo developers, or even political leaders, may very well be the strongest advocates of the social equity goals of community, diversity, and access.

Unfortunately, important decisions are already being made about highway widening, bridge replacement, density increases, and casino rebuilding in ways that run counter to social equity goals. This situation gives some weight to the argument that it is precisely the difficulty of implementation of equity goals that makes broad-based public participation essential. It is only with some groundswell of community interest that these proposals, which are at times radically counter to prevailing development patterns, have any hope of reversing the status quo. For example, the call to create a light-rail and pedestrian-oriented boulevard in place of a freeway along the coast means that previous decisions about land use and transportation will have to be reevaluated. In the rush to reinstate prior proposals, this interruption of pre-Katrina development inertia will only occur in the wake of strong leadership undergirded if not demanded by widespread public support.

The rebuilding of the bridge over Biloxi Bay is a compelling example of the need for widespread public support. The Mississippi Department of Transportation (MDOT) pushed through its standard approach, and absent a thorough dialogue on the wider social equity implications of conventional highway construction and bridge design, MDOT was unlikely to change. To some extent, the dismissal of social equity goals by a public entity is symptomatic of a failed participatory process on multiple fronts. One transportation engineer on the New Urbanist team commented that absent a more direct dialogue with MDOT, the New Urbanist plans may have been interpreted as threatening, both to MDOT's authority as well as to the jobs they were hoping to create.[5]

The planning process itself can have great value because it widens the horizons of people "beyond their own narrow self-interest to consider a broader public good" (Sirianni and Friedland 2001: 23). The simple fact of participation gives divergent people a chance to create some sense of shared goals. As many neighborhood organizers know, engagement in plan formation at the neighborhood level, like many aspects of civic organizing, provides an essential mechanism for coming together. Some neighborhood planning efforts have been deemed effective for exactly this reason. A study of Minneapolis's Neighborhood Revitalization Program found that one of the main results of the new program was that citizens were engaged and as a result, had increased feelings of "community and efficacy" (Fainstein and Hirst 1996: 108).

Some New Urbanists have complained that the conventional public participation process in the United States is dominated by self-interest and thus, will not necessarily result in the achievement of social equity goals (Duany et al. 2000). Some have pointed out that if minority interests are allowed to dominate, it is alarmingly easy to thwart a well-conceived plan (National Association of Home Builders 2003). This could be remedied by instituting a process that is not open-ended in the sense of allowing self-interest to dominate but is instead focused on establishing a clear connection between social goals and community plans. Political scientists and activists have long recognized the problem of "participatory ills," including "the tyranny of structurelessness, false consensus" and "lack of democratic accountability" (Sirianni and Friedland 2001: 24). The need, especially in the case of social equity, is to establish a participatory process that avoids these pitfalls.

As things currently stand, however, there is a substantial difference in emphasis between the New Urbanist approach to neighborhood planning and the approach taken by community organizations, CDCs, and neighborhood planning focused on equity and advocacy. Where New Urbanism consists of a normative statement of design principles, CDCs and neighborhood planning agencies are focused on supporting the collective, self-help capabilities of residents, helping them fight problems of poverty and blight by building organizational capacity. It is therefore not surprising that neighborhood planning is defined by many in the planning profession as a *process*. The recently released how-to manual of neighborhood planning, *Building Great Neighborhoods: A Citizens' Guide for Neighborhood Planning* (Townsend 2005), states emphatically that "in neighborhood planning, the process is just as important as the plan itself!" (p. 1.1). In this approach, what is considered most important about neighborhood planning is that it is resident driven and that form follows function. It is the professional planner that has the role of finding out how neighborhood planning goals are to be accomplished through design and physical improvement. The danger in this is not so much that design is seen as secondary but that it is not a process that has any particular interest in achieving social equity. There is no guarantee that resident viewpoints will not be committed, in fact, to exclusion.

What may be needed is a new kind of merger between New Urbanist social equity proposals and more conventional, bottom-up approaches to participatory neighborhood planning. This parallels the need to integrate plan and policy as discussed above. Design-based planning committed to social equity need not be in conflict with the idea of planning as process, but it may mean that new, more creative approaches to the participatory process are needed. On one hand, neighborhood residents are likely to be the strongest backers of elements such as community, diversity, and access, but on the other hand, these goals are particularly susceptible to narrow interests and could be derailed by a few outspoken adversaries. One idea is to use methods that ensure a more representative review of proposed change. Better use could be made of methods that rely on the random selection of participants, an approach that has been successfully implemented in other countries (Carson and Martin 1999; Gastil and Levine 2005).

Conclusion

This chapter has spelled out the basic requirements that will be needed to realize the social equity goals inherent in the New Urbanist Mississippi Gulf Coast plans. While conventional transportation, casino, and condo development is proceeding, the New Urbanists can claim some success in their ability to engender grassroots enthusiasm for the Katrina Cottage and the SmartCode. However, additional policy, programmatic, institutional, and process requirements will be needed if the New Urbanists hope to continue working toward the realization of social equity goals. Without policy assurances, civic institutional connections, and broad-based participation directed toward the realization of social equity, the New Urbanist plans may be legitimately critiqued as simply perpetuating an old social order. New Urbanists were outraged when their work in Mississippi was described by Mike Davis as being pleasing to "the slave masters" and reflective of the desire to keep people in their proper social roles (Hawthorne 2005). Such criticism would quickly lose validity in the face of design proposals effectively wedded to the institutions, policies, programs, and processes devoted to promoting social equity.

In *Region, Race and Cities: Interpreting the Urban South*, David Goldfield (1997) offered what he believed could constitute "the promise of the urban South" (p. 302). He argued that even in the midst of urban-rural, center-periphery imbalances, the urban South still retains the benefit of relatively small-scale cities and towns, a tradition of volunteerism, and a strong sense of history and identity. Given these assets, Goldfield asserts that the urban South can provide an example for the whole nation "in race relations, in economic development, in the value of traditions, and maybe even in equity" (p. 304).

The eleven towns along the Gulf of Mississippi offer just this kind of opportunity. Whether they will rise to the occasion and provide an example of model development based on equity is likely to vary widely among the communities. Biloxi's recent call for increased casino development—in bigger, more lavish form than before—and its embrace of an expanded bridge and freeway through town are signs that Biloxi is not prepared to achieve social equity in the terms spelled out in the plans. But other towns may take a different route. Towns that are now committed to rewriting their zoning codes to allow mixed use and mixed housing types, important physical planning components of social equity, are positive steps.

The New Urbanists should be applauded for providing some traction in the ongoing effort to promote social equity through city design. They have given the residents of the Mississippi Gulf Coast a physical representation of what social equity can be, articulated in a form consistent with the scale and vernacular traditions of the South. But the job is incomplete. Greatly needed are the merger of design, policy, and program and the integration of a more complete participatory process. While it is the responsibility of local residents and officials to work out this implementation, the New Urbanists can help by being vocal advocates of social equity and promoting a greater understanding of what it takes to achieve it.

Notes

1   Personal communication with Gayla Schmitt, resident of Pass Christian, MS, December 11, 2006.
2   Personal communication with Ann Daigle, Special Assistant to the Director, Mississippi Redevelopment Authority, December 3, 2006.
3   Personal communication with Phyllis Bleiweis, Executive Director of the Seaside Institute, November 16, 2006.
4   The property crime rate for Harrison County in 2000 was 6,642 crimes per 100,000 population, significantly higher than neighboring Jackson County (4,888 per 100,000), and the states of Mississippi (3,478 per 100,000) and Louisiana (4,410 per 100,000). The violent crime rate for Harrison County was not significantly higher. Source: U.S. Bureau of Justice Statistics (2004).
5   Personal communication with Rick Hall, civil engineer involved with the Mississippi Renewal Forum charrette, December 4, 2006.

References

Banerjee, Tridib, and William C. Baer (1984) *Beyond the neighborhood unit: Residential environments and public policy*, New York: Plenum.

Barbour, Haley (2005) The Mississippi Forum: From tragedy to opportunity. Opening statement presented at the Isle of Capri Hotel in Biloxi, MS, October 11. www.mississippirenewal .com/info/governor.html (accessed on May 1, 2007).

Barksdale, Jim (2005) "Mississippi Renewal Forum summary report: Recommendations for rebuilding the Gulf Coast," *The Town Paper*, Gaithersburg, MD, 2.

Bayer, Patrick (2000) "Tiebout sorting and discrete choices: A new explanation for socioeconomic differences in the consumption of school quality." Working paper available online. http://aida.econ.yale.edu/~pjb37/papers.htm (accessed on May 1, 2007).

Bernstein, Mark A., Julie Kim, Paul Sorensen, Mark Hanson, Adrian Overton, and Scott Hiromoto (2006) *Rebuilding housing along the Mississippi Coast: Ideas for ensuring an adequate supply of affordable housing*. Rand Gulf States Policy Institute, Jackson, MS. http:// www.rand.org/pubs/occasional_papers/OP162/ (accessed on May 1, 2007).

Berube, Alan, and Bruce Katz (2005) *Katrina's window: Confronting concentrated poverty across America*, Washington, DC: Brookings Institution, Metropolitan Policy Program.

Blackwell, Angela Glover, and Judith Bell (2005) "Equitable development for a stronger nation: Lessons from the field." In *The geography of opportunity: Race and housing choice in metropolitan America*. Xavier de Souza Briggs (ed), 289–309. Washington, DC: Brookings Institution.

Boarnet, Marlon, and Randall Crane (2001) *Travel by design: The influence of urban form on travel*, New York: Oxford University Press.

Briggs, Xavier de Souza (2005) "Conclusion: Desegregating the city." In *Desegregating the city: Ghettos, enclaves, and inequality*. David P. Varady (ed), 233–57. Albany: State University of New York Press.

Briggs, Xavier de Souza, Elizabeth J. Mueller, and Mercer L. Sullivan (1997) *From neighborhood to community: Evidence on the social effects of community development*, New York: New School for Social Research.

Burnham, Daniel H., and Edward H. Bennett [1909] (1970) *Plan of Chicago*, New York: Da Capo.

Burton, Elizabeth (2000) "The compact city: Just or just compact? A preliminary analysis," *Urban Studies* 37:11, 1969–2001.

Carson, Lyn, and Brian Martin (1999) *Random selection in politics*, Westport, CT: Praeger.

Cole, Ian, and Barry Goodchild (2001) "Social mix and the 'balanced community' in British housing policy: A tale of two epochs," *GeoJournal* 51:351–60.

Congress for the New Urbanism (2000) *Charter of the new urbanism*, edited by Michael Leccese and Kathleen McCormick, New York: McGraw-Hill.

Deitrick, Sabina, and Cliff Ellis (2001) The importance of design. *Shelterforce Online*. National Housing Institute. March/April. www.nhi.org (accessed on May 1, 2007).

Duany, Andres (2005) A Gulf Coast renaissance. In *Mississippi Renewal Forum summary report: Recommendations for rebuilding the Gulf Coast*, 3–5, Gaithersburg, MD: The Town Paper.

Duany, Andres, Elizabeth Plater-Zyberk, and Jeff Speck (2000) *Suburban nation: The rise of sprawl and the decline of the American dream*, New York: North Point.

Fainstein, Susan S., and Clifford Hirst (1996) "Neighborhood organizations and community planning: Minneapolis Neighborhood Revitalization Program." In *Revitalizing Urban Neighborhoods*, W. Dennis Keating, Norman Krumholz, and Philip Star (eds), 96–111. Lawrence: University Press of Kansas.

Galster, George C., Peter A. Tatian, Anna M. Santiago, Kathryn L.S. Pettit, Robin E. Smith (2003) *Why* not *in my backyard? Neighborhood impacts of deconcentrating assisted housing*, New Brunswick, NJ: Center for Urban Policy Research.

Gastil, John, and Peter Levine (eds) (2005) *The deliberative democracy handbook: Strategies for effective civic engagement in the twenty-first century*, San Francisco: Jossey-Bass.

Goetz, Edward G. (1996) *Clearing the way: Deconcentrating the poor in urban America*, Washington, DC: Urban Institute.

Goldfield, David (1982) *Cotton fields and skyscrapers: Southern city and region, 1607–1980*, Baton Rouge: Louisiana State University Press.

——— (1997) *Region, race and cities: Interpreting the urban South*, Baton Rouge, LA: Louisiana State University Press.

Grigsby, W., M. Baratz, G. Galster, and D. MacLennan (1987) *The dynamics of neighbourhood change and decline*, Oxford, UK: Pergamon.

Gulf Regional Planning Commission (2006) Gulf Coast Area Transportation Study (GCATS). www.grpc.com/gcats .html (accessed on May 1, 2007).

Habitat Permits Denied (2006) *South Mississippi SunHerald*. October 17. www.topix.net (accessed in February 2007).

Hall, Peter (1989) "The turbulent eighth decade: Challenges in American city planning," *Journal of the American Planning Association* 55:3, 275–82.

Handy, Susan (1996) "Urban form and pedestrian choices: A study of Austin neighborhoods," *Transportation Research Record 1552*, TRB, Washington, DC: National Research Council, 135–44.

Harmon, Tasha (2003) *Integrating social equity and growth management*, Springfield, MA: Institute for Community Economics.

——— (2004) *Integrating social equity and smart growth: an overview of tools*, Springfield, MA: Institute for Community Economics.

Hawthorne, Christopher (2005) "Critics notebook: In the rush to rebuild, a house divided," *Los Angeles Times*, December 4.

Jacobs, Jane (1961) *The death and life of great American cities*, New York: Vintage.

Janowitz, Morris (1952) *The community press in an urban setting*, Chicago: University of Chicago Press.

Karlinsky, Sarah (2000) *Community development corporations and smart growth: Putting policy into practice*, Cambridge, MA: Neighborhood Reinvestment Corporation and the Joint Center for Housing Studies of Harvard University.

Kretzmann, John P., and John L. McKnight (1997) *Building communities from the inside out: A path toward finding and mobilizing a community's assets*, Skokie, IL: ACTA.

Krumholz, Norman (1982) "A retrospective view of equity planning," *Journal of the American Planning Association* 41:3, 298–304.

Krumholz, Norman, and John Forester (1990) *Making equity planning work: Leadership in the public sector*, Philadelphia, PA: Temple University Press.

Kunstler, James Howard (1993) *Geography of nowhere: The rise and decline of America's man-made landscape*, New York: Free Press.

Langdon, Philip, and Steven Bodzin (2000) Mixed-income, mixed-use: More than a dream. *Congress for the New Urbanism Update.* www.cnu.org/cnu_updates/CNUpgsDec00.pdf (accessed in November 2006).

Levine, Jonathan (2005) *Zoned out: Regulation, markets, and choices in transportation and metropolitan land-use*, Washington, DC: Resources for the Future.

Lewis, Jim (2006) Battle for Biloxi, *New York Times Magazine*, May 21.

Living Cities and Goody Clancy (2006) "Moving forward: Recommendations for rebuilding East Biloxi." July 2006. http://www .biloxi.ms.us/Reviving_the_Renaissance/ (accessed on May 1, 2007).

Lucy, W. (1981) "Equity and planning for local services," *Journal of the American Planning Association* 47:447–57.

Marsh, M. T., and D. A. Schilling (1994) "Equity measurement in facility location analysis: A review and framework," *European Journal of Operational Research* 74:1–17.

Marshall, Alex (2000) *How cities work: Suburbs, sprawl, and the roads not taken*, Austin: University of Texas Press.

Mathews, Ricky (2006) "461 days after the storm. . . Editorial," *SunHerald*, Biloxi, MS, December 3. www.sunherald.com (accessed in November 2006).

Miller, Jason, Ben Brown, and Sandy Sorlien (2006) "Katrina plus one: At the one-year mark, groundwork is set for widespread renewal," www.mississippirenewal.com/info/dayAug23-06.html.

Mississippi Development Authority (2005) *Mississippi Consolidated Plan for Housing & Community Development, Final Report*, Jackson: Mississippi Development Authority.

Mississippi Renewal Forum (2006) www.mississippirenewal.com

National Association of Home Builders (NAHB) (2003) *The builder's guide to the APA's growing smart legislative guidebook*, Washington, DC: National Association of Home Builders.

National Public Radio (2005) Rebuilding, and Redesigning, New Orleans, *All Things Considered*, September 14.

Nechyba, Thomas (2001) "School finance, spatial segregation and the nature of communities: Lessons for developing countries?" Paper presented at the International seminar on Segregation in the City, Lincoln Institute of Land Policy, Cambridge, MA, July 26–28.

1000 Friends of Florida (2006) Creating balanced residential communities in Florida. Macromedia Flash presentation. www.1000friendsofflorida.org/housing/main.asp (accessed on May 1, 2007).

Pattillo-McCoy, Mary (1999) *Black picket fences: Privilege and peril among the black middle class*, Chicago: University of Chicago Press.

Peterson, Ruth D., Lauren J. Krivo, and Mark A. Harris (2000) "Disadvantage and neighborhood violent crime: Do local institutions matter?" *Journal of Research in Crime and Delinquency* 37:31–63.

Phillips, Steve (2006) East Biloxi neighbors organize to fight rise in crime. WLOX TV. September 21. www.wlox.com (accessed in November 2006).

Putnam, Robert (2000) *Social capital community benchmark survey*, Cambridge, MA: Saguaro Seminar of the John F. Kennedy School of Government, Harvard University.

Reviving the Renaissance Steering Committee (2006) *Coming back: An executive summary of the Reviving the Renaissance Steering Committee*, July 20. http://biloxi.ms.us/pdf/RTR07202006.pdf (accessed on May 1, 2007).

Risen, Clay (2005) "Op-ed: Wrong way home," *The Morning News*. www.themorning news.org/archives/oped/wrong_way_ home.php (accessed in December 2005).

Rivlin, Gary (2005) Bright spot on Gulf as casinos rush to rebuild, *New York Times*, December 14, 1.

Rosen, Jill (2005) "A new, improved New Orleans seen," *Baltimore Sun*, September 23.

Sarkissian, S. (1976) "The idea of social mix in town planning: An historical overview." *Urban Studies* 13:3, 231–46.

Silver, Christopher (1985) "Neighborhood planning in historical perspective," *Journal of the American Planning Association* 51:2, 161–74.

Sirianni, Carmen, and Lewis Friedland (2001) *Civic innovation in America: Community empowerment, public policy, and the movement for civic renewal*, Berkeley: University of California Press.

SmartCode (2006) www.placemakers.com.

Sorkin, Michael (2006) "Will new plans for the Gulf drown it again, this time in nostalgia?" *Architectural Record* February: 47–50.

Steffel, Jennifer (2002) "Affordable housing in New Urbanist communities: Challenges and solutions." Paper presented at the Association of Collegiate Schools of Planning conference, Baltimore, MD, November 22.

Suttles, Gerald (1972) *The social construction of communities*, Chicago: University of Chicago Press.

Talen, Emily (1996) "After the plans: Methods to evaluate the implementation success of plans," *Journal of Planning Education and Research* 16:2, 79–91.

—— (2000) "The problem with community in planning," *Journal of Planning Literature* 15:2, 171–83.

—— (2005) *New Urbanism and American planning: The conflict of cultures*, London: Routledge.

Thompson, Richard (2005) "Mississippi tackles a tough one," *Planning*, December: 6–11.

Townsend, Carol L. (2005) *Building great neighborhoods: A citizens' guide for neighborhood planning*, East Lansing: Michigan State University.

Truelove, M. (1993) "Measurement of spatial equity," *Environment and planning C* 11:19–34.

U.S. Bureau of Justice Statistics (2004). http://bjs.ojp.usdoj.gov/.

U.S. Census, 1990–2000.

van Kempen, Ronald (2002) "The academic formulations: Explanations for the partitioned city." In *Of state and cities: The partitioning of urban space*, Peter Marcuse and Ronald van Kempen (eds), 35–56, Oxford, UK: Oxford University Press.

Varady, D. P., J. A. Raffel, S. Sweeney, and L. Denson (2005) "Attracting middle-income families in the HOPE VI Public Housing Revitalization Program," *Journal of Urban Affairs* 27:2, 149–64.

Wassmer, Robert W. (2001) "The economics of the causes and benefits/costs of urban spatial segregation." Paper presented at the "International Seminar on Segregation in the City," July 26–28, 2001, at the Lincoln Institute of Land Policy, Cambridge, MA.

Winter, William (2005) "A top rebuilding priority is affordable housing," *SunHerald*, Biloxi, MS, October 30.

Chapter 10

Immature take-offs

Urbanization, industrialization and development
in twentieth-century Latin America

Arturo Almandoz

On the basis of some aspects of Rostow's theory that explore the relationship between industrialization, urbanization and development, this chapter aims to illustrate how the imbalance between urbanization and the production system helps to explain the issue of Latin America's economic "take-off" throughout the twentieth century, and the lack of an ensuing "drive to maturity". For that purpose, other political and social analyses are incorporated into a classification of development periods starting with the phase when Latin America first showed significant urbanization. Such a panoramic attempt is made from the historiographic and methodological premise that urbanization and development studies have often lost the historical perspective. It is therefore necessary to resume a panoramic, and to some extent comparative, review – both in historical and territorial terms. A possible contribution of this research is its periodization of Latin American development.

Introduction

According to functionalist sociology, the relationship amongst industrialization, urbanization and modernization was a sort of causal sequence. This paradigm was derived from the cases of countries that had successfully industrialized and urbanized throughout the nineteenth century and the first half of the twentieth (Reissman 1970; Davis 1982). It remained the dominant view until the mid-1960s. Such a sequence was adopted by some of the first comparative studies of Latin America's historical process, at least in relation to the phase of incipient industrialization and high urban growth experienced in several countries after the interwar *masificación*[1] (Hauser 1967; Harris 1971). The ideal modernizing sequence relied, to a great extent, on the mid-twentieth century theories of economic development, especially

---

Originally published in (2008) *Revista Eure* XXXIV: 102, 61–76.

By kind permission of the publisher.

on Walt Whitman Rostow's, which attached great importance to industrialization, growth and political stability within the equation.

By the mid-1960s, some Latin American countries were regarded by Rostow and other economists as examples that had technically initiated the "take-off" towards development. The indicator of this stage was that more than 25% of the Gross National Product (GNP) went towards investment. Venezuela had entered this progressive path after the 1950s, overtaking Brazil, Colombia, Chile and the Philippines (in decreasing order). If Mexico and Argentina had already taken off in the previous decade, the North American professor pointed to Venezuela and Brazil as the aeroplanes of the 1960s (Rostow 1990: 44, 127).

Development was not *guaranteed* by an irreversible moment of take-off, but required a long and deep process of economic, social and political changes, articulated in the phases explained by Rostow in *The Stages of Economic Growth* (1960). Combining elements of economic and political history with sociological descriptions, professors at the Massachusetts Institute of Technology and other Anglo-Saxon universities studied and grouped the cases of "traditional societies" that, from late-eighteenth century England onwards, had modernized their agrarian and industrial sectors. Those societies later increased the resources dedicated to investment, as required by the take-off, in order to eventually achieve a "drive to maturity" – not only economic but also social and political. This drive needed to be maintained throughout two generations of sustained prosperity, before finding the path towards permanent development. Based on the great diversity of take-offs throughout the nineteenth and twentieth centuries – France and the USA in the 1860s, followed by Germany in the 1870s, Russia and Canada after World War I, among the most spectacular examples – Rostow summarized the phases as follows:

> Here then, in an impressionistic rather than an analytic way, are the stages-of-growth which can be distinguished once a traditional society begins its modernization: the transitional period when the preconditions for take-off are created generally in response to the intrusion of a foreign power, converging with certain domestic forces making for modernization; the take-off itself; the sweep into maturity generally taking up the lives of about two further generations; and then, finally, if the rise of income has matched the spread of technological virtuosity . . . the diversion of the fully mature economy to the provision of durable consumer goods and services (as well as the welfare state) for its increasingly urban – and then suburban – population.
>
> (Rostow 1990: 12)

So, among other requirements of political stability and social modernization: for about fifty years after taking off, the above-referred to countries had to sustain a "drive to maturity" in the technological domain of those products whose industrialization had served to enlarge their economic bases, in order to secure development. Especially in the cases of a small country with relatively scarce natural resources (like Sweden or Switzerland), it was not necessary to be self-sufficient in industrial terms, but rather to demonstrate that it had "the technical and entrepreneurial skills to produce not everything, but anything that it chooses to produce" (Rostow 1990: 10). Until that maturity was complete (as evinced in the consolidation of an urbanized society with mass consumption and a welfare state), countries that had taken off were still considered to be "developing countries" – a category that had much resonance in different contexts, from Taiwan and South Korea to India and Turkey.[2]

Among the group of developing countries, Brazil, Venezuela, Chile and Colombia were considered to be outstanding Latin American examples. Even though by the 1960s, according to Rostow's stages, Argentina and Mexico had taken off, they had not reached maturity and in fact would not do so for the rest of the century. It is extremely complex to explain why such maturity was not obtained in Latin American countries, after several decades of the failed *despegues*.

As Rostow points out, there are (after the take-off) many decisions to make and balances to strike regarding development priorities. For instance, the diffusion of modern technology and the rise of the growth rate relative to the increase of per capita consumption, as well as the growth of expenditures on social welfare (without excessive growth of state bureaucracy) are all typical milestones on the path towards development. These choices are made more complex and particular by the political and social situations that each country must face (Rostow 1990: 14–16).

On the basis of Rostow's explanation of the relationship among industrialization, economic growth and development, this chapter aims to illustrate how the imbalance of urbanization in relation to the production sector is at the root of explaining immature take-offs in Latin America throughout the twentieth century. Such an explanatory attempt must pay attention to the factors that obstruct the application of Rostow's theory in the Latin American context. Political and social factors must be considered in the period when Latin America began to experience significant urbanization. At the same time, this chapter's panoramic stance assumes that studies of urbanization and development have often overlooked historical perspectives (Clichevski 1990; Drakakis-Smith 1990: 11–90; Potter and Lloyd-Evans 1998: 28). So it is necessary to adopt from urban historiography a general and comparative standpoint – both in historical and territorial terms – enabling us to periodize a long-term vision of Latin America's process as a possible contribution to this research (Almandoz 2006; 2008: 145–181).[3]

Early urbanization and massification

By the 1920s, Latin America's biggest countries boasted two urban dwellers for every peasant that had remained in the backward *pampas, llanos* or *sertôes* of their vast countryside. This is a gross indicator that camouflages sharply contrasting realities, as is often the case with this continent, in which Argentina and the Southern Cone had more than 50% of its population urbanized by 1914, whereas Andean and Central American countries were predominantly rural until the 1950s (Beyhaut 1985: 210–211). In spite of their relative simplification, demographic indicators do mirror an unmistakable reality: since the very beginning of the twentieth century in some republics, the process of urbanization accelerated in most of Latin America in the second third of the twentieth century. In a few decades a demographic cycle was completed that had taken more than one hundred years in Britain and other industrializing and urbanizing nations throughout the nineteenth century (Potter and Lloyd-Evans 1995: 9–11).

As in other regions of today's so-called Third World, the speedy pace of Latin America's twentieth-century urbanization was extreme in contrast to its rural backwardness and dispersion. Flooded with both foreign and rural-to-urban immigration, former colonial capitals and new centers rapidly reached magnitudes which rivaled European and North American metropolises. Buenos Aires jumped from 663,000 people in 1895 to 2,178,000 in 1932; Santiago from 333,000 in 1907 to 696,000 in 1930; and Mexico City from 328,000 in 1908 to 1,049,000 in 1933. As a dramatic case comparable to the growth of industrial cities like Manchester and Chicago, São Paulo expanded from 240,000 inhabitants in 1900 to 579,000 in 1920, and 1,075,000 in 1930, while the urban predominance of Rio was diminished, its population increasing from only 650,000 in 1895 to 811,433 in 1906 (Geisse 1987; Hardoy 1988).

The expansion of these capitals was partly due to an incipient process of industrialization that accelerated urbanization in Argentina, Uruguay, Chile and Cuba – which ranked among the world's most urbanized countries at the time of the Depression in 1930. Havana's population had jumped from 250,000 inhabitants by the turn of the century to 500,000 in 1925. Mainly fuelled by rural-to-urban migration, other capitals of the Andean countries also grew considerably: Bogotá went from 100,000 people in 1900 to 330,000 in 1930, while Lima increased from 104,000 in 1891 to 273,000 in 1930. Although Caracas only rose from 72,429 inhabitants in 1891 to 92,212 in 1920, the first effects of the oil boom pushed the population from 135,253 in 1926 to 203,342 in 1936 (Hardoy 1997; Almandoz 2002: 21).

Latin America's economies remained mainly agrarian or mineral-based, which was still evident by the 1929 Crash. The industrial share of the GNP was very low: 22.8% in Argentina, 14.2% in Mexico, 11.7% in Brazil and 7.9% in Chile – just to refer to the highest cases (Pozo 2002: 72–118). In the political arena, without reaching the dramatic fame of the Mexican Revolution in 1910 – which had been partly unleashed by resistance to the feudal-like countryside under the modernizing governments of Porfirio Díaz (1877–80, 1884–1911) – Latin American states could not replicate in the twentieth century the liberalism and positivism of the nineteenth, as the Porfiriato, and later Juan V. Gómez's regime in Venezuela (1909–35), tried to do. Challenged by demands of universal suffrage, constitutions of unions and other political rights, the governments of José Batlle y Ordóñez (1903–07, 1911–15) in Uruguay, Roque Sáenz Peña (1910–13) and Hipólito Yrigoyen (1916–22) in Argentina, followed by Arturo Alessandri's first tenure in Chile (1920–24, 1925), exemplified early attempts to adjust their emerging populism to the demands of rapid urbanization (Baer and Pineo 1998).

The relative economic bonanza and the overcoming of the conflicts between federalism and unionism, both catalyzed by the demands of international immigration from the late nineteenth century, enabled the Southern countries to shift their economic and political liberalism towards a socially oriented agenda. Though flimsy in terms of their economic bases – more commercial than industrial – national bourgeoisies led constitutional reforms that responded to part of the challenges posed by increasingly urban and heterogeneous populations. Thus, for instance, universal suffrage of male population was achieved in Argentina in 1912 and in Uruguay in 1919 – earlier than in some European countries (Halperin and Donghi 2005: 288–326). Similar reforms in Andean and Caribbean nations were delayed by several factors, including persistent internal conflicts, the weak attraction of those countries within international circuits of investment, along with the economically liberal yet politically conservative (or rather, autocratic) climate of long-lasting regimes such as Augusto Leguia's in Peru (1919–30), Gerardo Machado's in Cuba (1925–33), as well as the above-mentioned dictatorship of Gomez in Venezuela (Caballero 1994; Almandoz 2009).

Most of the claims of the heterodox *masa* – which, according to Romero, resulted from foreign and provincial immigrants mixed with traditional yet declining urban sectors (Romero 1984: 337) – were catalyzed by housing and sanitary conditions in volatile cities. For political and demographic reasons alike, their post-colonial shortages in services and infrastructure were no longer tolerated (Romero 1984; Pineo and Baer 1998). Public and private answers to those demands shaped the urban agenda for the first three decades of the twentieth century, especially in terms of health and housing reforms in the historic centers, complemented by residential suburbs for an increasingly cosmopolitan bourgeoisie (Almandoz 2002: 28–31).

Although industrialization and urban growth were still insufficient for national take-offs, the early agenda of urban *masificación* set up the social bases for an incipient welfare state that was necessary for the drive towards development. The most obvious demands for this were no longer posed by the bourgeoisie – as had been the case since the late nineteenth century – but by the expanding masses.

Imbalance between industrialization and urbanization

By 1950, urban centers contained more than half the population of Uruguay (78.0%), Argentina (65.3%), Chile (58.4%) and Venezuela (53.2%). While Latin America's average percentage of urban population was still 41.6, some other countries such as Brazil and Mexico were not demographically urban, due to their huge populations, yet boasted some of the world's greatest metropolises (United Nations 1996: 47). Mexico City and Rio de Janeiro were both about 3 million, while São Paulo had already grown to 2.5 million. The first rank of Latin America's metropolitan areas was still led by Greater Buenos Aires, with 4.7 million (Harris 1971: 167).

From the end of World War II until the mid-1960s, Latin America's biggest countries showed relative prosperity, marked by significant economic expansion amidst sustained urbanization. Fuelled by the massive markets targeted by the import substitution industrialization model (ISI), Brazil and Mexico reached yearly growth rates of 6%, which made them appear, by the 1950s and 1960s respectively, as economies on the eve of "taking off" towards development, in terms of Rostow's phases (Rostow 1990: 127). Even though the Southern Cone countries had been more dynamic in the inter-war period, they remained at a level of growth of about 4% (Clichevsky 1990: 22–23). Meanwhile, epitomized by the windfall of oil-producer Venezuela, the surplus yielded by the export of raw materials financed a second generation of ISI in Colombia and Peru also. In all of these countries, the rate of industrial growth almost doubled that of the primary sector (Williamson 1992: 334–335; Pozo 2002: 118).

The modernizing climate was penetrated by economic nationalism, shared by Latin America's socialism and liberalism, democracies and dictatorships alike. It ranged from the populist regimes of Mexico's Lázaro Cárdenas (1934–40), Argentina's Juan D. Perón (1946–55) and Brazil's Getúlio Vargas, to the economically progressive yet brutal dictatorships of Cuba's Fulgencio Batista (1940–44, 1952–59) and Venezuela's Marcos Pérez Jiménez (1952–58). Their common agenda of *desarrollismo* ("developmentalism") was backed since 1948 by the creation of international agencies such as the Organization of American States (OAS) and the Economic Commission for Latin America (ECLA), both sponsored by the United Nations (UN) and the USA, with its growing interest in the region. Headquartered in Santiago de Chile and led by Raúl Prebisch – former director general of Argentina's Central Bank – the ECLA was a cornerstone of Latin America's post-war developmentalism, aimed at implementing ISI and other economic policies that consolidated the "corporate state" in industrializing countries until the mid-1960s, when the "easy phase" of ISI was over (Franco 2007; Williamson 1992: 338–39).

Although they were regarded as promising examples of developing countries – a category that seemed to have great resonance until the 1960s – most of Latin America's industrializing societies were seemingly exemplars of the classic theory of modernization, as it was explained by the theory of economic growth and functionalist sociology. From the early 1960s, the connection between industrialization, urbanization and modernization was formulated, by Leonard Reissman and Kingsley Davis, from the standpoints of social change and demographic transition. Their (almost causal) theories relied on the examples of the North Atlantic countries that had industrialized in the nineteenth century (Reissman 1970; Davis 1982).[4] From that literature, it could be deduced that Latin America's developing nations seemed to be on a path towards urbanization and industrialization, but they actually suffered from profound distortions by comparison with successful experiences of modernization in Europe, North America and other parts of the world (Philip Hauser 1967).

Distortions can be categorized into two groups. On the one hand, a fledgling industrialization had not preceded, but rather followed, urbanization in Latin America, so the ISI was not the equivalent of an "industrial revolution" with its dynamic effects on the economic system and demographic transition (Drakakis-Smith 1990: 53–57; Williamson 1992: 333). As it happened in other parts of what was starting to be labeled as the "Third World", instead of waves of population being "*pulled*" to cities, which could be actually absorbed by manufacturing and other production sectors, most of Latin America's rural–urban migration was "*pushed*" by a countryside that had been abandoned after a long period of urban-focused policies carried on by corporate states (Potter and Lloyd-Evans 1998: 12–13). The adoption of ISI had aggravated the rural crisis in many countries that had not undergone land reforms. Not only did the labor force engaged in agriculture decline in the 1945–62 period, but also its productivity in terms of per capita GNP was (in the best of cases), less than one fourth of the USA's for the same period (Harris 1971: 74; Williamson 1992: 337–38).

On the other hand, levels of urbanization almost doubled industrial participation in the economies of Argentina, Chile, Venezuela, Colombia and Brazil, according to 1950s censuses (Harris 1971: 85). Such levels could not be absorbed by the production system, so in the long term would fuel "urban inflation" or "hyper-urbanization" – which happened in other parts of the Third World (Potter and Lloyd-Evans 1998: 14–15). In the decades to come, this surplus of unproductive population living in cities had to take refuge in slums, shanty-towns, and the informal economy. But it was already clear by the late 1960s that neither development nor modernization – understood as the outcome of ECLA-style developmentalism and functionalist sociology (Franco 2007) – would be possible anymore, given Latin America's imbalance between industrialization and urbanization. Not only being unable to reach the drive to maturity in Rostow's sense, this hyper-urbanization also raised questions about the applicability of his theory of late take-offs during the post-war era, at least in a region of high demographic mobility such as Latin America.

The failure of modernization and the Marxist response

Cuba's 1959 Revolution, which ousted Batista from power and installed the Marxist regime led by Fidel Castro, preconfigured Latin America's political and economic climate during the remainder of the Cold War. In order to forestall further leftist revolutions, the Kennedy administration decided to promote the so-called Alliance for Progress (AfP), a program aimed at consolidating the ISI, stepping up land reform and reducing social inequalities through the Americas to help new democratic governments in the region. Beneficiaries included Rómulo Betancourt (1959–64) in Venezuela; Arturo Frondizi (1958–62) in Argentina; Fernando Belaúnde Terry (1963–68) in Peru; Eduardo Frei (1964–70) in Chile; and, especially, Alberto Lleras Camargo (1958–62) and Carlos Lleras Restrepo (1966–70) in Colombia (Williamson 1992: 349; Halperin-Donghi 2005: 534–569).

In spite of the AfP aid and ISI's long presence in the biggest economies, by the late 1960s industrialization had neither diversified nor consolidated in Latin America – especially in terms of durable consumer goods and machinery. Factors argued to explain the ISI's structural and contextual constraints include the weakness of economic integration within the region, the small size of some of the national markets, and the disadvantage of most of the countries in competing with their manufactured goods in international circuits – already flooded with products made in Hong Kong, Taiwan and the rest of the Far East. But before its eventual failure, the "deepening" of the ISI from light to "intermediate" manufacturing and heavy machinery (which had been stepped up throughout the 1960s), aggravated the economic and social distortions of Latin America's underdevelopment.

Indeed, by the early 1970s it was already evident that, beyond the industrial bourgeoisie and middle classes, the "modernizing style of development" of the previous decades had not spread its effects to other strata of population, especially to the growing mass of "urban poor" which was amplified by rural–urban migration (Clichevsky 1990: 25). The failure of economic growth, developmentalism and modernization were worsened, after 1973, by the inflationary effects of the international oil crisis. In Latin America this effect was caused not only by the soaring prices of fuels, but also by unaffordable increases in the price of machinery imported from the industrialized world. Fuelled by the penetration of Cuban-like communism and guerrillas, which recreated both in the countryside and the cities the deep-rooted myth of the "good savage" – as it was then denounced by Carlos Rangel's best-seller *Del buen salvaje al buen revolucionario* (1976) (Rangel 2005). The economic and social malaise led some of Latin America's most stable democracies to embrace dictatorships or military juntas that would endure up to the 1980s, as dramatically epitomized in Augusto Pinochet's Chile (1973–90).

In relation to such a political reverse, it has been pointed out that the "optimistic" and functionalist assumption that more "socio-economic development will increase the probability of emergence and consolidation of political democracy", did not take into account that such development, at least in Latin America, tends to bring about "a process of greater political populism, which in a context of growing economic difficulties of peripheral countries, not properly inserted within the framework of global capitalism, generates in turn more intense 'constellations of problems', many of which become 'insoluble problems'". This would explain the "bureaucratic-authoritarian" regimes which appeared in Argentina, Brazil, Peru and Chile since the late 1960s. The case of Venezuela is different: the oil bonanza during those decades avoided the collapse that prompted military intervention in the Southern Cone (Romero 1999: 182–183).

The exhaustion of ISI fractured the fragile support that industrialization had provided in the post-war decades to the urbanization process. The urban share of population increased from 57.4% in 1970 to 65.4% in 1980 in Latin America overall (Clichevsky 1990: 42). Far above Africa and Asia – which were still 28.7% and 26.6% by 1980, respectively – Latin America was the most urbanized region of what started to become known as the "Third World" rather than the "developing world" (Drakakis-Smith 1990; Potter and Lloyd-Evans 1998: 24–25).

The Third World syndrome has been summed up in the following terms:

> As health and social welfare standards are generally so much better in the cities than in the rural areas, Third World cities exemplify par excellence the combination of pre-industrial fertility with post-industrial mortality. Contemporary cities in the developing world exhibit some of the highest rates of natural increase ever found in cities.
>
> (Potter and Lloyd-Evans 1998: 12)

Indeed, throughout most of the 1970s, Latin America's gross rates of urban growth were six times higher than the rural ones (Clichevsky 1990: 48), resulting from the massive flows arriving in cities from the countryside. On top of that, most of this population was highly concentrated in national territories: Latin America boasted three of the Third World's five megalopolises (above 8 million) by 1970 (Clark 2000: 46). More than a half of the national populations of Argentina, Brazil,

Mexico, Venezuela, Chile and Colombia lived in metropolitan areas above 100,000 inhabitants by 1980 (Clichevsky 1990: 54).

From the 1970s Latin America exhibited the most dramatic effects of hyper-urbanization, such as the excessive growth of the tertiary sector and the informal economy that camouflaged the surplus urban labor force, and led to the proliferation of squatter settlements and poverty. The failure of developmentalism and modernization and the ensuing syndrome of Third World urbanization challenged ECLA's functionalist approach in social sciences that had prevailed in Latin America up to the 1960s. The diversification and modernization of productive sectors implied by Rostowian maturity had been left behind, overcome by the urbanized (yet unqualified) masses that populated the administrative apparatuses. Although it had been prefigured since the previous decade, by the 1970s it was evident that Rostow's stages of development, as well as the modernization described by functionalist sociology, were components of an economic and social paradigm that was invalid for Latin America's political, technical and academic reality.

The theory of "Dependence" reinterpreted the centre/periphery antinomy as a structural hindrance that could only be overcome on the basis of state intervention, similar to the USA's Keynesianism (Williamson 1998: 334–335). This theory was conceived partly as an alternative to the liberal doctrine of comparative advantage, which had traditionally explained Latin America's historical sluggishness within the world economy since the late colonial period (Morse 1975). In this respect the Dependence approach was not originally opposed to ECLA's initiatives, including the ISI; but insofar as the latter proved to be exhausted, *dependentismo* became a predominantly Marxist response to capitalist developmentalism. With later contributions by Brazil's Celso Furtado, Fernando H. Cardoso and Enzo Faletto, among others, the theory turned into a Marxist school of social sciences, providing a historical matrix aimed at understanding Latin America's backwardness during the colonial and republican eras, including the economic, political and social dimensions of underdevelopment (Cardoso and Faletto 1969; Palma 1978).

Marking Latin America's intellectual climate until the early 1980s, the School of Dependence (SoD) also reinforced the nationalist orientations of governments opposed to the presence of foreign capital investment in the exploitation of raw materials and industrial processes (del Pozo 2002: 177–179). At the same time, the economic and urban effects of the exhausted "easy phase" of the ISI were summarized by SoD scholars in the following terms:

> . . . this first phase of the process of substitution, which was labelled as "easy" because it resulted in rapid and visible paces of industrial growth, led to shaping a basis of light manufacture and, therefore, to the emergence of an industrial bourgeoisie and a considerable proletariat mainly concentrated in urban nuclei.

Notwithstanding its quick growth, the system was unable to absorb the redundant population coming from rural areas and generated by the penetration of capitalist production in agriculture, which had led to the destruction of non-capitalist sectors. The result was the concentration in cities of a mass of people that has been empirically described as "marginal".

(Malavé Mata et al. 1979: 93)

The lost decade and neo-liberalism

The exhaustion of ISI, the 1970s oil crisis, the decrease of Latin America's exports on a global scale – having fallen from 13.5% in 1946 to 5.1% in 1970 (del Pozo 2002: 176) – aggravated foreign debt, which had plagued Latin American republics since political independence. In spite of the reluctance of nationalist-oriented developmentalism to accept foreign capital, the loans from international private banks soared (as in any other bloc in the Third World), reaching more than 50% of Latin America's GNP by 1980. The only exception amongst the region's biggest economies was Brazil. Only Jamaica, Colombia and Chile avoided a general decline in per capita income (United Nations Centre for Urban Settlements 1996: 45). At the same time, slower rates of GNP growth made the regional average of economic growth plummet from 4.5% in the 1970s to 1.3% in the following decade, while annual hyperinflation spiraled to more than 2,000% in Argentina, Brazil and Peru, amongst the most dramatic cases (del Pozo 2002: 179–180). This is why the 1980s was dubbed as Latin America's "lost decade", especially in contrast to the astounding development achieved by the Asian "tigers" – South Korea, Singapore, Malaysia, Thailand – as well as Spain and other new members of the former European Community (EC), whose evolution since the 1950s seemed to confirm Rostow's stages of development.

The lost decade marked the end of the so-called "Latin American model of development" whose features had been, from the 1930s, an unproductive agriculture that was subordinated to industry, as well as a strong presence of the corporate state (del Pozo 2002: 182).[5] It entailed a new and growing dependence on agencies such as the International Monetary Fund (IMF) and the World Bank (WB), which thereafter dictated economic and social recipes to be adopted by the debt-stricken countries. Imbued with the New Right of the Anglo-American axis,[6] the "plans of adjustments" prescribed from 1982 were in fact packages of neo-liberal policies, including reductions of the huge bureaucracies and privatization of many services and companies of Latin American corporate states. With the direct advice of Milton Friedman and the "Chicago School", including a renewal and orientation of agriculture towards exportation, Pinochet's Chile was an early success that demonstrated how reforms could be undertaken under

216 Arturo Almandoz

authoritarian regimes – as had been the case with South Korea and Franco's Spain in the preceding years.

The neo-liberal package was applicable yet unstable in the long term, as illustrated in the cases of Carlos Salinas de Gortari (1988–94) in México and Carlos Menem in Argentina (1989–95, 1995–99), especially during his first term. But reforms required by the IMF and WB proved to be disastrous when they were introduced too late and too drastically after a period of relative bonanza. This was illustrated by the second government of Carlos A. Pérez in Venezuela (1989–93), which was marked by riots, social unrest and military coups. Venezuela's climate of political and social violence was exported to other Latin American countries where, worsened by financial crashes in the mid-1990s, neo-liberal adjustments did not diminish social inequities but rather increased poverty and crime by the end of the decade (Rotker 2000).

The privatization and dismantling of public bureaucracies that had swollen the tertiary sector raised levels of unemployment above the 20% by the early 1980s, with an alarming growth of the informal economy – even worse than after the 1929 crisis. It is true that the burden of the debt repayment would eventually be reduced by the 1990s, when it amounted to less than 25% of export value in many of the economies, while the average growth of Gross Domestic Product (GDP) was above 3%. However, social inequities were worse in Latin America's major countries, where the income of the richest quintile of the population – which absorbed more than 40% of the national income – was 15 times higher than the poorest quintile by the late 1990s (del Pozo 2002: 181, 237–238, 242).

With 71.4% of its population living in urban settlements by 1990, Latin America had completed the cycle of urbanization, torn between a neo-liberalism that tried to palliate the recession of the lost decade and a leftist revival that opposed the inequities aggravated by liberal policies, but ended up being perceived as neo-imperialist (Irazábal 2008). Slower population growth was prompted by lower fertility rates and less rural–urban immigration, which resulted in a "smaller increase in the levels of urbanization and much smaller rates of growth for many of the region's larger cities" (United Nations Centre for Urban Settlements 1996: 42–43).[7] The completion of the cycle in countries like Argentina, Chile, Uruguay and Venezuela must not be mistaken (in spite of the Chilean "miracle"), for the correction of territorial imbalances or economic distortions. The fact that, with 29% by 1990, there were more people living in "million-cities" than living in rural areas can be reckoned as an indicator of Latin America's excessive urban concentration (United Nations Centre for Urban Settlements 1996: 47–48).

The stronger attraction of "million-cities" over larger major metropolises was due to the latter's exhaustion in terms of worn-out infrastructure and deterioration of living standards, which in turn led to a shift from rural–urban to inter-urban migration (Clichevsky 1990: 47). Medium-size cities of some of the region's most urbanized countries were favored by this turn, especially in the case of middle-class professionals looking for affordable housing and services. Although the relative loss of concentration can be regarded as a positive effect of the completion of the cycle of urbanization, such mobility actually mirrors the impoverishment of middle classes affected by the decrease of per capita income and even of real wages. By the late 1990s, unemployment rates neared 15% in Argentina, Colombia, Ecuador, Jamaica, Dominican Republic and Venezuela, while the proportion of poor households was 36% for the region altogether. This proportion rose to 40 during the 2001 crisis in Argentina, which had been the haven of Latin America's middle class (del Pozo 2002: 240–241).

The diversification of the informal economy and the aggravation of poverty have had dramatic effects on the urban scene, especially in terms of the invasion of public space by street vendors and the establishment of gated communities in both residential districts and squatter settlements. The segregation of the dual city has been accompanied by the deterioration of infrastructure and a general increase of urban poverty, which by 1990 amounted to 40% of the population in Colombia, 38 in Brazil, 28 in Venezuela and 23 in Mexico, among Latin America's most advanced countries (United Nations Centre for Urban Settlements 1996: 528). Crime, social unrest and the lack of governance remained as national problems, most dramatically evident in metropolitan areas under pressure from conflicting groups (Villasante 1994; Rotker 2000; Irazábal 2008). Thus economic and social malaise characterize a continent that has completed its urbanization after more than a century, an eventful process that can only be partially explained in terms of Rostow's stages of development.

Final comments

The failure of Latin American countries to reach the maturity after take-off predicted by Rostow can be attributed to numerous and diverse reasons, many of them particular to specific countries. However, some more general causes can be identified. Most obviously political instability was clearly responsible for deterring the consolidation of a welfare state aimed at disseminating social benefits. That was first apparent in Argentina – the earliest country to have taken off after the 1932 coup d'état that ousted Yrigoyen from power. His government had promoted social reforms that responded to urban *masificación*. In spite of its comprehensive progress during the three first decades of twentieth century, by the 1950s the Southern nation was seriously in debt due to the excesses of Perón and his populist ideology. Already at the brink of the forthcoming dictatorships, Argentina began to be perceived as Latin America's first immature take-off.

The exhaustion of ISI and other economic programs, such as land reform, also weighed in, as both causes and consequences at the same time, in what could be called the immaturity of Latin American development – a euphemism drawn from Rostowian stages that became part of the vocabulary of underdevelopment itself (Sunkel 1973). Indeed, as we have seen, by the late 1960s it was apparent that the region's industrialization had neither diversified nor consolidated, especially in terms of durable and capital goods. Thus the possibilities of maturity as defined by Rostow were eliminated. In spite of the relative advances in those decades – the constitution of the Central American Common Market (MCC, by its initials in Spanish), the Latin American Association of Free Trade (ALALC, later ALADI), as well as the Andean Pact of 1969 – the weakening of Latin America's industrialization was confirmed by the decreasing participation of its production in world markets until the 1970s.

Even though the periods distinguished in this chapter have attempted to span the twentieth century, it can be said that the imbalance between industrialization and urbanization after the former's exhaustion, followed by the urban inflation and the hypertrophy of the tertiary sector, had annihilated Latin America's possibilities of Rostowian development maturity by the last third of the century. As was demonstrated early by Pinochet's regime in Chile, Latin America governments thereafter revised the nature of industrialization formerly sponsored by the corporate state. They modernized the most internationally competitive sectors, while dismissing the others, and minimizing expenditures on public services. That was the turning point of the "Latin American model" that, since the 1930s *masificación*, could still be understood in terms of Rostow's stages of development, which are only achievable on the basis of industrialization.

But with emerging technological and geopolitical horizons, along with the new paradigm of sustainable development (Gabaldón 2008), other economic models had to be adopted from the 1980s, such as neo-liberalism until the end of the 1990s, or even the left-wing corporatism at the beginning of the twenty-first century, whose results remain to be assessed. Meanwhile, in spite of its shortcomings in explaining the immature Latin American take-offs, I have tried to show that, until the 1960s, Rostow's theory of development provided a useful perspective for understanding Latin America's economic and urban process, and for setting it in a historical perspective.

Notes

1   The sense of *masa* as a heterogeneous social body in process of urbanization is drawn from the Argentine historian José Luis Romero (1984: 337): "It was the amalgamation between immigrant groups and popular and low-middle-class sectors of the traditional society what informed the mass of Latin American cities from the

World War I years. More frequently than 'crowd', that name acquired a restricted and precise sense. The *masa* was that heterogeneous ensemble, marginally located beside the normal society, by contrast to which it appeared as an informal and outside sector. The *masa* was an urban group, though urbanized differently, considering that it comprised long-standing urban people as well as people of rural background that started to urbanize. But the *masa*'s face was decidedly urban, as it was its behaviour . . .."

2 For the examples of developing countries, I rely on Davis (1982).

3 Given the panoramic and comparative purpose of the chapter – in addition to its relatively reduced length – the bibliography quoted or referred throughout the text will give priority to general or comparative sources about Latin America, rather than national or urban case studies.

4 Davis's well-known interpretation was included in the popular volume *The City* (1965), which was translated into Spanish in 1967. Reissman's *The Urban Process. Cities in Industrial Societies* (1964) took a bit longer, but reached a wide audience through the Escuela Técnica de Arquitectura de Barcelona (ETSAB), which also translated many other titles from English (Reissman 1972).

5 As it is summarized by del Pozo (2002: 182, author's translation): "All these elements led, by the late-1980s, to questioning whether the Latin American model of development – which had been created between 1930 and 1950 on the basis of the priority of an industrialization sponsored behind the walls of protectionism, with the subordination of agriculture and an important presence of the state had come to dead end. This had undoubtedly been the perception of the military and entrepreneurs that devised the dictatorships' economic policies, especially in the Southern Cone from the 1970s, which led, in countries such as Argentina and Chile, to a reorientation of the economy that gave priority to the export of primary over industrial products, while emphasizing the privatization of public enterprises, along with a relentless repression of unions and any kind of workers' organization."

6 We refer, of course, to the axis defined by the neo-liberal reforms during the governments of Ronald Reagan in the US (1981–89) and Margaret Thatcher in the UK (1979–90).

7 It is convenient to finally differentiate this general panorama according to the countries' level of urbanization by the mid-1990s (United Nations Centre for Urban Settlements 1996: 48): "Although the accuracy of comparisons between countries in their level of urbanization is always limited by the differences in the criteria used to define urban centres, it is possible to identify three groups of nations. The first, the most urbanized with more than 80% of their population living in urban areas, includes the three nations in the Southern Cone and Venezuela. The second, with between 50 and 80% in urban areas, includes most of the countries that had rapid and industrial development during the period 1950–90 – Dominican Republic, Mexico, Brazil, Ecuador and Colombia – and also Cuba (that was already one of the most urbanized nations in the region in 1950), Bolivia, Peru and Nicaragua and Jamaica and Trinidad and Tobago. The third, with less than 50% of the population in urban areas, includes only one in South America (Paraguay) and one in the Caribbean (Haiti) along with a group of countries in Central America (Costa Rica, El Salvador, Guatemala and Honduras) . . .".

References

Almandoz, A. (2002) "Urbanization and urbanism in Latin America: from Haussmann to CIAM," In A. Almandoz (Ed.), *Planning Latin America's capital cities*, 1850–1950 (1st ed., pp. 13–44), London and New York: Routledge.

Almandoz, A. (2006) "Urban planning and historiography in Latin America. *Progress in Planning*, 65: 2, 81–123.

Almandoz, A. (2008) *Entre libros de historia urbana. Para una historiografía de la ciudad y el urbanismo en América Latina* (1st ed.), Caracas: Equinoccio, Editorial de la Universidad Simón Bolívar.

Almandoz, A. (2009) "Demandas políticas y reformas sociales en la masificación urbana latinoamericana," 1900–1930. In F. Aguiar, F. Lara, and N. Lara (Eds.), *Decidir en sociedad. Homenaje a Julia Barragán* (1st ed., pp. 329–343), Caracas: Ediciones Chiryme-kp.

Baer, R. and Pineo, J. A. (1998) "Introduction" In R. Pineo and J. A. Baer (Eds.), *Cities of hope. People, protests and progress in urbanizing Latin America*, 1870–1930 (1st ed., pp. 7–14), Boulder: Westview Press.

Beyhaut, G. H. (1985) *Historia universal siglo XXI. América Latina. III. De la independencia a la segunda guerra mundial* (1st ed., Vol. 23), México: Siglo Veintiuno Editores.

Caballero, M. (1994) *Gómez, el tirano liberal* (2nd ed.), Caracas: Monte Avila Editores Latinoamericana.

Cardoso, F. H. and Faletto, E. (1969) *Dependencia y desarrollo en América Latina* (1st ed.), México: Siglo XXI.

Clark, D. (2000) *Urban world / Global city* (1st ed.), London: Routledge.

Clichevsky, N. (1990) *Construcción y administración de la ciudad latinoamericana* (1st ed.), Buenos Aires: Instituto Internacional de Medio Ambiente y Desarrollo (IIED-América Latina), Grupo Editor Latinoamericano (GEL).

Davis, K. (1982) "La urbanización de la población mundial", *In La ciudad* (4th ed., pp. 11–36). Madrid: Scientific American, Alianza Editorial.

De Mattos, C. (2006) "Modernización capitalista y transformación metropolitana en América Latina: cinco tendencias constitutivas," In *A. I. G. Lemos*, M. Arroyo and M. L. Silveira (Eds.), *America Latina: cidade, campo e turismo* (1st ed., pp. 41–73), Buenos Aires: CLACSO Universidad de San Pablo.

Del Pozo, J. (2002) *Historia de América Latina y del Caribe 1825–2001* (1st ed.), Santiago: LOM Ediciones.

Drakakis-Smith, D. (1990) *The third world city* (1st ed.), London: Routledge.

Franco, R. (2007) *La FLACSO clásica (1957–1963): vicisitudes de las ciencias sociales latinoamericanas* (1st ed), Santiago de Chile: Catalonia.

Gabaldón, A. J. (2008) *Desarrollo sustentable. La salida de América Latina* (1st ed.), Caracas: Grijalbo.

Geisse, G. (1987) "Tres momentos históricos en la ciudad hispanoamericana del siglo XIX", In G. Alomar (Coord.), *De Teotihuacán a Brasilia. Estudios de historia urbana iberoamericana y filipina* (1st ed., pp. 397–433), Madrid: Instituto de Estudios de Administración Local (IEAL).

Halperin Dongui, T. (2005) *Historia contemporánea de América Latina* (6th ed.), Madrid y Buenos Aires: Alianza Editorial.

Hardoy, J. E. (1988) "Teorías y prácticas urbanísticas en Europa entre 1850 y 1930. Su traslado a América Latina," In J. E. Hardoy and R. M. Morse (Eds.), *Repensando la ciudad de América Latina* (1st ed., pp. 97–126), Buenos Aires: Grupo Editor Latinoamericano (GEL).

Hardoy, J. E. (1997) "Las ciudades de América Latina a partir de 1900", In *La ciudad hispanoamericana, El sueño de un orden* (2nd ed., pp. 267–274). Madrid: Centro de Estudios Históricos de Obras Públicas y Urbanismo (CEHOPU), Centro de Estudios y Experimentación de Obras Públicas (CEDEX), Ministerio de Fomento.

Harris, W. D. Jr. (1971) *The growth of Latin American cities* (1st. ed.), Athens, Ohio: Ohio University Press.

Hauser, P. M. (Ed.) (1967) *La urbanización en América Latina* (2nd. ed.), Buenos Aires: Solar, Hachette.

Irazábal, C. (2008) "Citizenship, democracy and public space in Latin America," In C. Irazábal (Ed.), *Ordinary places, extraordinary events. Citizenship, democracy and public space in Latin America* (1st ed., pp. 11–34), London and New York: Routledge.

Malavé Mata, H., Silva Michelena, H. and Sonntag, H. R. (1979) "El contenido conflictivo del actual proceso económico-social de América Latina" (pp. 92–105), In *Ensayos venezolanos* (1st ed., pp. 92–122), Caracas: Ateneo de Caracas.

Morse, R. M. (1975) "El desarrollo de los sistemas urbanos en las Américas durante el siglo XIX," In J. E. Hardoy and R. P. Schaedel (Eds.), *Las ciudades de América Latina y sus áreas de influencia a través de la historia* (1st ed., pp. 263–290), Buenos Aires: Sociedad Interamericana de Planificación (SIAP).

Palma, G. (1978) "Dependency: a formal theory of underdevelopment or a methodology for the analysis of concrete situations ofunderdevelopment," *World Development*, 7:8, 881–920.

Pineo, R. and Baer, J. A. (1998) "Urbanization, the working class and reform," In R. Pineo & J. A. Baer (Eds.), *Cities of hope. People, protests and progress in urbanizing Latin America, 1870–1930* (1st. ed., pp. 258–274), Boulder: Westview Press.

Potter, R. B. and Lloyd-Evans, S. (1998), *The city in the developing world* (1st ed.), London: Longman.

Rangel, C. (2005) *Del buen salvaje al buen revolucionario. Mitos y realidades de América Latina* (4th ed.), Caracas: Criteria.

Reissman, L. (1970), *The urban process. Cities in industrial societies* (2nd ed.), New York: The Free Press.

Romero, A. (1999) *Decadencia y crisis de la democracia* (3rd ed.), Caracas: Editorial Panapo.

Romero, J. L. (1984) *Latinoamérica: las ciudades y las ideas* (3rd ed.), México: Siglo Veintiuno.

Rostow, W. W. (1990), *The stages of economic growth. A non-communist manifesto* (2nd ed.), New York, NY: Cambridge University Press.

Rotker, S. (Ed.) (2000) *Ciudadanías del miedo* (1st ed.), Caracas: Nueva Sociedad, Rutgers University.

Sunkel, O. (1973) *El subdesarrollo latinoamericano y la teori_a del desarrollo* (6th ed.), México: Siglo Veintiuno Editores.

United Nations Centre for Human Settlements (HABITAT) (1996) An urbanizing world. Global report on human settlements (1st ed.). Oxford: Oxford University Press.

Villasante, T. R. (Ed.) (1994) *Las ciudades hablan. Identidades y movimientos sociales en seis metrópolis latinoamericanas* (1st ed.) Caracas: Nueva Sociedad.

Williamson, E. (1992) *The Penguin history of Latin America* (1st ed.), London: Penguin Books.

Chapter 11

Policy and planning for large-infrastructure projects

Problems, causes, and curses

Bent Flyvbjerg

This chapter focuses on problems and their causes and cures in policy and planning for large-infrastructure projects. First, it identifies as the main problem in major infrastructure developments pervasive misinformation about the costs, benefits, and risks involved. A consequence of misinformation is cost overruns, benefit shortfalls, and waste. Second, it explores the causes of misinformation and finds that political-economic explanations best account for the available evidence: planners and promoters deliberately misrepresent costs, benefits, and risks in order to increase the likelihood that their projects, and not those of their competition, that gain approval and funding. This results in the 'survival of the unfittest', in which often it is not the best projects that are built, but the most misrepresented ones. Finally, it presents measures for reforming policy and planning for large-infrastructure projects with a focus on better planning methods and changed governance structures, the latter being more important.

Introduction

For a number of years the research group on large infrastructure at Aalborg University, Denmark, has explored different aspects of the planning of large-infrastructure projects (Flyvbjerg 2005a; 2005b; Flyvberg and COWI 2004; Flyvbjerg et al. 2002; 2003; 2004; 2005).[1] In this chapter I take stock of what may be learned from the research so far.

First, I will argue that a major problem in the planning of large-infrastructure projects is the high level of misinformation about costs and benefits that decision-makers face in deciding whether to build, and the high risks such misinformation generates. Second, I will explore the causes of misinformation and risk, mainly in the guise of optimism bias and strategic misrepresentation. Finally, I will present a

---

Originally published in (2007) *Environment and Planning B: Planning and Design*, 34: 578–597.

By kind permission of Pion Ltd Publishers. www.pion.co.uk.

number of measures aimed at improved planning and decision-making, including changed incentive structures and better planning methods. Thus the chapter is organized as a simple triptych consisting of problems, causes, and cures.

The emphasis will be on transportation infrastructure projects. I would like to mention at the outset, however, that comparative research shows that the problems, causes, and cures we identify for transportation apply to a wide range of other project types including power plants, dams, water projects, concert halls, museums, sports arenas, convention centers, information technology systems, oil and gas extraction projects, aerospace projects, and weapons systems (Altshuler and Luberoff 2003; Flyvbjerg 2005a; Flyvbjerg et al. 2002: 286; 2003: 18–19).

Problems

Large-infrastructure projects, and planning for such projects, generally have the following characteristics (Flyvbjerg and COWI 2004):

1) First: Such projects are inherently risky because of long planning horizons and complex interfaces.
2) Technology is often not standard.
3) Decision-making and planning are often multi-actor processes with conflicting interests.
4) Often the project scope or ambition level will change significantly over time.
5) Statistical evidence shows that such unplanned events are often unaccounted for, leaving budget contingencies sorely inadequate.
6) As a consequence, misinformation about costs, benefits, and risks is the norm.
7) The results are cost overruns and/or benefit shortfalls for the majority of projects.

The size of cost overruns and benefit shortfalls

For transportation infrastructure projects, Table 11.1 shows the inaccuracy of construction-cost estimates measured as the size of cost overrun. The cost study covers 258 projects in twenty nations on five continents. All projects for which data were obtainable were included in the study.[2] For rail, average cost overrun is 44.7% measured in constant prices. For bridges and tunnels, the equivalent figure is 33.8%, and for roads 20.4%. The difference in cost overrun between the three project types is statistically significant, indicating that each type should be treated separately (Flyvbjerg et al. 2002).

Table 11.1 Inaccuracy of transportation-project cost estimates by type of project, in constant prices

Type of project	Number of cases (N)	Average cost overrun (%)	Standard deviation
Rail	58	44.7	38.4
Bridges and tunnels	33	33.8	62.4
Road	167	20.4	29.9

The large standard deviations shown in Table 11.1 are as interesting as the large average cost overruns. The size of the standard deviations demonstrates that uncertainty and risk regarding cost overruns are large, indeed.

The following key observations pertain to cost overruns in transportation infrastructure projects:

1) Nine out of ten projects have cost overrun.
2) Overrun is found across the twenty nations and five continents covered by the study.
3) Overrun is constant for the seventy-year period covered by the study—estimates have not improved over time.

Table 11.2 shows the inaccuracy of travel demand forecasts for rail and road projects. The demand study covers 208 projects in fourteen nations on five continents. All projects for which data were obtainable were included in the study.[3] For rail, actual passenger traffic is 51.4% lower than estimated traffic on average. This is equivalent to an average overestimate in rail-passenger forecasts of no less than 105.6%. The result is large benefit shortfalls for rail. For roads, actual vehicle traffic is on average 9.5% higher than forecast traffic. We see that rail-passenger forecasts are biased, whereas this is not the case for road-traffic forecasts. The difference between rail and road is statistically significant at a high level. Again the standard deviations are large, indicating that forecasting errors vary widely across projects (Flyvbjerg 2005b; Flyvbjerg et al. 2005). The following observations hold for traffic-demand forecasts:

1) 84% of rail passenger forecasts are wrong by more than ±20%.
2) Nine out of ten rail projects have overestimated traffic.
3) 50% of road-traffic forecasts are wrong by more than ±20%.
4) The number of roads with overestimated and underestimated traffic, respectively, is about the same.
5) Inaccuracy in traffic forecasts is found in the fourteen nations and five continents covered by the study.
6) Inaccuracy is constant for the thirty-year period covered by the study—forecasts have not improved over time.

226 Bent Flyvbjerg

Table 11.2 Inaccuracy in forecasts of rail passenger and road-vehicle traffic

Type of project	Number of cases (N)	Average inaccuracy (%)	Standard deviation
Rail	25	–51.4	28.1
Road	183	9.5	44.3

We conclude that, if techniques and skills for arriving at accurate cost and traffic forecasts have improved over time, these improvements have not resulted in an increase in the accuracy of forecasts.

If we combine the data in Tables 11.1 and 11.2, we see that for rail an average cost overrun of 44.7% combines with an average traffic shortfall of 51.4%.[4] For roads, an average cost overrun of 20.4% combines with a fifty-fifty chance that traffic is also wrong by more than 20%. As a consequence, cost–benefit analyses and social-impact and environmental-impact assessments based on cost and traffic forecasts like those described above will typically be highly misleading.

Examples of cost overruns and benefit shortfalls

The list of examples of projects with cost overruns and/or benefit shortfalls is seemingly endless (Flyvberg 2005a). Boston's Big Dig, otherwise known as the Central Artery/Tunnel Project, was 275% or US $11 billion over budget in constant dollars when it opened, and further overruns are accruing because of faulty construction. Actual costs for Denver's $5 billion International Airport were close to 200% higher than estimated costs. The overrun on the San Francisco–Oakland Bay Bridge retrofit was $2.5 billion, or more than 100%, even before construction started. The Copenhagen metro and many other urban rail projects worldwide have had similar overruns. The Channel tunnel between the United Kingdom and France came in 80% over budget for construction and 140% over for financing. At the initial public offering, Eurotunnel, the private owner of the tunnel, lured investors by telling them that 10% "would be a reasonable allowance for the possible impact of unforeseen circumstances on construction costs" (The Economist 1989). Outside of transportation, the $4 billion cost overrun for the Pentagon spy-satellite program and the over $5 billion overrun on the International Space Station are typical of defense and aerospace projects. Our studies show that large infrastructure and technology projects tend statistically to follow a pattern of cost underestimation and overrun. Many such projects end up financial disasters. Unfortunately, the consequences are not always only financial, as is illustrated by the NASA space shuttle. Here, the cooking of budgets to make this underperforming project look good on paper has been linked with

Policy and planning for large-infrastructure projects **227**

shortchanged safety upgrades related to the deaths of seven astronauts aboard the Columbia shuttle in 2003 (Flyvbjerg 2004).

As for benefit shortfalls, consider Bangkok's US $2 billion Skytrain, a two-track elevated urban rail system designed to service some of the most densely populated areas from the air. The system is greatly oversized, with station platforms too long for its shortened trains; many trains and cars sit in the garage, because there is no need for them; terminals are too large; and so on. The reason is that actual traffic turned out to be less than half that forecast (Flyvbjerg et al. 2005: 132). Every effort has been made to market and promote the train, but the project company has ended up in financial trouble. Even though urban rail is probably a good idea for a dense, congested, and air-polluted city like Bangkok, overinvesting in idle capacity is hardly the best way to use resources, especially in a developing nation in which capital for investment is particularly scarce. Such benefit shortfalls are common and have also haunted the Channel tunnel, the Los Angeles and Copenhagen metros, and Denver's International Airport.

Other projects with cost overruns and/or benefit shortfalls are, in North America: the F/A-22 fighter aircraft; the FBI's Trilogy information system; Ontario's Pickering nuclear plant; subways in numerous cities, including Miami and Mexico City; convention centers in Houston, Los Angeles, and other cities; the Animas–La Plata water project; the Sacramento regional sewer-system renewal; the Quebec Olympic stadium; Toronto's Sky Dome; the Washington Public Power Supply System; and the Iraq reconstruction effort. In Europe: the Euro-fighter military jet, the new British Library, the Millennium Dome, the Nimrod maritime patrol plane, the UK West Coast rail upgrade and the related Railtrack fiscal collapse, the Astute attack submarine, the Humber Bridge, the Tyne metro system, the Scottish parliament building, the French Paris Nord TGV, the Berlin–Hamburg maglev train, Hanover's Expo 2000, Athens' 2004 Olympics, Russia's Sakhalin-1 oil and gas project, Norway's Gardermo airport train, the Øresund Bridge between Sweden and Denmark, and the Great Belt rail tunnel linking Scandinavia with continental Europe. In Australasia: Sydney's Olympic stadiums, Japan's Joetsu Shinkansen high-speed rail line, India's Sardar Sarovar dams, the Surat–Manor tollway project, Calcutta's metro, and Malaysia's Pergau dam. I end the list here only for reasons of space.

This is not to say that projects do not exist for which costs and/or benefits were on or better than the budget, even if they are harder to find. For instance, costs for the Paris Southeast and Atlantic TGV lines were on budget, as was the Brooklyn Battery tunnel. The Third Dartford Crossing in the United Kingdom, the Pont de Normandie in France, and the Great Belt road bridge in Denmark all had higher traffic and revenues than projected. Finally, the Bilbao Guggenheim Museum is an example of that rare breed of projects, the cash cow, with costs on budget and with revenues much higher than expected.[5]

Why cost overruns and benefit shortfalls are a problem

Cost overruns and benefit shortfalls of the frequency and size described above are a problem for the following reasons:

1) They lead to a Pareto-inefficient allocation of resources—that is, waste.
2) They lead to delays and further cost overruns and benefit shortfalls.
3) They destabilize policy, planning, implementation, and operations of projects.
4) The problem is getting bigger, because projects get bigger.

Let us consider each point in turn. First, an argument often heard in the planning of large-infrastructure projects is that cost and benefit forecasts at the planning stage may be wrong, but if one assumes that forecasts are wrong by the same margin across projects, cost–benefit analysis would still identify the best projects for implementation. The ranking of projects would not be affected by the forecasting errors, according to this argument. However, the large standard deviations shown in Tables 11.1 and 11.2 falsify this argument. The standard deviations show that cost and benefit estimates are not wrong by the same margin across projects; errors vary extensively and this will affect the ranking of projects. Thus we see that misinformation about costs and benefits at the planning stage is likely to lead to Pareto-inefficiency, because in terms of standard cost–benefit analysis decision-makers are likely to implement inferior projects.

Second, cost overruns of the size described above typically lead to delays, because securing additional funding to cover overruns often takes time. In addition, projects may need to be renegotiated or reapproved when overruns are large, as the data show they often are (Flyvbjerg 2005a). In a separate study, my colleagues and I demonstrated that delays in transportation-infrastructure implementation are very costly, increasing the percentage construction-cost overrun measured in constant prices by 4.64% per year of delay incurred after the time of decision to build (Flyvbjerg et al. 2004). For a project of, say, US $8 billion—that is the size range of the Channel Tunnel and about half the size of Boston's Big Dig—the expected average cost of delay would be approximately $370 million per year, or about $1 million per day. Benefit shortfalls are an additional consequence of delays, because delays result in later opening dates and thus extra months or years without revenues. Because many large-infrastructure projects are loan financed and have long construction periods, they are particularly sensitive to delays, as delays result in increased debt, increased interest payments, and longer payback periods.

Third, large cost overruns and benefit shortfalls tend to destabilize policy, planning, implementation, and operations. For example, after several overruns in the initial phase of the Sydney Opera House, the Parliament of New South Wales decided that every further 10% increase in the budget would need their approval. After this decision, the Opera House became a political football needing constant reapproval. Every overrun set off an increasingly menacing debate about the project, in Parliament and outside, with total cost overruns ending at 1400%. The unrest drove the architect off the project, destroyed his career and oeuvre, and produced an Opera House unsuited for opera. Many other projects have experienced similar, if less spectacular, unrest, including the Channel Tunnel, Boston's Big Dig, and Copenhagen's metro.

Finally, as projects grow bigger, the problems with cost overruns and benefit shortfalls also grow bigger and more consequential (Flyvbjerg et al. 2004: 12). Some megaprojects are becoming so large in relation to national economies that cost overruns and benefit shortfalls from even a single project may destabilize the finances of a whole country or region. This occurred when the billion-dollar cost overrun on the 2004 Athens Olympics affected the credit rating of Greece and when benefit shortfalls hit Hong Kong's new $20 billion Chek Lap Kok airport after it opened in 1998. The desire to avoid national fiscal distress has recently become an important driver in attempts at reforming the planning of large-infrastructure projects, as we will see later.

Policy implications

The policy implications of the results presented above are clear:

1) Lawmakers, investors, and the public cannot trust information about costs, benefits, and risks of large-infrastructure projects produced by promoters and planners of such projects.
2) The current way of planning large-infrastructure projects is ineffective in conventional economic terms, that is, it leads to Pareto-inefficient investments.
3) There is a strong need for reform in policy and planning for large infrastructure projects.

Before depicting what reform may look like in this expensive and consequential policy area, we will examine the causes of cost overruns and benefit shortfalls.

Causes

Three main types of explanation exist that claim to account for inaccuracy in forecasts of costs and benefits: technical, psychological, and political-economic explanations.

Three explanations of cost overruns and benefit shortfalls

Technical explanations account for cost overruns and benefit shortfalls in terms of imperfect forecasting techniques, inadequate data, honest mistakes, inherent problems in predicting the future, lack of experience on the part of forecasters, and so on. This is the most common type of explanation of inaccuracy in forecasts (Ascher 1978; Flyvbjerg et al. 2002; 2005; Morris and Hough 1987; Wachs 1990). Technical error may be reduced or eliminated by developing better forecasting models, better data, and more experienced forecasters, according to this explanation.

Psychological explanations account for cost overruns and benefit shortfalls in terms of what psychologists call the planning fallacy and optimism bias. Such explanations have been developed by Kahneman and Tversky (1979), Kahneman and Lovallo (1993), and Lovallo and Kahneman (2003). In the grip of the planning fallacy, planners and project promoters make decisions based on delusional optimism rather than on a rational weighting of gains, losses, and probabilities. They overestimate benefits and underestimate costs. They involuntarily spin scenarios of success and overlook the potential for mistakes and miscalculations. As a result, planners and promoters pursue initiatives that are unlikely to come in on budget or on time, or to ever deliver the expected returns. Overoptimism can be traced to cognitive biases, that is, errors in the way the mind processes information. These biases are thought to be ubiquitous, but their effects can be tempered by simple reality checks, thus reducing the odds that people and organizations will rush blindly into unprofitable investments of money and time.

Political-economic explanations see planners and promoters as deliberately and strategically overestimating benefits and underestimating costs when forecasting the outcomes of projects. They do this in order to increase the likelihood that it is their projects, and not those of the competition, that gain approval and funding. Political-economic explanations have been set forth by Flyvbjerg et al. (2002; 2005) and Wachs (1989; 1990). According to such explanations planners and promoters purposely spin scenarios of success and gloss over the potential for failure. Again, this results in the pursuit of ventures that are unlikely to come in on budget or on time, or to deliver the promised benefits. Strategic misrepresentation can be traced to political and organizational pressures, for instance competition for scarce funds or jockeying for position, and it is rational in this sense. If we now define a lie in the conventional fashion as making a statement intended to deceive others (Bok 1979: 14; Cliffe et al. 2000: 3), we see that deliberate misrepresentation of costs and benefits is lying, and we arrive at one of the most basic explanations of lying that exists: lying pays off, or at least political and economic agents believe it does. Where there is political pressure there is misrepresentation and lying, according to this explanation, but misrepresentation and lying can be moderated by measures of accountability.

How valid are explanations?

How well does each of the three explanations of forecasting inaccuracy—technical, psychological, and political-economic—account for the data on cost overruns and benefit shortfalls presented earlier? This is the question to be answered in this section. Technical explanations have, as mentioned, gained widespread credence among forecasters and planners (Ascher 1978; Flyvbjerg et al. 2002; 2005). It turns out, however, that such credence could mainly be upheld because until now samples have been too small to allow tests by statistical methods. The data presented above, which come from the first large-sample study in the field, lead us to reject technical explanations of forecasting inaccuracy. Such explanations do not fit the data well. First, if misleading forecasts were truly caused by technical inadequacies, simple mistakes, and inherent problems with predicting the future, we would expect a less biased distribution of errors in forecasts around zero. In fact, we have found with high statistical significance that for four out of five distributions of forecasting errors, the distributions have a mean statistically different from zero. Only the data for inaccuracy in road-traffic forecasts have a statistical distribution that seems to fit with explanations in terms of technical forecasting error. Second, if imperfect techniques, inadequate data, and lack of experience were main explanations of inaccuracies, we would expect an improvement in accuracy over time, as in a professional setting errors and their sources would be recognized and addressed through the refinement of data collection, forecasting methods, and so on. Substantial resources have in fact been spent over several decades on improving data and methods. Still our data show that this has had no effect on the accuracy of forecasts. Technical factors, therefore, do not appear to explain the data. It is not so-called forecasting 'errors' or their causes that need explaining. It is the fact that, in a large majority of cases, costs are underestimated and benefits overestimated. We may agree with proponents of technical explanations that it is, for example, impossible to predict for the individual project exactly which geological, environmental, or safety problems will appear and make costs soar. But we maintain that it is possible to predict the risk, based on experience from other projects, that some such problems will haunt a project and how this will affect costs. We also maintain that such risk can and should be accounted for in forecasts of costs, but typically is not. For technical explanations to be valid, they would have to explain why forecasts are so consistent in ignoring cost and benefit risks over time, location, and project type.

Psychological explanations better fit the data. The existence of optimism bias in planners and promoters would result in actual costs being higher and actual benefits being lower than those forecast. Consequently, the existence of optimism bias would be able to account, in whole or in part, for the peculiar bias found in most of our data. Interestingly, however, when you ask forecasters about causes for forecasting inaccuracies in actual forecasts, they do not mention optimism bias as a main cause of inaccuracy (Flyvbjerg et al. 2005: 138–140). This could of course be because optimism bias is unconscious and thus not reflected by forecasters. After all, there is a body of experimental evidence for the existence of optimism bias (Buehler et al. 1994; 1997; Newby-Clark et al. 2002). However, the experimental data are mainly from simple, nonprofessional settings. This is a problem for psychological explanations, because it remains an open question whether they are general and apply beyond such simple settings. Optimism bias would be an important and credible explanation of underestimated costs and overestimated benefits in infrastructure forecasting if estimates were produced by inexperienced forecasters, that is, persons who were estimating costs and benefits for the first or second time and who were thus unknowing about the realities of infrastructure building and were not drawing on the knowledge and skills of more experienced colleagues. Such situations may exist and may explain individual cases of inaccuracy. But given the fact that in modern society it is a defining characteristic of professional expertise that possible optimism bias would constantly be tested—through scientific analysis, critical assessment, and peer review—in order to root out bias and error, it seems unlikely that a whole profession of forecasting experts would continue to make the same mistakes decade after decade instead of learning from their actions. Learning would result in the reduction, if not elimination, of optimism bias, which would then result in estimates becoming more accurate over time. But our data clearly show that this has not happened. The profession of forecasters would indeed have to be an optimistic—and nonprofessional—group to keep their optimism bias throughout the seventy-year period our study covers for costs, and the thirty-year period covered for patronage, and not learn that they were deceiving themselves and others by underestimating costs and overestimating benefits. This would account for the data, but is not a credible explanation. Therefore, on the basis of our data, we are led to reject optimism bias as a primary cause of cost underestimation and benefit overestimation.

Political-economic explanations and strategic misrepresentation account well for the systematic underestimation of costs and overestimation of benefits found in the data. A strategic estimate of costs would be low, resulting in cost overrun, whereas a strategic estimate of benefits would be high, resulting in benefit shortfalls. A key question for explanations in terms of strategic misrepresentation is whether estimates of costs and benefits are intentionally biased to serve the interests of promoters in getting projects started. This question raises the difficult issue of lying. Questions of lying are notoriously hard to answer, because a lie is making a statement intended to deceive others, and in order to establish whether lying has taken place, one must therefore know the intentions of actors. For legal, economic, moral, and other reasons, if promoters and planners have intentionally cooked estimates of costs and benefits to get a project started, they are unlikely to formally

Policy and planning for large-infrastructure projects **233**

tell researchers or others that this is the case. Despite such problems, two studies exist that succeeded in getting forecasters to talk about strategic misrepresentation (Flyvbjerg and COWI 2004; Wachs 1990).

Flyvbjerg and COWI (2004) interviewed public officials, planners, and consultants who had been involved in the development of large UK transportation infrastructure projects. A planner with a local transportation authority is typical of how respondents explained the basic mechanism of cost underestimation: "You will often as a planner know the real costs. You know that the budget is too low but it is difficult to pass such a message to the counsellors [politicians] and the private actors. They know that high costs reduce the chances of national funding" (Flyvbjerg and COWI 2004: 44). Experienced professionals like the interviewee know that outturn costs will be higher than estimated costs, but because of political pressure to secure funding for projects they hold back this knowledge, which is seen as detrimental to the objective of obtaining funding.

Similarly, an interviewee explained the basic mechanism of benefit over-estimation: "The system encourages people to focus on the benefits—because until now there has not been much focus on the quality of risk analysis and the robustness [of projects]. It is therefore important for project promoters to demonstrate all the benefits, also because the project promoters know that their project is up against other projects and competing for scarce resources" (Flyvbjerg and COWI 2004: 50).

Such a focus on benefits and disregard of risks and robustness may consist, for instance, in the discounting of spatial assimilation problems described by Priemus (2007) elsewhere in this issue. Competition between projects and authorities creates political and organizational pressures that in turn create an incentive structure that makes it rational for project promoters to emphasize benefits and deemphasize costs and risks. A project that looks highly beneficial on paper is more likely to get funded than one that does not.

Specialized private consultancy companies are typically engaged to help develop project proposals. In general, the interviewees found that consultants showed high professional standard and integrity. But interviewees also found that consultants appeared to focus on justifying projects rather than critically scrutinizing them. A project manager explained:

> Most decent consultants will write off obviously bad projects but there is a grey zone and I think many consultants in reality have an incentive to try to prolong the life of projects which means to get them through the business case. It is in line with their need to make a profit.
>
> (Flyvbjerg and COWI 2004: 46–47)

The consultants interviewed confirmed that appraisals often focused more on benefits than on costs. But they said this was at the request of clients and that for

234 Bent Flyvbjerg

specific projects discussed "there was an incredible rush to see projects realized" (Flyvbjerg and COWI 2004: 47).

One typical interviewee saw project approval as 'passing the test' and precisely summed up the rules of the game like this: "It's all about passing the test [of project approval]. You are in, when you are in. It means that there is so much focus on showing the project at its best at this stage" (Flyvbjerg and COWI 2004: 50).

In sum, the UK study shows that strong interests and strong incentives exist at the project approval stage to present projects as favorably as possible, that is, with benefits emphasized and costs and risks deemphasized. Local authorities, local developers and landowners, local labor unions, local politicians, local officials, local members of parliament, and consultants all stand to benefit from a project that looks favorable on paper and they have little incentive to actively avoid bias in estimates of benefits, costs, and risks. National bodies, like certain parts of the Department for Transport and the Ministry of Finance, who fund and oversee projects may have an interest in more-realistic appraisals, but so far they have had little success in achieving such realism, although the situation may be changing with the initiatives to curb bias set out in HM Treasury (2003) and Flyvbjerg and COWI (2004).

The second study was carried out as case-study research by Wachs (1986; 1990). Wachs spoke with public officials, consultants, and planners who had been involved in transit planning cases in the United States. He found that a pattern of highly misleading forecasts of costs and patronage could not be explained by technical errors, honest mistakes, or inadequate methods. In case after case, planners, engineers, and economists told Wachs that they had had to 'revise' their forecasts many times because they failed to satisfy their superiors. The forecasts had to be cooked in order to produce numbers that were dramatic enough to gain federal support for the projects whether or not they could be fully justified on technical grounds. Wachs (1990: 144) recounts from his conversations with those involved:

> One young planner tearfully explained to me that an elected county supervisor had asked her to estimate the patronage of a possible extension of a light-rail (streetcar) line to the downtown Amtrak station. When she carefully estimated that the route might carry two to three thousand passengers per day, the supervisor directed her to redo her calculations in order to show that the route would carry twelve to fifteen thousand riders per day because he thought that number necessary to justify a federal grant for system construction. When she refused, he asked her superior to remove her from the project, and to get someone else to 'revise' her estimates.

In another typical case of cost underestimation and benefit overestimation, Wachs (1990: 144–145) gives the following account:

a planner admitted to me that he had reluctantly but repeatedly adjusted the patronage figures upward, and the cost figures downward to satisfy a local elected official who wanted to compete successfully for a federal grant. Ironically, and to the chagrin of that planner, when the project was later built, and the patronage proved lower and the costs higher than the published estimates, the same local politician was asked by the press to explain the outcome. The official's response was to say, 'It's not my fault; I had to rely on the forecasts made by our staff, and they seem to have made a big mistake here'.

As in the UK study above, Wachs specifically spoke with consultants. He found, as one consultant put it, that "success in the consulting business requires the forecaster to adjust results to conform with the wishes of the client", and clients typically wish to see costs underestimated and benefits overestimated (Wachs 1990: 151–152).

On the basis of his pioneering work, Wachs (1990: 145) concludes that forecasts of costs and benefits are presented to the public as instruments for deciding whether or not a project is to be undertaken, but they are actually instruments for getting public funds committed to a favored project. Wachs (1986: 28; 1990: 146) talks of "nearly universal abuse" of forecasting in this context, and he finds no indication that it takes place only in transit planning: it is common in all sectors of the economy where forecasting routinely plays an important role in policy debates, according to Wachs.

In conclusion, the results of the UK and US studies are basically similar. Both studies account well for existing data on cost underestimation and benefit overestimation. Both studies falsify the notion that in situations with high political and organizational pressure the lowballing of costs and highballing of benefits is caused by non-intentional technical error or optimism bias. Both studies support the view that in such situations promoters and forecasters intentionally use the following formula in order to secure approval and funding for their projects:

underestimated costs + overestimated benefits = project approval.

Using this formula, and thus "showing the project at its best" as one interviewee said above, results in an inverted Darwinism, that is, the 'survival of the unfittest'. It is not the best projects that get implemented, but the projects that look best on paper. And the projects that look best on paper are the projects with the largest cost underestimates and benefit overestimates, other things being equal. But these are the worst, or 'unfittest', projects in the sense that they are the very projects that will encounter most problems during construction and operations in terms of the largest cost overruns, benefit shortfalls, and risks of non-viability. They have been designed like that.

Cures

As should be clear, the planning and implementation of large-infrastructure projects stand in need of reform. Less deception and more honesty are needed in the estimation of costs and benefits if better projects are to be implemented. This is not to say that costs and benefits are or should be the only basis for deciding whether to build large-infrastructure projects. Clearly, forms of rationality other than economic rationality are at work in most projects and are balanced in the broader frame of public decision-making. But the costs and benefits of large-infrastructure projects often run in the hundreds of millions of dollars, with risks correspondingly high. Without knowledge of such risks, decisions are likely to be flawed.

When contemplating what planners can do to help reform come about, we need to distinguish between two fundamentally different situations: (1) planners and promoters consider it important to get forecasts of costs, benefits, and risks right, and (2) planners and promoters do not consider it important to get forecasts right, because optimistic forecasts are seen as a necessary means to getting projects started. The first situation is the easier one to deal with and here better methodology will go a long way in improving planning and decision-making. The second situation is more difficult, and more common as we saw above. Here changed incentives are essential in order to reward honesty and punish deception, whereas today's incentives often do the exact opposite.

Thus two main measures of reform are (1) better forecasting methods, and (2) improved incentive structures, with the latter being the more important.

Better methods: reference-class forecasting

If planners genuinely consider it important to get forecasts right, we recommend they use a new forecasting method called 'reference-class forecasting' to reduce inaccuracy and bias. This method was originally developed to compensate for the type of cognitive bias in human forecasting that Princeton psychologist Daniel Kahneman found in his Nobel prize-winning work on bias in economic forecasting (Kahneman 1994; Kahneman and Tversky 1979). Reference-class forecasting has proven more accurate than conventional forecasting. In April 2005, on the basis of a study by Flyvbjerg et al. (2005), the American Planning Association (2005) officially endorsed reference-class forecasting:

> APA encourages planners to use reference class forecasting in addition to traditional methods as a way to improve accuracy. The reference class forecasting method is beneficial for non-routine projects such as stadiums, museums, exhibit centers, and other local one-off projects. Planners should

never rely solely on civil engineering technology as a way to generate project forecasts.

For reasons of space, here we present only an outline of the method, based mainly on Lovallo and Kahneman (2003) and Flyvbjerg (2003). In a different context, we are currently developing what is, to our knowledge, the first instance of practical reference-class forecasting in planning (Flyvbjerg and COWI 2004).

Reference-class forecasting consists in taking a so-called 'outside view' on the particular project being forecast. The outside view is established on the basis of information from a class of similar projects. Use of the outside view does not involve trying to forecast the specific uncertain events that will affect the particular project, but instead involves placing the project in a statistical distribution of outcomes from this class of reference projects. Reference-class forecasting requires the following three steps for the individual project:

1) Identification of a relevant reference class of past projects. The class must be broad enough to be statistically meaningful but narrow enough to be truly comparable with the specific project.
2) Establishing a probability distribution of outcomes for the selected reference class. This requires access to credible, empirical data for a sufficient number of projects within the reference class to make statistically meaningful conclusions.
3) Compare the specific project with the reference-class distribution in order to establish the most likely outcome for the specific project.

Daniel Kahneman relates the following story about curriculum planning to illustrate reference-class forecasting in practice (Lovallo and Kahneman 2003: 61). We use this example, because similar examples do not exist as yet in the field of infrastructure planning. Some years ago, Kahneman was involved in a project to develop a curriculum for a new subject area for high schools in Israel. The project was carried out by a team of academics and teachers. In time, the team began to discuss how long the project would take to complete. Everyone on the team was asked to write on a slip of paper the number of months needed to finish and report the project. The estimates ranged from 18 to 30 months. One of the team members—a distinguished expert in curriculum development—was then posed a challenge by another team member to recall as many projects similar to theirs as possible and to think of these projects as if they were in a stage comparable with their project. "How long did it take them at that point to reach completion?", the expert was asked. After a while he answered, with some discomfort, that not all the comparable teams he could think of ever did complete their task. About 40% of them eventually gave up. Of those remaining, the expert could not think of any that completed their task in less than seven years, nor of any that took more than ten. The expert was then asked if he had reason to believe that the present team was more skilled in curriculum development than the earlier ones had been. The expert said no, he did not see any relevant factor that distinguished this team favorably from the teams he had been thinking about. His impression was that the present team was slightly below average in terms of resources and potential. The wise decision at this point would probably have been for the team to break up, according to Kahneman. Instead, the members ignored the pessimistic information and proceeded with the project. They finally completed the project eight years later, and their efforts went largely wasted—the resulting curriculum was rarely used.

In this example, the curriculum expert made two forecasts for the same problem and arrived at very different answers. The first forecast was the inside view; the second was the outside view, or the reference-class forecast. The inside view is the one that the expert and the other team members adopted. They made forecasts by focusing tightly on the case at hand, considering its objective, the resources they brought to it, and the obstacles to its completion. They constructed in their minds scenarios of their coming progress and extrapolated current trends into the future. The resulting forecasts, even the most conservative ones, were overly optimistic. The outside view is the one provoked by the question to the curriculum expert. In the outside view the details of the project at hand were completely ignored, and there was no attempt at forecasting the events that would influence the future course of the project. Instead, the experiences of a class of similar projects were examined, a rough distribution of outcomes for this reference class were laid out, and then the current project was positioned in that distribution. The resulting forecast, as it turned out, was much more accurate.

Similarly—to take an example from our work with developing reference-class forecasting for practical infrastructure planning—planners in a city preparing to build a new subway would, first, establish a reference class of comparable projects. This could be the relevant rail projects from the sample used for this chapter. Through analyses the planners would establish that the projects included in the reference class were indeed comparable. Second, if the planners were concerned, for example, with getting construction-cost estimates right, they would then establish the distribution of outcomes for the reference class regarding the accuracy of construction-cost forecasts. Figure 11.1 shows what this distribution looks like for a reference class relevant to building subways in the United Kingdom, developed by Flyvbjerg and COWI (2004: 23) for the UK Department for Transport. Third, the planners would compare their subway project with the reference-class distribution. This would make it clear to the planners that unless they have reason to believe they are substantially better forecasters and planners than their colleagues who did the forecasts and planning for projects in the reference class, they are likely to grossly underestimate construction costs. Finally, planners would then use this knowledge to adjust their forecasts for more realism. Figure 11.2 shows what such adjustments are for the UK situation. More specifically, Figure 11.2 shows that for a forecast of construction costs for a rail project, which has been planned in the manner that such projects are usually planned, that is, like the projects in the reference class, this forecast would have to be adjusted upwards by 40% if investors were willing to accept a risk of cost overrun of 50%. If investors were willing to accept a risk of overrun of only 10%, the uplift would have to be 68%. For a rail project initially estimated at, say £4 billion, the uplifts for the 50% and 10% levels of risk of cost overrun would be £1.6 billion and £2.7 billion, respectively.

The contrast between inside and outside views has been confirmed by systematic research (Gilovich et al. 2002). The research shows that, when people are asked simple questions requiring them to take an outside view, their forecasts become significantly more accurate. However, most individuals and organizations are inclined to adopt the inside view in planning major initiatives. This is the conventional and intuitive approach. The traditional way to think about a complex project is to focus on the project itself and its details, to bring to bear what one

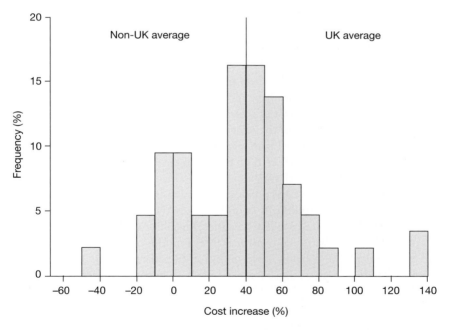

11.1 Shows the inaccuracy of construction-cost forecasts for rail projects in reference class. The histogram shows the inaccuracy of forecasts for forty-three non-UK rail projects. The points below the histogram indicate the inaccuracy of forecasts for three UK rail projects. Average cost increase is indicated for non-UK and UK projects, separately.

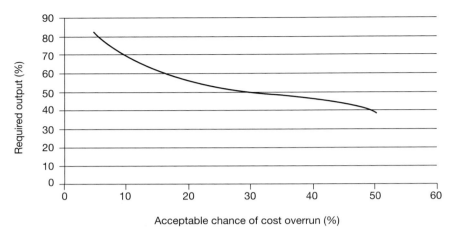

11.2 Required adjustments to cost estimates for UK rail projects as function of the maximum acceptable level of risk for cost overrun. Constant prices.

knows about it, paying special attention to its unique or unusual features, trying to predict the events that will influence its future. The thought of going out and gathering simple statistics about related cases seldom enters a planner's mind. This is the case in general, according to Lovallo and Kahneman (2003: 61–62). And it is certainly the case for cost and benefit forecasting in large-infrastructure projects. Despite the many forecasts we have reviewed, we have not come across a single genuine reference-class forecast of costs and benefits and neither has Kahneman, who first conceived the idea of the reference-class forecast.[6]

Although understandable, planners' preference for the inside view over the outside view is unfortunate. When both forecasting methods are applied with equal skill, the outside view is much more likely to produce a realistic estimate. That is because it bypasses cognitive and political biases such as optimism bias and strategic misrepresentation and cuts directly to outcomes. In the outside view, planners and forecasters are not required to make scenarios, imagine events, or gauge their own and others' levels of ability and control, so they cannot get all these things wrong. Surely the outside view, being based on historical precedent, may fail to predict extreme outcomes, that is, those that lie outside all historical precedents. But for most projects, the outside view will produce more accurate results. In contrast, a focus on inside details is the road to inaccuracy.

The comparative advantage of the outside view is most pronounced for non-routine projects, understood as projects that planners and decision-makers in a certain locale have never attempted before—like building an urban rail system in a city for the first time, or a new major bridge or tunnel where none existed before. It is in the planning of such new efforts that the biases toward optimism and

Policy and planning for large-infrastructure projects **241**

strategic misrepresentation are likely to be largest. To be sure, choosing the right reference class of comparative past projects becomes more difficult when planners are forecasting initiatives for which precedents are not easily found: for instance, the introduction of new and unfamiliar technologies. However, most large-infrastructure projects are both non-routine locally and use well-known technologies. Such projects are, therefore, particularly likely to benefit from the outside view and reference-class forecasting. The same holds for concert halls, museums, stadiums, exhibition centers, and other local one-off projects.

Improved incentives: public-sector and private-sector accountability

In this section I consider the situation where planners and other influential actors do not find it important to get forecasts right and where planners, therefore, do not help to clarify and mitigate risk but, instead, generate and exacerbate it. Here planners are part of the problem, not the solution. This situation may need some explication, because it might sound to many like an unlikely state of affairs. After all, it may be agreed that planners ought to be interested in being accurate and unbiased in forecasting. It is even stated as an explicit requirement in the American Institute of Certified Planners (AICP) Code of Ethics and Professional Conduct that "a planner must strive to provide full, clear and accurate information on planning issues to citizens and governmental decision-makers" (American Planning Association 1991: A.3). The British Royal Town Planning Institute (RTPI) has laid down similar obligations for its members (RTPI 2001).

However, the literature is replete with things planners and planning 'must' strive to do, but which they do not. Planning must be open and communicative, but often it is closed. Planning must be participatory and democratic, but often it is an instrument of domination and control. Planning must be about rationality, but often it is about power (Flyvbjerg 1998; Watson 2003). This is the 'dark side' of planning and planners identified by Flyvbjerg (1996) and Yiftachel (1998), which is remarkably underexplored by planning researchers and theorists.

Forecasting, too, has its dark side. It is here that "planners lie with numbers", as Wachs (1989: 476) has aptly put it. Planners on the dark side are busy not with getting forecasts right and following the AICP code of ethics but with getting projects funded and built. And accurate forecasts are often not an effective means for achieving this objective. Indeed, accurate forecasts may be counterproductive, whereas biased forecasts may be effective in competing for funds and securing the go-ahead for construction. "The most effective planner", says Wachs (1989: 477), "is sometimes the one who can cloak advocacy in the guise of scientific or technical rationality." Such advocacy would stand in direct opposition to AICP's ruling that "the planner's primary obligation [is] to the public interest" (American Planning Association 1991: B.2). Nevertheless, seemingly rational forecasts that underestimate costs and overestimate benefits have long been an established formula for project approval as we saw above. Forecasting is here mainly another kind of rent-seeking behavior, resulting in a make-believe world of misrepresentation which makes it extremely difficult to decide which projects deserve undertaking and which do not. The consequence is, as even one of the industry's own organs, the Oxford-based Major Projects Association, acknowledges, that too many projects proceed that should not. One might add that many projects do not proceed that probably should, as a result of losing out to projects with 'better' misrepresentation (Flyvbjerg et al. 2002).

In this situation, the question is not so much what planners can do to reduce inaccuracy and risk in forecasting, but what others can do to impose on planners the checks and balances that would give planners the incentive to stop producing biased forecasts and begin to work according to their code of ethics. The challenge is to change the power relations that govern forecasting and project development. Better forecasting techniques and appeals to ethics will not do here; institutional change with a focus on transparency and accountability is necessary.

As argued in Flyvbjerg et al. (2003), two basic types of accountability define liberal democracies: (1) public-sector accountability through transparency and public control, and (2) private-sector accountability via competition and market control. Both types of accountability may be effective tools to curb planners' misrepresentation in forecasting and to promote a culture which acknowledges and deals effectively with risk. In order to achieve accountability through transparency and public control, the following would be required as practices embedded in the relevant institutions [the full argument for the measures may be found in Flyvbjerg et al. (2003: chapters 9–11)].

1) National-level government should not offer discretionary grants to local infrastructure agencies for the sole purpose of building a specific type of infrastructure. Such grants create perverse incentives. Instead, national government should simply offer 'infrastructure grants' or 'transportation grants' to local governments, and let local political officials spend the funds however they choose to, but make sure that every dollar they spend on one type of infrastructure reduces their ability to fund another.

2) Forecasts should be made subject to independent peer review. Where large amounts of taxpayers' money are at stake, such review may be carried out by national or state accounting and auditing offices, like the General Accounting Office in the United States or the National Audit Office in the United Kingdom, who have the independence and expertise to produce such reviews. Other types of independent-review bodies may be established,

Policy and planning for large-infrastructure projects **243**

for instance within national departments of finance or with relevant professional bodies.

3) Forecasts should be benchmarked against comparable forecasts: for instance, using reference-class forecasting as described in the previous section.

4) Forecasts, peer reviews, and bench-markings should be made available to the public as they are produced, including all relevant documentation.

5) Public hearings, citizen juries, and the like should be organized to allow stakeholders and civil society to voice criticism and support of forecasts. Knowledge generated in this way should be integrated in planning and decision-making.

6) Scientific and professional conferences should be organized where forecasters would present and defend their forecasts in the face of colleagues' scrutiny and criticism.

7) Projects with inflated benefit–cost ratios should be reconsidered and stopped if recalculated costs and benefits do not warrant implementation. Projects with realistic estimates of benefits and costs should be rewarded.

8) Professional and occasionally even criminal penalties should be enforced for planners and forecasters who consistently and foreseeably produce deceptive forecasts. An example of a professional penalty would be the exclusion from one's professional organization if one violates its code of ethics. An example of a criminal penalty would be punishment as the result of prosecution before a court or similar legal setup: for instance, cases in which deceptive forecasts have led to substantial mismanagement of public funds (Garett and Wachs 1996). Malpractice in planning should be taken as seriously as it is in other professions. Failing to do this amounts to not taking the profession of planning seriously.

In order to achieve accountability in forecasting via competition and market control, the following would be required, again as practices that are both embedded in and enforced by the relevant institutions:

1) The decision to go ahead with a project should, where at all possible, be made contingent on the willingness of private financiers to participate without a sovereign guarantee for at least one third of the total capital needs.[7] This should be required whether projects pass the market test or not, that is, whether projects are subsidized or not or provided for social-justice reasons or not. Private lenders, shareholders, and stock-market analysts would produce their own forecasts or would critically monitor existing ones. If they were wrong about the forecasts, they and their organizations would be hurt. The result would be more realistic forecasts and reduced risk.

2) Full public financing or full financing with a sovereign guarantee should be avoided.

3) Forecasters and their organizations must share financial responsibility for covering cost overruns and benefit shortfalls resulting from misrepresentation and bias in forecasting.
4) The participation of risk capital should not mean that government gives up or reduces control of the project. On the contrary, it means that government can more effectively play the role it should be playing, namely as the ordinary citizen's guarantor for ensuring concerns about safety, environment, risk, and a proper use of public funds. Whether projects are public, private, or public and private, they should be vested in one and only one project organization with a strong governance framework. The project organization may be a company or not, public or private, or a mixture. What is important is that this organization enforces accountability vis-à-vis contractors, operators, and so on, and that, in turn, the directors of the organization are held accountable for any cost overruns, benefits shortfall, faulty designs, unmitigated risks, and so on, that may occur during project planning, implementation, and operations.

If the institutions with responsibility for developing and building major infrastructure projects would effectively implement, embed, and enforce such measures of accountability, then the misrepresentation in cost, benefit, and risk estimates, which is widespread today, may be mitigated. If this is not done, misrepresentation is likely to continue, and the allocation of funds for infrastructure is likely to continue to be wasteful and undemocratic.

Towards better practice

Fortunately, after decades of widespread mismanagement of the planning and design of large-infrastructure projects, signs of improvement have recently appeared. The conventional consensus that deception is an acceptable way of getting projects started is under attack, as will be apparent from the examples below. This is in part because democratic governance is generally getting stronger around the world. The Enron scandal and its successors have triggered a war on corporate deception that is spilling over into government with the same objective: to curb financial waste and promote good governance. Although progress is slow, good governance is gaining a foothold even in large-infrastructure project development. The conventional consensus is also under attack for the practical reason mentioned earlier that the largest projects are now so big in relation to national economies that cost overruns, benefit shortfalls, and risks from even a single project may destabilize the finances of a whole country or region, as happened in Greece and Hong Kong. Lawmakers and governments begin to see that national fiscal distress is too high a price to pay for the conventional way of planning and designing large projects. The main drive for reform comes from outside the agencies and industries conventionally involved in infrastructure development, which increases the likelihood of success.

In 2003 the Treasury of the United Kingdom required, for the first time, that all ministries develop and implement procedures for large public projects that will curb what it calls—with true British civility—'optimism bias'. Funding will be unavailable for projects that do not take into account this bias, and methods have been developed for how to do this (Flyvbjerg and COWI 2004; HM Treasury 2003; Mott MacDonald 2002). In the Netherlands in 2004 the Parliamentary Committee on Infrastructure Projects conducted for the first time extensive public hearings to identify measures that will limit the misinformation about large-infrastructure projects given to the Parliament, public, and media (Tijdelijke Commissie Infrastructuurprojecten 2004). In Boston, the government sued to recoup funds from contractor overcharges for the Big Dig related to cost overruns. More governments and parliaments are likely to follow the lead of the United Kingdom, the Netherlands, and Boston in coming years. It is too early to tell whether the measures they implement will ultimately be effective. It seems unlikely, however, that the forces that have triggered the measures will be reversed, and it is those forces that reform-minded groups need to support and work with in order to curb deception and waste. This is the 'tension point' where convention meets reform, power balances change, and new things may happen.

The key weapons in the war on deception and waste are accountability and critical questioning. The professional expertise of planners, engineers, architects, economists, and administrators is certainly indispensable for constructing the infrastructures that make society work. Our studies show, however, that the claims about costs, benefits, and risks made by these groups usually cannot be trusted and should be carefully examined by independent specialists and organizations. The same holds for claims made by project-promoting politicians and officials. Institutional checks and balances including financial, professional, or even criminal penalties for consistent and unjustifiable biases in claims and estimates of costs, benefits, and risk should be developed and employed. The key principle is that the cost of making a wrong forecast should fall on those making the forecast, a principle often violated today.

Many of the public and private partnerships currently emerging in large-infrastructure projects contain more and better checks and balances than previous institutional setups, as has been demonstrated by the UK National Audit Office (2003). This is a step in the right direction but should be no cause for repose. All available measures for improvement must be employed. The conventional mode of planning and designing infrastructure has long historical roots and is deeply ingrained in professional and institutional practices. It would be naive to think it is easily toppled. Given the stakes involved—saving taxpayers from billions of dollars of waste, protecting citizens' trust in democracy and the rule of law, avoiding the destruction of spatial and environmental assets—this should not deter us from trying.

Notes

1   By 'large-infrastructure projects' I here mean the most expensive infrastructure projects that are built in the world today, typically at costs per project from around a hundred million to several billion dollars.

2   The data are from the largest database of its kind. All costs are construction costs measured in constant prices. Cost overrun, also sometimes called 'cost increase' or 'cost escalation', is measured according to international convention as actual out-turn costs minus estimated costs as a percentage of estimated costs. Actual costs are defined as real, accounted construction costs determined at the time of project completion. Estimated costs are defined as budgeted, or forecasted, construction costs at the time of decision to build. For reasons explained in Flyvbjerg et al. (2002) the figures for cost overrun presented here must be considered conservative. Ideally, financing costs, operating costs, and maintenance costs would also be included in a study of costs. It is difficult, however, to find valid, reliable, and comparable data on these types of costs across large numbers of projects. For details on methodology, see Flyvbjerg et al. (2002).

3   Following international convention, inaccuracy is measured as actual traffic minus estimated traffic as a percentage of estimated traffic. Rail traffic is measured as number of passengers; road traffic as number of vehicles. The base year for estimated traffic is the year of decision to build. The forecasting year is the first full year of operations. Two statistical outliers are not included here. For details on methodology, see Flyvbjerg (2005b).

4   For each of twelve urban rail projects, we have data for both cost overrun and traffic shortfall. For these projects average cost overrun is 40.3%; average traffic shortfall is 47.8%.

5   For an explanation of the success of the Bilbao Guggenheim Museum, see Flyvbjerg (2005a).

6   Personal communication, author's archives. The closest we have come to an outside view in large-infrastructure forecasting is Gordon and Wilson's (1984) use of regression analysis on an international cross-section of light-rail projects to forecast patronage in a number of light-rail schemes in North America. At the time of writing this paper, several reference-class forecasts were in progress in the United Kingdom and the Netherlands, based on Flyvberg and COWI (2004).

7   The lower limit of a one-third share of private-risk capital for such capital to effectively influence accountability is based on practical experience. See more in Flyvbjerg et al. (2003: 120–123).

Policy and planning for large-infrastructure projects **247**

References

Altshuler, A., Luberoff, D. (2003) *Mega-projects: The Changing Politics of Urban Public Investment*, Washington, DC: Brookings Institution.

American Planning Association (1991) "AICP Code of Ethics and Professional Conduct," adopted October 1978, as amended October 1991, www.planning.org

American Planning Association (2005) "JAPA article calls on planners to help end inaccuracies in public project revenue forecasting," www.planning.org/news releases/2005/ftp040705.htm

Ascher, W. (1978) *Forecasting: An Appraisal for Policy-makers and Planners*, Baltimore, MD: Johns Hopkins University Press.

Bok, S. (1979) *Lying: Moral Choice in Public and Private Life*, New York, NY: Vintage.

Buehler, R., Griffin, D., Ross, M. (1994) "Exploring the 'planning fallacy': why people underestimate their task completion times," *Journal of Personality and Social Psychology*, 67: 366–381.

Buehler, R., Griffin, D., MacDonald, H. (1997) "The role of motivated reasoning in optimistic timepredictions," *Personality and Social Psychology Bulletin*, 23: 238–247.

Cliffe, L., Ramsey, M., Bartlett, D. (2000) *The Politics of Lying: Implications for Democracy*, London: Macmillan.

Flyvbjerg, B. (1996) "The dark side of planning: rationality and Realrationalität", in S. Mandelbaum, L. Mazza, R. Burchell (Eds) *Explorations in Planning Theory*, Center for Urban Policy Research Press, New Brunswick, NJ, 383–394.

Flyvbjerg, B. (1998) *Rationality and Power: Democracy in Practice*, Chicago, IL: University of Chicago Press.

Flyvbjerg, B. (2003) "Delusions of success: comment on Dan Lovallo and Daniel Kahneman," *Harvard Business Review*, December: 121–122.

Flyvbjerg, B. (2004) "Megaprojects and risk: a conversation with Bent Flyvbjerg," *Critical Planning*, 11: 51–63.

Flyvbjerg, B. (2005a) "Design by deception: the politics of megaproject approval" *Harvard Design Magazine*, Number 22, Spring/Summer: 50–59.

Flyvbjerg, B. (2005b) "Measuring inaccuracy in travel demand forecasting: methodological considerations regarding ramp up and sampling," *Transportation Research*, A 39: 522–530.

Flyvbjerg, B., COWI (2004) Procedures for Dealing with Optimism Bias in Transport Planning: Guidance Document, UK Department for Transport, London.

Flyvbjerg, B., Holm, M. K. S., Buhl, S. L. (2002) "Underestimating costs in public works projects: error or lie?," *Journal of the American* Planning Association, 68: 279–295.

Flyvbjerg, B., Bruzelius, N., Rothengatter, W. (2003) *Megaprojects and Risk: An Anatomy of Ambition*, Cambridge, MA: Cambridge University Press.

Flyvbjerg, B., Holm, M. K. S., Buhl, S. L. (2004) "What causes cost overrun in transport infrastructure projects?," *Transport Reviews*, 24: 3–18

Flyvbjerg, B., Holm, M. K. S., Buhl, S. (2005) "How (in)accurate are demand forecasts in public works projects? The case of transportation," *Journal of the American Planning Association*, 71: 131–146.

248 Bent Flyvbjerg

Garett, M., Wachs, M. (1996) *Transportation Planning on Trial: The Clean Air Act and Travel Forecasting*, Thousand Oaks, CA: Sage.

Gilovich, T., Griffin, D., Kahneman, D. (2002) (Eds) *Heuristics and Biases: The Psychology of Intuitive Judgment*, Cambridge, MA: Cambridge University Press.

Gordon, P., Wilson, R. (1984) "The determinants of light-rail transit demand: an international cross-sectional comparison," *Transportation Research A*, 18: 135–140.

HM Treasury (2003) *The Green Book: Appraisal and Evaluation in Central Government, Treasury Guidance*, London: The Stationery Office.

Kahneman, D. (1994) "New challenges to the rationality assumption," *Journal of Institutional and Theoretical Economics*, 150: 18–36.

Kahneman, D., Lovallo, D. (1993) "Timid choices and bold forecasts: a cognitive perspective on risk taking," *Management Science*, 39: 17–31.

Kahneman, D., Tversky, A. (1979) "Prospect theory: an analysis of decisions under risk," *Econometrica*, 47: 313–327.

Lovallo, D., Kahneman, D. (2003) "Delusions of success: how optimism undermines executives' decisions," *Harvard Business Review*, July: 56–63

Morris, P. W. G., Hough, G. H. (1987) *The Anatomy of Major Projects: A Study of the Reality of Project Management* , New York, NY: John Wiley.

Mott MacDonald (2002) Review of Large Public Procurement in the UK study for HM Treasury, www.hm-treasury.gov.uk/media/A00/D3/greenbook mott.pdf

National Audit Office (2003) *PFI: Construction Performance*, The Stationery Office, London.

Newby-Clark, I. R., McGregor, I., Zanna, M. P. (2002) "Thinking and caring about cognitive inconsistency: when and for whom does attitudinal ambivalence feel uncomfortable?," *Journal of Personality and Social Psychology*, 82: 157–166.

Priemus, H. (2007) "Design of large infrastructure projects: disregarded alternatives and issues of spatial planning," *Environment and Planning B: Planning and Design*.

RTPI (2001) "Code of professional conduct: as last amended by the Council on 17 January 2001." Royal Town Planning, Institute, www.rtpi.org.uk/about-the-rtpi/codecond.pdf

The Economist (1989) "Under water over budget," October: 37–38.

Tijdelijke Commissie Infrastructuurprojecten (2004) Grote Projecten Uitvergroot: Een Infrastructuur voor Besluitvorming [Large projects become larger: an infrastructure for decision making] Tweede Kamer der Staten-Generaal, The Hague.

Wachs, M. (1986) "Technique vs. advocacy in forecasting: a study of rail rapid transit, " *Urban Resources*, 4: 23–30.

Wachs, M. (1989) "When planners lie with numbers," *Journal of the American Planning Association*, 55: 476–479.

Wachs, M. (1990) "Ethics and advocacy in forecasting for public policy," *Business and Professional Ethics Journal*, 9: 141–157.

Watson, V., (2003) "Conflicting rationalities: implications for planning theory and ethics," *Planning Theory and Practice* 4: 395–408.

Yiftachel, O. (1998) "Planning and social control: exploring the dark side," *Journal of Planning Literature*, 12: 395–406.

Chapter 12

Socio-political analysis of French transport policies

The state of practice

Vincent Kaufmann, Christophe Jemelin, Géraldine Pflieger, and Luca Pattaroni

The chapter takes stock of transport policy research which was carried out in France between 1995 and 2002 or which deals specifically with French transport policy. It is divided into four parts. The first concerns the quantitative analysis of the body of research, which enables us to pinpoint those studies which deal specifically with our chosen topic. The second part consists of an analysis of the content of the selected research according to the sequential model used to analyse public policy, i.e. starting with the construction of the problem and ending with an evaluation of the measures introduced to solve the problem. The third provides a discussion on the overall studies compared to the transport policy in France. Finally, the last section deals with the three central questions raised in this study.

Introduction

Between the mid-1990s and early 2000, a body of research carried out in France adopted a highly original approach to the analysis of transport policy, one which incorporated political science and sociology of action. In a field of research traditionally dominated by transport economics and its concomitant models, this trend owed much to the particular situation in which France found itself at that time, namely the introduction of a series of new urban transport legislation and regulations with an environmental focus, and the coordination of local transport policies and territorial development. The adoption of these laws led to the re-definition of local public action in terms of transport and prompted a number of studies on transport policy development.

The present chapter provides an overview of the scientific contribution of this research. However, the primary aim is not to set an agenda for future research, but rather: (1) to capitalise on existing knowledge with regard to the implementation

---

Submitted by the Association for Development of Planning Education and Research.

Originally published in *Transport Policy*, 15: 12–22. ©2008. By kind permission of Elsevier Ltd.

of French transport policies during the 1990s, (2) to identify the contributions made by the public policy analysis and the sociology of public action to the analysis and socio-economic evaluation of transport policy, and (3) to illustrate how this research is of interest to the debates that drive international research in the field.[1]

Public policy can be defined as "a collection of coherent decisions taken at various legal and project levels, which focus on objectives decided upon or applied by the relevant actors and which are aimed at resolving a particular social problem" (Knoepfel and Bussmann 1998). Basing our study on this definition and in order to impose a clear structure on our work, we have retained the sequential model of public policy analysis developed by the American political scientist Jones (1970) and used widely in both the United States and Europe. According to Jones, public policy refers to a process which begins with the formulation of the problem to be solved (or programme identification). These stages are themselves linked to various events or to changes in the power play between the actors involved. The programme identification also covers the inclusion of the problem in the political agenda. The next stage of the process involves the design of solutions and the legitimisation of the as yet unimplemented public policy (programme development). The final two phases relate to the implementation of the policy, namely the use of resources and the enforcement of decisions at different institutional and territorial levels. There may also be an evaluation phase, concerning the definition of criteria, data gathering and analyses.

As a general rule, the analysis of this process breaks down into two main phases. First, there is the setting of an agenda, in other words the social construction of the problem to be solved and the development of the legal and procedural frameworks within which the problem lies. The second phase concerns the implementation or application of the measures chosen to tackle the problem and their impact in terms of *outputs* and *outcomes*.

These phases allow us to deconstruct the analysis of transport policies and the research studies dealing with these policies, their origins and their evaluation. However, the systematic nature of the sequences described by Charles O. Jones requires further qualification (Muller and Surel 1998). Indeed, the order of the phases may be inversed or upset, as illustrated in the *garbage can model* (Cohen et al. 1972), which shows that the invention of a solution can precede or even occur concomitantly with the identification of the problem. This is an inversion of the sequence frequently observed in the development of policies that involve important technical or organisational innovations. Even if the sequence can vary, yet the categories in which problem setting, problem solving, implementation and evaluation are grouped continue to dominate public policy analysis.

On this basis, we have adopted a mixed method methodology (Bergman 2007) separated in four phases:

1. An analysis of an established corpus comprised of multiple studies. The corpus contains research led by French and other European programmes concerning France between 1995 and 2002. This refers to all research carried out on a subject related to transport policy in France and which include an analysis of public policies or the sociology of public action. About 100 studies were identified. These constitute the bases for the present analysis.
2. A quantitative analysis of the corpus, which allows to localise the research in terms of subject areas (urban, regional, passenger traffic, freight, etc.) and in terms of the process of developing a policy from its integration in the political agenda to its evaluation.
3. A systematic content analysis of the 100 researches identified. This analysis was organised according to topic and is based on the sequential model for public policy analysis. As a result, we are able to identify each distinct phase in the development of a public policy.
4. An evaluation of the findings in relation to the central issues raised in this study: (1) greater knowledge about French transport policy development in the 1990s, (2) the contributions of this research to more traditional research, which focuses on the socio-economic analysis and evaluation of transport policies, and (3) the contribution this research could make to the debates which drive international research.

The present chapter is divided into four parts. The first concerns the quantitative analysis of the body of research, which enables us to pinpoint those studies which deal specifically with our chosen topic. The second part consists of an analysis of the content of the selected research according to the sequential model used to analyse public policy, i.e. starting with the construction of the problem and ending with an evaluation of the measures introduced to solve the problem. The third provides a discussion on the overall studies compared to the transport policy in France. Finally, the last section deals with the three central questions raised in this study.

Transport policy analysis: the body of research and quantitative benchmarks

A two-step procedure was used to compile an inventory of the research to be examined. The first phase focused on research financed by French research programmes, while the second looked at research on French transport policy which received European funding or which was carried out as part of foreign research programmes.

The survey of the research carried out in France identified 69 studies. The survey of European comparative research, including French case studies, identified 31 research projects, the majority of which (19) were conducted as part of the transport programme of the Fourth and Fifth Framework Programmes of the European Union.

A first obvious result is that the whole set of research focus only on the appraisal methods in transport policy and the socio-economic expected effects of transport. Therefore, two main research topics and orientations are clearly missing:

- Research on the earliest phases of the decision-making process, such as the frames of reference for initiatives by the public authorities, and the structure of political opportunities.
- Research on the latter stages of public action, namely their evaluation.

It should also be noted that the selected research focuses heavily on public and on urban transport. Passenger rail transport is largely overlooked (apart from analyses of the decision-making process in relation to new TGV Méditerranée routes); references to road networks and freight are virtually non-existent. There are two reasons for this absence. The first reason is cyclical. In the 1990s, France introduced a series of new laws on urban transport, which raised awareness of this issue, either because of the debates which the legislation generated, or because of the projects and infrastructural developments which they produced (tramway networks, traffic plans). A second reason can be attributed to the scientific disciplines which are traditionally involved in transport research in France. While the analysis of transport demand, logistics and the rail transport sector is dominated by the disciplines of economics and civil engineering, the disciplines of urban planning, geography, political science and sociology are very active in areas of transport research that touch on the urban environment. It is normal, therefore, that we found an over-representation of subjects dealing with the urban environment in the body of research on which the present study is based.

Insights from the research and issues under discussion

The following section sets out the central issues, which we have identified in our selected body of research. These are grouped according to six topics, and account for more than 80% of the research.

Frame of reference for initiatives by the public authorities

What transport-related problems remain unsolved? How far is there agreement on these problems? Shared values, which provide the foundations for initiatives by the public authorities, are particularly important when it comes to setting the policy agenda. In France, further development of these research topics came from the world of political science, where researchers took the notion of frame of references, in other words all arguments which can provide a basis for consensus-building, as the starting point of their work. Several studies were carried out on this topic, in the knowledge that the main obstacle to developing transport policy is the lack of a shared cognitive corpus (Jouve 2002).

While this is a generally accepted fact (Ollivier-Trigalo and Piéchaczyk 2002; Offner 2003b), transport research rarely includes in-depth analysis of the frames of reference. This is all the more surprising given that this field of research a priori offers rich pickings. The trans-sectoral nature of transport policy leads to friction due to the contradictory objectives it creates. This friction can be observed in the many different areas affected by transport-related initiatives that were undertaken by the public authorities. An in-depth analysis of frames of reference is not only clearly absent in our research inventory but also in the public policy development process. For example, scant regard is paid to local frames of reference when developing local urban transport policies. Any consideration of these references merely amounts to an assertion of a few central principles, such as environmental protection as well as the interlinking of transport and urban planning (Offner 2003a).

The Mobiscopie research project (Mermoud et al. 2001)—an audit of public surveys on urban transport—sees the limited impact of transport policies on modal shift as a reflection of the many different frames of reference in this field and their inherent tensions. From a socio-cultural perspective, mobility is highly valued in our societies. From a socio-political perspective, however, awareness of environmental issues which has grown significantly in France over recent years is not viewed as a welcome development. Consequently, policies must continue to take account of this contradiction, while at the same time conciliation strategies are introduced to deal with it.

Decision-making procedures and processes

The decision-making process has an important bearing on the actual content and shape of any decision taken. With this in mind, a series of studies looked at the development of "Plans de déplacements urbains" (PDU), a multimodal instrument of urban transport planning, which became compulsory in 1996 for all French cities with more than 100,000 inhabitants. Its aim is to contain rising car use. In their summary report on the summative evaluation of the PDU development process, Jean-Marc Offner unequivocally states: "Opponents who claim that the PDU are vague, wishy-washy, timid and conservative are not entirely mistaken. However, they fail to see what the true aim of this instrument is. If we follow the teachings of political science, the effectiveness of planning procedures lies less in their intrinsic content than in their ability to procure the ingredients for future local policies and decisions. PDU are thus not an end in themselves but are an active part of 'constitutive' policies" (Offner 2003a: 12).

The starting point for most of the studies in our body of research on decision-making procedures and processes was the crisis of legitimacy within government (Fourniau 1996). Philippe Blancher singles out the highly formal interplay between participatory dynamics and public enquiries (Blancher 2001: 59). These tensions lie at the very heart of research on transport-related decision-making procedures and processes. Such research goes in two distinct directions: concentrating on the importance of procedures or concentrating on the actors and their dynamics.

The first set of studies deals with institutional innovative measures to enhance coordination. If transport policies are to meet their objectives (particularly with regard to sustainable development), they require seamless intermodal and inter-sectoral coordination. Most cities which have embarked on ambitious transport policies already had a working transport authority (Jouve 2002). Nevertheless, there is still room for improvement due to the dearth of instruments to coordinate transport and urban planning and due to the fact that projects are funded on a sectoral or even modal basis. The legislative apparatus recently introduced in France—LAURE (law on air and rational energy use) and the SRU (law on solidarity and urban renewal)—promotes greater coordination between sectoral policies, and as such offers a promising area of research. To what extent do institutional innovations (the obligation to develop urban transport plans, or a "schéma de cohérence territoriale"—a long-term urban planning concept in France), potentially aided by this legislation, enable the development of original policies? In their analysis of the Marseille PDU which resulted from LAURE, Yerpez and Hernandez (2000) show that this instrument created an opportunity for professionals to hold discussions with other partners (such as associations and neighbourhood committees) and opened up areas for consultation, which in turn could bring its influence to bear on the policy-makers' decisions.

The second set of research focuses on the network of actors who are favoured or disadvantaged by the institutional architecture, and on how these networks are conducive to the development of new political configurations. In their analyses of the PDU of Lille and Valenciennes, Frère et al. (2000) show that the main advantage of the PDU lies less with its capacity to resolve environmental problems associated with mobility but rather with its capacity for all concerned to embark on a collective learning process.

Dialogue and transport policies

The size and recurrence of movements which oppose the construction of transport infrastructures reflect the crisis of state legitimacy, particularly with regard to its role as the guardian of public interest. As a result, some countries, including France, have introduced changes to their decision-making instruments over the last 20 years to meet growing calls for greater transparency and more public participation in the decision-making process (Fourniau 1996). In France, these calls led to changes in the instruments which regulate initiatives by the public authorities. In the transport sector, a legislative apparatus which institutionalises dialogue was introduced incrementally.

A number of studies examined these new citizen participation tools. They reveal that calls for consultation on transport issues give rise to two questions:

- The first is resolutely pragmatic; given the rising opposition to transport projects (infrastructures or traffic management), does consultation help boost consensus and thus optimise the chances of these projects being implemented (MATE 1996)?
- The second is motivated by democratic concerns and the qualitative optimisation of projects. Here consultation is viewed as a means of involving citizens in initiatives by the public authorities against the backdrop of their crisis of legitimacy as guardian of the public interest and as a valuable project participant (Fourniau 1996; Louvet 2002).

More precisely, a number of studies developed a so-called "ingénierie de la concertation" (a set of guiding principles for the consultation process) which suggests that if consultation is to have a positive influence on a project and is not to exacerbate the NIMBY phenomenon, it must take place long before the implementation phase to allow community actors to exert a degree of control (Lolive 1999). However, implicit within such an early participatory procedure is the choice between contradictory options. Furthermore, consultation does mean a simple juxtaposition of more or less consensual opinions. The consistency of the different aims pursued within one project depends largely on the ability of the authorities to organise and steer debate, while at the same time having different strategies at their disposal (Roy and Damart 2002; Jouve 2002).

Despite the legislative apparatus which aims to increase citizen participation, the power which the public authorities and the administration in France exert on decision-making processes is unique in Europe (Newman and Thornley 1996; Ollivier-Trigalo and Piéchaczyk 2002). Compared to other European countries, local associations in France have only a limited role to play. The analysis by Cuel of the Grenoble north bypass project echoes this observation: "The north bypass highlights the fact that such major projects with a high economic and social impact are too often carried out under the unilateral rule of the public authorities, who keep a tight rein on their control of public policy. In terms of the appraisal of decision-making processes, there is a lack of independent instruments that are governed by a truly ethical code of conduct. Furthermore, there is little regard for public opinion" (Cuel 2002: 112).

Appraisals

Several authors show that a feature which is unique to transport is the central importance of "non-human" agents, i.e. the technical aspects of the transport system (Lolive 1999). Traditionally, the handling of these nonhuman agents falls to the experts, who define the impact that these agents are likely to have. The expert, therefore, is a central figure in transport-related decision-making. Yet, at the present time, obedience to the techno-economic reality is undergoing a serious crisis of legitimacy. According to Durand and Thoenig (1996), this is reflected in the growing gap between transport policies and initiatives undertaken by the public authorities. This type of appraisal, which largely dominates transport policy research, produces sectoral diagnostics that are ill-suited to participation and take little account of the trans-sectoral nature of transport (e.g. in his study of the PDU in Lyons, Bernard Jouve reveals that fewer than 15 professions are involved in the process (Jouve 2002: 100)). As Rui et al. (2001) show, appraisals become a communication and evaluation tool during the consultation process. Given these new roles, the research recommends an overhaul of appraisal methods. This research branches out in three complementary directions:

A review of forecasting methods

According to Bernard Jouve, "If one wants to adopt a global perspective on the issue of urban travel, we should perhaps begin by thinking of new types of appraisals which enable us to shift our focus" (Jouve et al. 2002: 104). In his comparative study on urban transport policies in Europe, Jouve shows that in the absence of alternative, legitimate forms of appraisal, both politicians and professionals remain the prisoners of technical know-how, which in turn hampers innovation.

The report by the Boiteux group (2001) proposes placing a monetary value on the non-market advantages and disadvantages of transport projects as an aid during the evaluation and selection process. By doing so, and in particular by integrating nuisance costs, this approach basically amounts to a type of "cost-benefit" analysis. The report also goes as far as recommending placing monetary values on the impact of other aspects such as atmospheric pollution, the number of human lives saved, noise, etc.

Research by Roy and Damart (2002) shows that cost-benefit analyses alone cannot boost the legitimacy of any decisions taken. In fact, the idea of seeking the optimum and the all-important "black box" feature, which are an integral part of such analyses, are in direct opposition to the participatory procedure. This leads the researchers to conclude that new instruments should be developed which enable information to be presented in an accessible way in order to facilitate both debate and reflection on the issues in terms of the actual process. Research by Barthélemy (2002) on the democratisation of the transport debate in France and the Netherlands also highlights the weakness of planning methods as well as the significant restrictions placed on public debate by decision-making tools that are based on cost-benefit analyses. By and large these conclusions are in line with the observations made by Michel Rousselot on the Boiteux group's study "To enter into a participatory process is to accept that decisions are generally not taken a priori, only to be justified to the public ex post, possibly accompanied by minor changes. It is to accept the need for debate when several solutions to a particular set of problems exist. It is to accept that the public interest cannot be calculated mathematically but is the result of the entire set of interests of the different population groups concerned" (Rousselot 2002: 20).

Favourable ground for the overhaul of appraisal methods
In their analysis of Munich and its transport programme, Mobinet and Jouve (2002) suggest that local policy initiatives could offer the basis for an overhaul of appraisal methods. Baye and Debizet (2002) also arrived at the same conclusions, stating that the driving force of innovative planning is the interaction between consultants and local stakeholders: where urban communities in effect become the managers of a transport project, demanding innovative policies which, given their controversial nature, require an approach that involves greater consultation. This gives rise to a creative emulation of new planning methods. As a result, planning departments acquire more interdisciplinary skills and pay greater consideration to environmental concerns, to the coordination of transport objectives and to issues of controlling urbanisation (Baye and Debizet 2002: 53).

Still on the subject of the overhaul of appraisal methods, Baye (2002) shows that the regionalisation of rail services in Europe goes hand in hand with calls for innovative appraisal methods. In several European countries, the train is an essential means of transport in suburban and periurban areas (particularly in Germany and the UK). Furthermore, in several European countries regional institutions are placed in charge of the rail transport system, but without having the necessary internal competences (Germany, Spain and France) (Baye 2002).

Innovative appraisal methods

The third focus of this research is the contribution of new appraisal methods.

Lolive and Tricot (2002) highlighted the emergence of a new form of environmental appraisal which has its origins in the opposition of local communities to the construction of the TGV Méditerranée line (high-speed rail link between Paris and Marseille) and to motorway projects in the Southern Alps. The aim of this new form of environmental appraisal is the adoption of new project evaluation criteria. Its international slant, involving Switzerland and Austria (two countries with a well-developed tradition of environmental appraisals of major transport infrastructures due to their alpine topography), was a conscious decision, as it enables the acquisition of greater technical expertise and legitimacy.

Guillaume Faburel shows that the contingent valuation method[2] is a traditional appraisal tool, sometimes applied in an innovative way (such as measuring noise pollution from planes—Faburel 2002), and is well accepted by institutional and local actors. It therefore has undeniable advantages in terms of linking the evaluation of variants with decision-making.

Other research studies deal with the contribution of methods borrowed from experimental economics (Denant-Boèmont 2002) on the evaluation of disputed risks (Kast and Lapied 2002) or of information systems (Houée 2002). European research has also produced a series of studies on new appraisal methods, such as the INTRAMUROS project, which developed an IT multi-criteria tool to facilitate the decision-making process, thereby enabling a quantitative comparison of the impact of local strategies on urban transport management and the identification of areas where enhanced coordination between the relevant actors would optimise the effectiveness of transport measures.

Socio-economic effects of transport

Transport is often considered as a lever of economic and social change. This touches on two distinct ideas: that the quality of service provision changes the modal behaviour of actors, and that good accessibility is synonymous with economic development. Yet, analyses reveal the limits of this approach. On the one hand, strategies to develop public transport services have had little impact on modal shifts, either in terms of urban or interurban travel or in terms of combined traffic (TGV traffic which had little impact on road traffic (Klein 2002), or the development of rail traffic through the Swiss Alps in line with the Alptransit are two good examples).

On the other hand, several studies have cast doubt on the structuring effect of accessibility on territorial development (Offner 1997).

Research on the socio-economic effects of transport go along two main lines: (1) the search for the most effective measures to resolve transport-related "problems", and (2) the re-definition of the notion of "effect".

The search for "good practices" to tackle transport-related problems

The search for the most effective measures to resolve transport-related problems has long resorted to the notion of "good practices". The basic idea which underpins a large part of European comparative transport research is the identification of policies that can resolve the "problem" effectively, and then to apply these policies to other contexts. Although this idea is well-meaning, it rarely creates the necessary transferability conditions. Two recent European research programmes which analysed coordination between transport management and urban planning in local policies, TRANSLAND and LEDA, conclude that the transferability of policies is greatly restricted by the simple fact that the vertical distribution of decision-making responsibilities varies across countries, and that often urban regimes and specific spatial morphologies which are highly context-specific mean that one policy has to be chosen over another. As a result, it is unclear whether the same measures can be taken and would have the same impact in other contexts. It now seems that the notion of simply importing "good practices" has taken a battering, thus leaving the way clear for comparative analyses which have a greater focus on the origins of policies.

Re-defining the notion of "effect"

Undoubtedly, Europe is the setting for the most sustained discussions on the impact of major transport infrastructures (Joignaux 1997). Based on an analysis of common transport policy documents that look at the development of major transport infrastructures, Adam suggests reviewing the importance which European transport policy attributes to speed. It is universally valued and is used to justify the expansion of the different transport sectors. In addition, speed is inextricably linked to economic competitiveness and profit, with the compression of distance and time viewed as a prerequisite of economic growth (Adam 2001).

A series of critical reviews on the impact of transport were developed by social scientists based on the following evidence: "Analyses of the links between transport and society often borrow from determinism. The current overhaul considers transport services first and foremost as a social production, and only afterwards as an instrument of social change" (Klein 2002).

Francois Plassard stresses that research on the effects of transport is a field which has a tendency towards complexification. Appearing for the first time in the 1960s to evaluate the consequences of new transport services on the main socio-economic variables, transport research progressively covered the spatial changes arising from transport infrastructures, and recently expanded to cover social changes based on the observed insufficiency of economic and spatial approaches. Plassard thus asks the question: "Does the limited impact of infrastructures have its roots in the inadequate manner in which this impact is measured?" (Plassard 1997).

Measuring the impact of transport infrastructures raises the issue of methodology and the areas in which they can be found.

- With regard to methodology, Chaplain shows that "To take account of the relation between transport and space, it is no longer suitable to start with transport and its effects but rather with the processes, namely the sequence of public action, which gave rise to the infrastructure in the first place, and which integrated it in the public space" (Chaplain 1997: 127). Re-read in these terms by Chaplain, the project of the high-speed rail link through the Channel Tunnel thus assumes the function of affirming and positioning the institutional actors.
- With regard to the area, several studies carried out as part of the PUCA research programme "Travel and Inequalities" highlight the effects which had received little attention in the existing research. The work of Dupuy et al. (2001) on the mobility of poor households in France and UK show that transport policies and territorial planning in these countries have very different effects on their disadvantaged populations. In France, urban planning tends to house these population groups in segregated though well-serviced areas. In the UK, on the other hand, the lack of planning leads to car dependency among the disadvantaged.

Evaluation ex post

An ex post evaluation enables us to link the analysis of public policy development procedures to their institutional framework, to their architecture, to their implementation and to their effects. Furthermore, it has the advantage of replying in a more in-depth fashion to the difficulties of translating intention into action. "To evaluate both ex ante and ex post is to acquire information on the consequences of decisions of each actor: the road policies adopted by each Département and their influence on the development of the periurban habitat, commercial urban planning choices and the concomitant transport demands, the actions of mayors in outlying communes faced with a new bypass . . ." (Offner 1997: 27).

Transport research has pioneered ex post evaluations, with work such as the monitoring of the underground transport system in Lyons, the evaluation of local transport policies by Pierre Lassave, and finally the evaluation programme of the French Ministry of Transport led by the Conseil Général des Ponts et Chaussées (French Public Works Department) (Offner 2003b). We can differentiate between two types of ex post evaluations of transport systems: (1) supervisory evaluations as a governance tool (transport observatories found in several cities) and (2) summative evaluations, which judge the effectiveness of initiatives by the public authorities.

Despite their promise, ex post evaluations are largely absent in our inventory of research. There is research on the evaluation of local urban transport policies by Offner (2003a), and an evaluation of Article 14 of LOTI (Loi d'Orientation des Transports Intérieurs) by Affichard et al. (1997). However, these amount to summative evaluations of public policy. There is a total absence of comparative evaluations.

Discussion of French transport policies

The six central topics which we have identified highlight the two basic features of transport policies implemented in France in the 1990s: tensions between contradictory principles of public action in the decision-making processes and an ambivalent political will. We shall look at both aspects in more detail in the following section.

Tensions between contradictory principles of public action in the decision-making process

The results of the analysed researches highlight several tensions within the various transport policies in France. With regard to policy content, the absence of an integrated transport policy in France (Bernat and Ollivier-Trigalo 1997) is both corroborated and detailed in the research. France currently does not have a transport policy in the sense of a collection of consistent decisions and initiatives which are targeted at specific objectives. Transport policy appears to be ridden with tensions, such as:

Tension between obvious tranversality and de facto sectorisation

The integration of different modes of transport within a veritable transport system, intermodality and interoperability, as well as the coordination between transport and urban planning are objectives which are constantly and forcefully re-affirmed. Yet, beyond the expression of these major principles, we have observed that practices remain essentially sectoral, as witnessed by the pursuit of motorway construction projects, the development of high-speed rail links or segregated public transport corridors. The starting point for most of these projects is new modal infrastructures. In addition, the planning of these projects takes little account of multimodal issues.

Tension between participation and decision-making

Most of the research on the overhaul of decision-making processes as well as on the contribution of consultation to decision-making highlights the inconsistencies between procedures and decisions. Several different mindsets meet head on, such as the state and its apparatus—as guarantors of the public interest—which makes decisions on a unilateral basis; planning where decisions are dictated by the demands of political feasibility and thus requires the involvement of other actors; and finally pragmatism, where broad-based consultation systems are introduced to respond to the crisis of citizen representation mechanisms, with the aim of avoiding legal impasses. In concrete terms, the convergence of these differing mindsets creates conflict situations, such as a participatory process which ends up in an awkward decision, taken as the result of a public enquiry wiping out the input from the consultation process.

The gulf which divides decision-making procedures often reflects the absence of a consensus on the objectives which transport policy should reach. To overcome this, experiments in participatory democracy and procedural innovations have been attempted, some of which have been the subject of in-depth research. Their aim is generally to boost the legitimacy of decisions and to appease opponents. However, we observed that the problem is more structural than procedural, and therefore innovations and other experiments are only partially successful.

Tension between contradictory objectives

Transport policy replies to sectoral objectives of economic promotion (support of industrial innovation, support of the construction sector), territorial planning (combating spatial isolation and interregional disparities), social policy (mobility rights), and respect for the environment (reducing atmospheric pollution and greenhouse gas emissions). Each of these areas is characterised by their own specific frame of reference. This implicitly leads to contradictory decisions on transport policy. Les Plans de déplacements urbains (PDU) are once again a good example: their aim is to guarantee consistency between different modal policies in a given territory and to link them with urban planning policy. Such general objectives presuppose that the measures to be taken concern the use of the networks and the coordination between the different sectors involved in initiatives undertaken by the public authorities. Yet, there are several PDU which simply amount to an instrument to lobby for the construction of a tram network and/or new road links, and do not propose any other type of transport-related measures (Offner 2003a). In these cases, the reconciliation of the demands of both economic promotion and environmental conservation leads to a focus on the creation of infrastructures to the detriment of measures to optimise the running of existing transport services. More generally, these many objectives lead to the adoption of a combination of measures that are ill-suited to creating the desired effects.

Several research studies show that these tensions must be interpreted through the prism of the crisis of state legitimacy, which France is currently experiencing (Jouve 2002; Fourniau 1996). In fact, this situation is not only a crisis of legitimacy, but also a "crisis of effectiveness".

- Crisis of legitimacy: As highlighted in several studies, the state is no longer the guarantor of "the public interest" and therefore the legitimacy of its decisions is undermined. This is further reflected in the weak acceptability of many economic regulatory instruments, such as congestion charges, the introduction of tolls in city spaces or restricted paid parking.
- A crisis of effectiveness: Several studies show that state appraisal has become obsolete, at least in terms of the methods used and the competences which the state can mobilise to this end. With its adherence to "cost benefit" analyses, which several researchers consider to be incapable of ensuring cohesion between the actors, state appraisals are increasingly limited by a double reasoning which links infrastructure services realisation with an engineering monoculture. As a result, these appraisals are impervious to the growing importance of transport management and service issues.

Limits of public initiatives: the ambivalence of political will

Although most of the research is applied and was commissioned by the public authorities, it reveals the transport interests and disinterests of the public authorities. In particular, they appear to concentrate on urban transport and the construction of new infrastructures. There is an almost total absence of research on interurban transport, road traffic in general and freight traffic. Undoubtedly, this is a reflection of the desire on the part of the public authorities to protect urban public transport and railways against car and air travel. In other words, the public authorities are concerned with defending their own interests, given that many parts of the French transport system are under state control. These studies also show the absence of a proactive stance in France in relation to setting ambitious transport policy objectives.

The search for "magic" innovations

Several measures to regulate mobility have long been identified and validated, yet a great deal of research remains to be conducted on the effectiveness of these measures and on the search for new instruments (TRANS-PRICE 2000; CONCERT 2000). However, it would appear that this process concerns discovering new regulatory instruments which would allow the state to avoid making unpopular decisions. This is particularly true for France, where the constant search for new instruments is not undertaken only after the potential of existing measures has been exhausted. Parking management, for example, avails of a broad range of measures, such as paid parking in "blue zones"[3] or norms linked to the construction of new buildings. In conjunction with good public transport services, these instruments have been successful in several European cities, but are only rarely applied on a systematic basis in France (with the exception of a few cities).

Research tenders issued by the public authorities are symptomatic of this state of affairs. Does the first phase of the "Agora 2020" workshops (organised by the Centre de Prospective et de Veille Scientifique, part of the French Ministry of Town and Country Planning), which focused on omissions and weakness in existing research, not reflect the fact that the public authorities simply wish to increase their room for manoeuvre in order to win back their influence?

A degree of disinterest in other countries' experiences

The weakness of international comparisons on public transport policy analyses illustrates the scant interest France pays to the transport-related experiences of other countries. The public authorities often cite foreign examples as role models of best practices, but use them merely to illustrate their general stance on the given issue. They appear rather disinterested in the origins of these policies, their frames of reference, procedures and institutions, and in their evaluation and transferability potential. Several studies refer to the "tram–train" in the German town of Karlsruhe as a transport system to which France aspires. However, this aspiration is rarely substantiated, given the absence of assessments of the impact such a measure would have on the modal share. The tram–train therefore is reduced to an abstract idea, a catchphrase, a fashion. The same applies to the BRT (bus rapid transit system) in the Brazilian city of Curitiba or the road–rail transport system in Switzerland.

This phenomenon has also been observed in policy areas which receive much less media attention. Take the example of regular hourly scheduling of train services. Adopted by the majority of large European rail networks in the 1970s and 1980s, this operating model boosts capacity and productivity. However, it was derided by the SNCF (the company which operates the French rail network). Until recently, only the RFF (the agency which manages the French railway infrastructure) has given serious thought to the introduction of regular hourly scheduling, having capitalised on the experience of foreign networks in this regard.

Dearth of research on acceptability

More generally speaking, transport policy research should give greater attention to the acceptability of transport policy measures. At the most, existing research merely takes note of instances of acceptability, but not why such measures are deemed acceptable. The acceptance of parking restriction measures and tolls in city areas are linked to the frames of reference specific to the initiative undertaken by the public authorities, as well as those specific to the interplay between user and citizen. According to research carried out by Brög since the 1980s: "As a citizen, I may agree to restrict my car use voluntarily, on the condition that there are credible alternatives and insofar as my behaviour will help resolve a public problem which I deem important." However, the credible alternative and the urban transport problem which could incite popular support have yet to be defined. To this end, it is essential that in-depth analyses are carried out on the frames of references of initiatives taken by the public authorities, what informs them and how they come about. Admittedly, several in-depth studies do cite these frames of reference, albeit less than systematically, as an important element in any analysis of transport policy. Nevertheless, they are rarely subject to close examination.

Absence of evaluations

There is an unquestionable dearth of evaluations in the body of research which we analysed, even in areas where important measures have been introduced, such as segregated public transport corridors (TCSP) or high-speed rail links. At the most, the research shows that the share of car journeys between the cities of Paris and Lille has remained high even after the introduction of the TGV Nord line. They also show that the introduction of segregated public transport corridors in provincial transport networks has not curbed the rising modal share of car journeys (with the exception of Grenoble), even though this was the main reason behind the project. Clearly, research tenders appear to have little interest in evaluations. There may be several reasons for this. Perhaps the public authorities fear that an evaluation would call into question the elements involved in achieving a political objective? Or perhaps the authorities view such an undertaking as pointless given the total absence of any real transport policy objectives.

Conclusion

What can the research tell us about (1) transport policy in France during the 1990s, (2) the contribution of the analysis of public policy and sociology of public action to transport policy analyses. What does it bring to (3) the international scientific debate?

Three central points emerge:

First of all, French transport policies in the 1990s had to wrestle with two specific problems: a decision-making crisis and a certain lack of political will. While new legislation should be both proactive and based on sustainable development, subsequent policies have remained focused on the construction of infrastructures and lack innovative flair. The opportunities to produce something new based on these legal frameworks are rarely exploited. If the decision-making crisis in France raises the more general issue of the reform of public institutions, the fact remains that the decision-making process curbs the development of truly innovative transport policies.

Second, the public policy analysis and the sociology of public action as applied in the evaluation of transport policies highlights the limitations of a cost-benefit approach. Given the dominance of such an approach (even within the legal framework), several research studies show that cost-benefit considerations present a double obstacle to decision-making. On the one hand, their application is debatable because they are based on non-consensual ideologies, such as the idea that a new transport infrastructure can generate benefits which can then be measured in terms of gains in speed, which itself is synonymous with economic growth. On the other hand, they prevent the participation of actors in the decision-making process, as they are characterised by a "black box" that is unquestioned by experts. A number of research studies are a reminder that in a decision-making process or as a user making choices, actors are not solely moved by economic interests, but also by ideas of what is good and fair, which cannot always be reduced to cost-benefit equations. This trend is more broadly inspired by the "pragmatic turn" of French sociology which stresses the need to "take moral seriously" (Boltanski and Thévenot 2000).

Third, the research shows that an interdisciplinary approach to the analysis and evaluation of transport policies holds significant promise. Econometric modelling is manifestly dominant in the study of transport policy, both in terms of it epistemologies, its approaches and its subjects. Yet, the research illustrates the importance of factors which are difficult to monetise and which go beyond utility maximisation. Even if this body of research has enjoyed only a limited following in France, the new perspectives that it offers for a socio-economic study of transport is undoubtedly its most important contribution to international research.

Notes

1   The present chapter is part of a study carried out within "PREDIT 3", Groupe Opérationnel 11 "Politiques des transports" a programme of research, experimentation and innovation in land transport by the relevant French authorities.
2   A contingent valuation involves conducting a survey among a representative sample of the population concerned by asking them questions on how much they would be willing to pay for services based on hypothetical scenarios.
3   "Blue zones" are parking spaces which are painted blue and indicate that the maximum parking time allowed is 90 min. A disc must be placed in the windscreen of the parked vehicle, indicating the time of arrival.

References

Adam, B. (2001) "The value of time in transport." In: Giorgi, L., Pohoryles, R. (Eds.), *Transport Policy: What Future?*, Aldershot: Ashgate.

Affichard, J., Champeil-Desplats, V., Lyon-Caen, A. (1997) "Définir le service public, réguler un secteur concurrentiel: genèse de la loi d'orientation des transports intérieurs." Paris: Institut International de Paris La Défense.

Barthélemy, J.-R. (2002) "De la complicité à la vertu: démocratisation du débat sur les transports, France/Pays-Bas" [From collusion to inclusion: democratisation of the transport debate]. In: *Métropolis*, 108/109: 46–51.

Baye, E. (2002) "Elus, experts, électeurs-usagers: qui décide?" [Elected representatives, experts and the voter-user: who decides?]. In: *Economie et Humanisme* no. 359: 55–58.

Baye, E., Debizet, G. (2002) "Innovation et bureaux d'études dans la planification des transports urbains" [Innovation and engineering consultancies in urban transport planning]. In: *Métropolis*, 108/109: 52–57.

Bergman, M.M. (2007) *Mixed Methods Research*. London: Sage.

Bernat, V., Ollivier-Trigalo, M. (1997) "Politique des transports en France. Processus de décision: discours et pratiques," [French Transport Policy. Decision Making—Discourse and Practices] TENASSESS (DG VII) Project, Work Package 6.

Blancher, Ph. (2001) "Débat public, concertation et politique des déplacements" [Public debate, consultation and travel policy]. In: *Economie et Humanisme*, December: 58–61.

Boiteux group (2001) "Transports: choix des investissements et coût des nuisances," Paris: Commissariat Général du Plan.

Boltanski, L., Thévenot, L. (2000) "The reality of moral expectations: a sociology of situated judgement," *Philosophical Explorations* III: 3, 208–231.

Chaplain, C. (1997) "Le Nord Pas de Calais face au transmanche: mobilisations différentiées et structure de relations" [Nord Pas de Calais and cross-channel links: mobilisation and relationship structures]. In: Burmeister, A., Joignaux, G. (Eds.), *Infrastructures de transport et territories—Approches de quelques grands projets*, Paris: L'harmattan.

Cohen, M.D., James, G., March, J. Olsen, P. (1972) "A garbage can model of organizational choice," *Administrative Science Quarterly* 17: 1, 1–25.

CONCERT, (2000) *Cooperation for novel city electronic regulating tools.* Final Report, Barcelona: Tecnologia.

Cuel, S. (2002) "Le processus d'élaboration du projet de Rocade Nord à Grenoble" [North Ringroad in Grenoble: the project development process]. Grenoble: Institut d'Etude Politique de Grenoble.

Denant-Boèmont, L. (2002) "Une approche par l'économie expérimentale" [An experimental economics approach]. In: *Métropolis*, 108/109: 36–40.

Dupuy, G., et al. (2001) *Les pauvres entre dépendance automobile et assignation territoriale : comparaison France/Royaume-Uni* [The poor: car dependency and territory ascription—a comparison of France and the UK]. Paris: Latts.

Durand, P., Thoenig, J.-C. (1996) "L'Etat et la gestion territoriale" [The State and territorial management]. In: *Revue Française de Science Politique* 46: 4, 580–623.

Faburel, G., 2002. "La Méthode d'évaluation contingente appliquée au bruit des avions: acceptabilité institutionnelle et sociale" [Contingent evaluation method applied to airplane noise pollution: institutional and social acceptability]. In: *Métropolis*, 108/109: 26–31.

Fourniau, J.-M. (1996) "Transparence des décisions et participation des citoyens" [Transparancy of decision-making and citizen participation]. In: "Projet d'infrastructures et débat public," Proceedings of the CPVS/DRAST meeting, 25 January 1996, Ministère de l'Aménagement du Territoire de l'Equipement et des Transports, CPVS, Techniques, territoires et sociétés, 31pp.

Frère, S., Menerault, Ph., Roussel, I. (2000) "PDU et dynamique des institutions à Lille et Valenciennes" [PDU and institutional dynamic in Lille and Valenciennes]. In: *RTS*, 69: 22–34.

Houée, M. (2002) "Systèmes d'information et evaluation" [Information systems and evaluation procedures]. In: *Métropolis*, 108–109: 41–45.

Joignaux, G. (1997) "L'approche des relations entre infrastructure et territoires: retour sur la théorie et les methods" [Relationships between infrastructure and territories: a review of theories and methods]. In: Burmeister, A., Joignaux, G. (Eds.), *Infrastructures de transport et territories—approche de quelques grands projets*, Paris: L'Harmattan.

Jones, C.O. (1970) *An introduction to the study of public policy*, Belmont: Duxbury Press.

Jouve, B. (2002) "Le Politique contre le Savant, tout contre: le PDU de Lyon" [Politics versus science, so close: the Lyon PDU]. In: *Métropolis*, 108/109: 99–104.

Jouve, B., Kaufmann, V., Di Ciommo, F., Falthauser, O., Schreiner, M., Wolfram, M. (2002) *Les politiques de déplacements urbains en quête d'innovations* [Urban Mobility Policies towards innovations], 2001 Paris: plus.

Kast, R., Lapied, A., 2002 "Modèles d'évaluation des risques controversies" [Models to assess controversial risks]. In: *Métropolis*, 108/109: 32–35.

Klein, O. (2002) "Les horizons de la grande vitesse: le TGV, une innovation lue à travers les mutations de son époque" [The future of high-speed rail travel: TGV—an innovation product of its times]. Doctoral thesis, ENTPE, Lyon.

Knoepfel, P., Bussmann, W. (1998) "Les politiques publiques comme objet d'évaluation" [Public polices as target of evaluation]. In: Bussmann, W., Klöti, U., Khoepfel, P. (Eds.), *Politiques publiques*, Paris: Economica.

Lolive, J. (1999) *Les contestations du TGV Méditerranée* [Controversies around the TGV Méditerranée], Paris: L'Harmattan.

Lolive, J., Tricot, A. (2002) "La constitution d'un réseau d'expertise environnementale" [Creation of a specialist environment network]. In: *Métropolis*, 108/109: 62–69.

Louvet, N. (2002) *Les plans de déplacements urbains et la concertation* [Urban travel planning and the consultation process]. Doctoral thesis, ENPC-LATTS.

MATE—Ministère de l'Aménagement du Territoire de l'Equipement et des Transports (1996) Projet d'infrastructures et débat public [Infrastructure projects and public debate]. CPVS, Techniques, territoires et sociétés, no. 31.

Mermoud, F., Walther, L., Kaufmann, V. (2001) *Mobiscopie—Opinions des Français sur les déplacements urbains* [Mobiscopie—French Public Survey on Urban Travel], Paris: GART.

Muller, P., Surel, Y. (1998) *L'analyse des politiques publiques* [Analysis of Public Policies], Paris: Montchrestien.

Newman, P., Thornley, A. (1996) *Urban Planning in Europe*, London: Routledge.

Offner, J.-M. (1997) "Transports et urbanisme : un régime matrimonial ambigu" [Transport and urban planning: an ambiguous marriage]. In: *Les cahiers du génie urbain*, June 1997: 25–27.

Offner, J.-M. (2003a) "L'élaboration des Plans de Déplacements Urbains de la loi sur l'air de 1996." *Latts Research Report*, Ecole Nationale des Ponts et Chaussées.

Offner, J.-M. (2003b) *Les nouvelles modalités de l'action publique—synthèse bibliographique* [New Forms of Public Action—A Literature Review], Document Prepared by IAURIF. Paris: Latts.

Ollivier-Trigalo, M., Piéchaczyk, X. (2002) "Le débat public en amont des projets d'aménagement : genèse et codifications" [Public debates on territorial planning projects: its origins and codifications]. In: *Métropolis*, 108/109: 72–76.

Plassard, F. (1997) "Effets des infrastructures de transport: modèles et paradigmes" [Effects of transport infrastructures: models and paradigms]. In: Bermeister, A., Joignaux, G. (Eds.), *Infrastructures de transport et territoiresapproche de quelques grands projets* [Territories and Transport Infrastructures—Examples from Selected Large-Scale Projects], Paris: L'Harmattan: 39–54.

Rousselot, M. (2002) "La monétarisation des effets externes des transports" [Monetization of transport's external effects]. In: *Métropolis*, 108/109: 17–20.

Roy, B., Damart, S. (2002) L'analyse Coûts-Avantages, outil de concertation et de légitimation? [Cost-benefit analysis, an instrument of consultation or legitimization?]. In: *Métropolis*, 108/109: 7–16.

Rui, S., Ollivier-Trigalo, M., Fourniau, J.-M. (2001) "L'expérience de la mise en discussion publique des projets: identités et légitimités plurielles" [Experiences gained from public debates on transport projects: multiple identities and legitimacy principles]. *Ateliers de bilan du débat public*, INRETS, Report No. 240, Arcueil.

TRANS-PRICE (2000) "Modal integrated urban transport pricing for optimal modal split," *Final Report*, London: Eurotrans.

Yerpez, J., Hernandez, F. (2000) "Les contraintes temporelles dans le processus d'élaboration des PDU—le cas de Marseille" [Marseille—time constraints in the PDU development process]. In: *RTS*, 69: 8–21.

Chapter 13
Institutional impediments to planning professionalism in Victoria, Australia

Alan March

It is common for urban spatial planning professionalism to be understood primarily as the expertise and skills required of practitioners in their assigned roles, putting aside the importance of the governance mechanisms within which planners work. However, urban spatial planning is located across a range of governance mechanisms, requiring particular understandings of the possibilities and problems of professionalism in this setting. The case of Victoria, Australia, is used to demonstrate how the institutional roles of planners strongly influence the exercise of their professional judgement and action. The concept of 'mediatization' is used to show how particular governance arrangements can result in a fragmented professional knowledge base that erodes planners' ability to act as a meaningful force for collective change. It is argued that planners, when being 'professional', need to consciously acknowledge and take on roles as democratic facilitators of knowledge development via planning processes. This enlarged role would begin with an appraisal of the existing institutional impediments to the process of democratic planning in a given setting.

Introduction

The proper role and education of urban planners is again under review in Australia. This review occurs against the backdrop of an international reappraisal of planners' credibility and skills (Hague et al. 2006; Hague 2007). These reappraisals consider planners' skills, but are also concerned with the governance structures in which planners work (Taylor 2007). However, it would seem that in practice, the majority of effort is currently directed towards individual planners' skills, rather than governance structures. For instance, the Planning Institute of Australia (PIA)

Submitted by Australia and New Zealand Association of Planning Schools.

Originally published in (2007) *International Planning Studies*, 12:4, 367–389. By kind permission of Taylor & Francis Ltd. www.informaworld.com.

recently introduced a certification scheme for planners dependent on qualifications and ongoing professional training, representing stricter membership requirements (Planning Institute of Australia 2006). In parallel is a recent 'widening' of the professional body via the introduction of chapters, dividing planners into specialists in social, economic, urban design, urban and regional, transport, environmental, and legal planning. This certification and its emphasis on skills is a variation of other schemes such as the certification required of American planners (American Planning Association 2007), or the Continuing Professional Development required of the British Royal Town Planning Institute members (Royal Town Planning Institute 2007).

This chapter contends that more thorough examination of the governance structures within which planners' work is required to reveal the difficulties faced by planners in applying their skills to address the many challenges of cities and regions. Urban planning, due to its unique nature as spatial, temporal and multi-expertise practices, combined with its location within governance, requires a particular understanding of professionalism. This understanding is not one where planning is idealized and unified by the possibility of a single shared substantive professional knowledge base. Nor is it a collection of separate, equally valid, but ultimately untestable viewpoints. Rather, it is contended here that planners' primary purpose (as individuals and collectively) should be to inclusively develop knowledge and facilitate collective action based upon these knowledges, through the economic, social and environmental processes of spatial planning (March and Low 2004). It is argued that, in contrast to this ideal, Victorian planning is 'hollow', within a shell of unquestioned and routinized processes and roles. This 'hollowness' is a lack in ways of developing collective knowledge and turning this to action – eroding planning's ability to determine and realize new and improved collective futures. The chapter uses an institutional approach to operationalize collaborative or communicative ideals (Healey 1999), allowing grounded analysis of planners in the context of practice.

The chapter begins by examining the difficulties of using a traditional view of professionalism to consider the nature of planning and its location within governance. To explore this, the case of Victorian urban planning is introduced, drawing on a number of case studies and interviews with planning professionals. These studies, and the roles played by professionals within them, are used to open out and problematize planning professionalism. The knowledge base used by planning professionals in these studies is then examined. This allows determination of actual, rather than claimed, knowledge used by professionals in the exercise of judgement. To explain and resolve the contrast between a traditional view of a relatively unified professional knowledge base, with the actual, highly disparate, knowledge and action of Victorian planners, the chapter then introduces a critical democratic view of the planning professional. The Habermasian concept of 'mediatization' is employed to explain how urban planners, in the range of roles they play, are strongly influenced by multiple systemic forces cutting across the possibilities for the development of professional knowledge. This calls to question the value of traditional ideals of professional knowledge and expertise, but further, suggests the need to examine the influence of governance systems upon planners. The concept of mediatization is then used to analyse the institutional causes of disparate planning knowledge in the Victorian case. This analysis is then turned to considering the system-wide implications of planning knowledge and action being so disparate.

It is concluded that a critical view of planning as democratic governance, using the critical lens of mediatization, exposes the institutional barriers to professional planners being able to fully play their roles – causing both the knowledge base and subsequent actions of planners to be 'hollow'. To combat this, urban spatial planners must develop a modified understanding of professionalism; as the practice of inclusively developing and acting on knowledge, through governance structures, including the economic, social, and environmental processes of spatial planning.

Professional ideals

It is useful to first briefly consider the core traditional ideals of professionalism. A fundamental feature of professionalism has always been the knowledge and practice of specialized skills. While many claim to act 'professionally', it is those who belong to professional organizations that have the strongest claim to call themselves professionals in this sense. In this view, professionalism is 'a set of institutions which permit the members of an occupation to make a living while controlling their own work' (Friedson 2001: 1). Control here refers to the self-regulation (although it is never absolute) professional bodies exercise over membership, conduct, and expertise.

However, professionalism's dedication to higher order skills and specialized knowledge were also founded upon their deployment in the public interest. Perkin (1989) argues that professionalism emerged as the development of a societal 'fourth class' in the seventeenth century. Professionals came to represent a new group defined by merit – knowledge and applied expertise that achieved outcomes perceived as achieving overall societal benefit (Perkin 1989). In return, professionals have traditionally received privileges or status, assuming that professional morality can be aligned with public morality (Koehn 1994: 147). Hence, professions are also characterized by codes of ethics and by having particular roles in the community, alongside their knowledge and expertise (Macdonald 1995: 167)

> The ideology of professionalism asserts above all else devotion to the use of disciplined knowledge and skill for the public good . . . proponents must

necessarily exercise a strong, principled voice both in policy-making forums and in the communities where practice takes place . . . from a collegial body that . . . expresses forcefully the collective opinion of the discipline.

(Friedson 2001: 217)

Professionals themselves depend on a particular concept of property – that their expertise is valuable – and that the rewards and privileges they receive are necessary to the continued deliverance of that value to society (Parsons 1968; Perkin 1989: 377; Eraut 1994). Conventionally, this exclusivity was seen as balanced by the professional bodies acting as defences against 'threats' to society, whether they be ill health, mercantile discord, or lawlessness (Durkheim 1957). For planners, this view has long been problematic, given the complex causes of the 'problems' they seek to address (Rittel and Webber 1973). Further, planning has a diverse 'client' base (Campbell and Marshall 2002), variously including paying clients, diverse 'publics', politicians, interest groups, government agencies, or even future generations. In addition, planners simply cannot know all there is to know about planning (Evans and Rydin 1997), evidenced by the ever multiplying planning specializations.

In terms of actual work, professionals were traditionally defined by intellectual 'skilled' (non-routine) thinking requiring high levels of discretionary consideration. The exercise of discretion is said to be based upon superior levels of abstract knowledge, typically based upon formal learning (Friedson 2001: 34–35). To this, some would add the ability to 'reflect in action' (Schon 1983) as the ongoing practical application of higher thinking. Broadbent et al. (1997: 10) contend that 'professionals cannot be simply told what to do', meaning the essence of professionalism is independent evaluation of the means and ends of work, by individuals and professional bodies.

However, from the late 1960s, the diffusion of planning expertise and action across multiple roles, agencies, including politicians, the public, and private consultants, made it debateable whether a planning 'profession' should actually exist, in favour of learned societies seeking to further knowledge regarding substantive problems (McLoughlin 1973: 90; Evans 1993; Healey 1993). This view questions an understanding of the planning profession as those 'who know', as distinct from those 'who do not' (Reade 1987), and leaves unclear who should be entrusted with exercising judgement on the public's behalf.

Completely independent evaluation of work and the exercise of discretion are clearly ideals that are never fully met by any profession. However, that planning is so far removed from this ideal is relevant. A fundamental and defining feature of planning is that it is always a form of collective governing, whether this occurs in central government bureaucracies, council offices, community meetings, or consultants' offices (Campbell and Marshall 2005). The sort of planning we are concerned with here is inherently an act above the level of the individual, even while individuals act within, and are affected by, planning. In this sense, planning diverges significantly from conventional professional models, which view professions as separate, but complementary to, government, such as doctors (Durkheim 1957).

To summarize the broad introductory sketch of professionalism above, it can be argued that several interrelated problems face the profession of planning if it is to be a useful force for change within modern regulatory institutions. The inability to meaningfully identify 'clients', its immersion in regulation and politics, and tensions between the profession being an exclusive controlling body versus a diffuse knowledge-based learned society, confound the very idea of what the planning profession represents from traditional perspectives. Campbell and Marshall, echoing an earlier call by Reade (1987), suggest that a way forward is to critically examine 'how practitioners perceive their roles, and how this actually matches with what they actually do in their working practices' (Reade 1987; Campbell and Marshall 2005: 211). The following case studies of planners in Victoria, Australia seek to allow critical examination of the professional ideals and roles of actual planners practising within governance institutions. It is primarily intended to demonstrate the influences of institutional contexts upon the professionalism of planners, in terms of their development and application of knowledge, as a basis for suggesting a modified view of professional urban spatial planning.

The Victorian planning study

The state of Victoria is at the south-east corner of mainland Australia. It has a population of 4.9 million, with some 3.6 million in the capital, Melbourne (ABS 2006). At state level, urban planning is undertaken by the Department of Sustainability and Environment (DSE), under Ministerial control (Planning and Environment Act 1987: 446, passim). At local level, municipal authorities prepare and administer planning schemes under DSE and state Ministerial jurisdiction.

The force of Victorian planning derives from development control – all land is affected by a local planning scheme, and all development must comply with the relevant scheme (March and Low 2004). Local planning authorities are required to use the Victoria Planning Provisions (VPP) as the basis of their respective schemes' format and content (Planning and Environment Act 1987: Part 1A). The VPPs provide predetermined zones, overlays, decision criteria, and state policy, with some limited allowance for locally determined policy. These schemes include some scope for the potential exercise of discretion by professional planners, local authorities, and the Victorian Civil and Administrative Tribunal (VCAT) and state-level appeal body for permitting decisions.

To characterize typical Victorian planning processes, standard planning processes in each of three separate authorities were examined during the period January 2001 to February 2004. These included:

- five plan-preparation processes; and
- fifteen completed plan-implementation processes.

Detailed examination of these cases allowed the day-to-day operations which shape public and professional understandings to be characterized. These plan-making and implementation processes constitute part of the institutional apparatus through which wider public and professional understandings of matters relating to planning occur – what can described as the way society understands itself, defines problems, and collectively acts on them (Giddens 1984: 376 and passim). Analysis of the 60 events in total indicated considerable similarity in the key procedural stages followed between separate cases of plan-making and plan-implementation. The similarity is easily explained by the requirements of the Planning and Environment Act (1987), which specifies highly routine and uniform procedures. While not representative of all planning in Victoria, it is contended that the characteristics of these repeated processes provide considerable insight into professionalism in Victorian planning. Plan-preparation occurs via the scheme amendment process, in which local authority planning schemes are modified. Within the framework established by the Planning and Environment Act (1987) amendments can be characterized as having eight main stages, as set out below in Table 13.1.

Implementation is almost always the responsibility of local government, however, contested matters are taken to the appeal hearing body VCAT, as shown in Table 13.2 below.

To consider the impacts of this institutional setting upon the knowledge and decisions of professionals, interviews were undertaken with 72 persons, including 36 planners, in 12 detailed case studies. Seven detailed cases of implementation, and five cases of plan-making were studied in depth. The studies were chosen with the intent of including cases that represented the 'typical types' identified in the

Table 13.1 Plan-preparation (scheme amendments)

1 Initiation and decision to commence scheme amendment
2 Preparation of documentation
3 Exhibition of documentation and opportunity for public submissions
4 Independent Panel hearing, allowing submissions
5 Report of Panel to Council
6 Council decision
7 Submission of proposed amendment to Minister
8 Decision of Minister

Table 13.2 Implementation (development control)

1	Trichotomy (VPP 31.01-1)
	'No permit required' uses or developments
	'Permit required'
	'Prohibited'
2	Application and initial appraisal
3	Advertising and objections
4	Professional appraisal and recommendation
5	Council meeting and decision
6	Appeal lodgement
7	Appeal hearing and determination

first part of the study. Additionally, cases were selected so that they included inner city, middle, and edge/rural municipalities. The detailed interviews with participants were undertaken to complement the longitudinal first hand observation and documentation of each case studied by the researcher.

Interviews were carried out with planners (and all other main actors involved in each case, such as objectors, councillors, applicants, and lawyers) after the observation and documentation of key events throughout the planning process of each case study. The confidential one and a half-hour interviews involved 53 questions which sought to determine why professionals acted in particular ways throughout the specific cases observed by the researcher. Each respondent was informed that the researcher had been observing events in the case study (although most had already seen the researcher observing). An indicative summary of the questions asked is included in Table 13.3 below.

Table 13.3 Indicative summary of interview questions

What would have represented 'success' for you at each stage of the process?

How did you try to be successful at each stage?

To what extent do you think understanding the others' motivations and expectations could have assisted a better outcome?

Why do you think the outcome/s (specify) occurred as it did?

What key professional standards did you seek to maintain in this process?

What actions did you personally take to be a good professional in this process?

What result/s would have represented professional success for you?

Does 'being professional' in this process require finding agreement with others?

Does 'being professional' require or encourage questioning the planning process and its rules?

How would you rate the influence of professionalism on your actions?

Influence: None———Little———Moderate———Strong———Very strong

The central theme of the questions and studies was to determine the knowledge and expertise actually deployed by planners in exercising their professional discretion. This provides a starting point for problematizing the value of a traditional view of professionalism as the use of expertise based on expert knowledge for the public good. Accordingly, the studies sought to determine why planners, while considering themselves to be professional, acted in particular ways as individuals within the institutional setting of the Victorian planning system. The questions asked planners what represented their own professional success, at various stages of the processes they used, and how this understanding of success informed their goal setting and decision-making. In parallel, separate questions also examined the influences that legal, political, bureaucratic, and financial factors had upon planners' success criteria, and upon their subsequent decision-making. This allowed the rationales and knowledge of individual planners to be understood, alongside the manner in which this knowledge is exercised and turned to influencing collective action. Further, the influences of institutional settings upon professional action can be examined, providing understandings of the systemic constraints to professional expertise being directed to collective goals.

Victorian professionalism as disparate knowledges

The current study examined the self-reported professional justifications of planners, coupled with independent observation of their actions over time. It revealed views so disparate (described in more detail below) as to render futile attempts to determine any sort of 'true' single substantive professional opinion on any of the matters studied. These initial findings begin to demonstrate the contradictions and tensions within planning professionalism as it is practised. Of course, there has been much debate over time regarding the many non-technical influences upon planning, often revolving around power and politics. However, in terms of professionalism, these examinations vary between artificially separating planners from decision-makers as technical experts (Faludi 1973), to discovering that power trumps professional planning reason (Flyvbjerg 1998), to seeing individual practitioners as actors inside (Krumholz and Forester 1990), alongside (Davidoff 1973) or outside (Friedmann 1973; Sandercock 1998) 'the system', but without really considering the system itself. Understandings drawing upon political economy have used inter alia Marxist, neoliberal, neocorporatist, and privatization perspectives to alternatively critique planning or its opponents. The institutional approach of Healey (1999) does locate planning within governance, but rests uneasily with ideal perspectives of collaborative action. The current study seeks to demonstrate a means for understanding and improving planners' professional roles within a working planning system, by directly examining their motives as professionals within specific institutional settings.

Analysis of the case studies demonstrated that planners are strongly influenced by their institutional settings. The planners observed and interviewed all drew their objectives and decision-making criteria from the role they played, rather than any abstract professional ideals. Of the 37 planners interviewed, 31 rated professionalism as a 'very strong' influence on their actions, five rated it as 'strong', and one as 'moderate'. All strongly believed the specific arguments and decisions they made were 'professional', suggesting professionalism has considerable meaning for them as individuals. These findings loosely concur with Howe's (1994) study of individual planners' ethics which identified a number of types of personal motivations in terms of planners' professional roles: as a force for social change; as servants of elected officials; or as coordinators and facilitators of planning process (Howe 1994: 91, 112–147). Howe found that planners' ethics were driven by a range of factors depending on the individual and the role they played (Howe 1994: 148–180). However, there is a vast difference between planners professing their ethics in abstract settings, and determining how (and why) they act in reality.

If planning 'knowledge' is disparate and contested, a key traditional ideal of professionals as a group based upon an exclusive knowledge base (Parsons 1968) is questionable. The diversity of opinion and action is, of course, easily explained by considering planners' locations in a complex of planning governance encompassing state agencies, local government, consultants, elected councillors, developers, activists, and residents. Traditionally, acknowledging the location of planning within and of governance has been uncomfortable for many – it has often been ostensibly convenient to practise it as if planning is apolitical. Lindblom and Cohen questioned delegation of decision-making powers regarding 'interactive matters' (requiring value judgements) to professionals or technicians, as allocation of political problem-solving tasks to 'experts' (Lindblom and Cohen 1979). Davidoff and Reiner emphasized the number of value choices made in supposedly neutral processes (Davidoff and Reiner 1962). Admitting the value laden nature of planning leads to the corollary that planning is an element of 'governance' – the wider mechanisms by which collective affairs are managed (Healey 1997: 207). Governance extends beyond formal institutions, political parties and lobby groups, to the wider shaping of rules and processes (March and Olsen 1995: 28; Healey 1999). The following sections examine the knowledge and decision making of planners within their various governance roles to demonstrate the influence of institutional roles upon professional action.

Despite their diverse opinions, the planners studied played regularized procedural roles in the institutional 'spaces' created by the regulatory framework. The main work of planners was ostensibly a range of communicative acts supported by technical arguments. Planners advanced opinions, made decisions, gave advice,

Institutional impediments to planning professionalism **279**

and managed processes, appearing much like Forester's deliberative practitioners. However, Forester suggests that before problems can be solved, they have to be 'constructed' (Forester 1999: 37). Closer examination of the Victorian planners within their institutional settings showed that the scope to actually 'construct' problems and viewpoints was severely limited by the 'facts' of the processes they worked within. Local planners decided on minor applications, while in more important matters could only make recommendations to councillors. They acted as councils' representatives in any subsequent panel (plan-making review processes) and VCAT hearings, having to put essentially political views sanctioned by council. This type of role matches, superficially at least, Faludi's rational planner separated from politics, while accepting decision-makers' goals (Faludi 1973: 87–103, 225). Consulting planners' roles revolved around managing or influencing processes, strongly oriented to achieving clients' desired outcomes. Consultant planners' goals ranged from seeking development approvals, opposing approvals, to modifying others' proposals to suit client desires, or even to acting for councils in planning appeals.

State-level DSE planners were limited to the assessment of amendments to local authority planning schemes. DSE planners assessed the suitability of councils' amendments against the overall policy matrix of the Planning Act, state and local policy. They made recommendations to the minister (almost always followed). Planning panel members heard and considered submissions regarding the appropriateness of amendments to planning schemes. They made recommendations about amendments, based on tests provided by the existing 'policy matrix'. Tribunal members considered planning applications that were disputed. Using the planning scheme alone as their decision criteria, they determined whether applications should be granted permits or refusals, and by virtue of their statutory role, their decisions were final.

Of course, a range of viewpoints might ultimately assist in determining appropriate action, if there are ways of determining which view is ultimately the most desirable. In the Victorian system decisions on any vaguely important matter are made by councillors, tribunal and panel members, or the minister. This might suggest that the opinions of all the different planners mean little, when ultimate decision-makers determine outcomes anyway. This is a vastly different scenario from Forester's deliberative practitioners (Forester 1999). The problem with planners representing a number of disparate professional views is essentially that action in the planning realm (if one accepts that it is governance) must always ultimately be oriented to collective substantive concerns and action, as governing. Does this mean that professional planning opinion is essentially unreliable when it comes to the substantive concerns of the wide range of decisions affected by, or made by, planning and planners?

A defining characteristic of professionalism is membership, and compliance with the rules, of a professional organization (Eraut 1994). Professional organizations are traditionally conceived as separate, but complementary, to government. However, to plan for urban and regional areas, one is inherently within government, in at least some sense. Further, it is common for planners to successfully fulfil their roles without professional membership. In contrast, members work within bureaucratic and political instruments of government, through to private land development companies. To deal with this, planners have often claimed that they can adopt a politically neutral position, even while advising politicians on deeply political matters (Hague 1984: 99). This argument, while flawed, has allowed planners to establish a domain of ostensible professional neutrality, even while acting within governmental complexes fraught with clear political and economic values (Long 1959; Davidoff and Reiner 1962; Tribe 1972; Lindblom and Cohen 1979). This position sets ideals of individual and professional autonomy, identity, and integrity directly against governmental roles, control, and identity. If one looks to the PIA, little substantive guidance is provided for the knowledge and behaviour of planners beyond generic triple bottom line statements regarding economic growth, social justice, and ecological sustainability. However, this level of generality is perhaps necessary to allowing the diversity of planners' views.

In the context of the wider 'knowledge society' setting, the public now accords little deference to urban planning – it is simply another knowledge work group (Marquand 1997: 144). Further, the incessant self-criticism of planners has rendered the planning profession itself practically incapable of coherently describing its own essence (Eversley 1973: 300; Tugwell 1974/1948; Hague 1984: 109; Fischer and Forester 1993; Hoch 1994; Geddes 1998/1915; Yiftachel 1999; Fainstein 2000; Birch 2001; Roberts 2002). This lack of self-identity and determination of the ends of work reduces planners from professionals to experts. This distinction can probably best be understood using Weber's distinction between a 'craft occupation' in which neophytes learn on-the-job skills of 'how to do things', versus professional learning (usually including at least some abilities learnt away from practice) which allows and encourages questioning of the ways things are done, oriented to improvement (Weber 1978/1914: 784–787). 'If professionals are only experts, they have no moral obligation to help people or to further the public good' (Koehn 1994: 22). Ironically, it would seem that being embedded in governance (itself ultimately intended for the collective good) is the primary impediment to planners developing independent goals to improve our cities and regions. It is to the effects of this institutional paradox of planning professionalism and collective governing that the chapter now turns.

Critical democratic professionalism

To meaningfully place professionalism within governance requires considering the wider context of democracy. Despite general acceptance that democracy of some type is generally desirable, the ideal of 'a people ruling itself' (Mayo 1960) is never fully achieved. 'Complete' democracy is always prevented by the size, diversity, and complexity of modern collective activity. Necessity requires various imperfectly democratic mechanisms to facilitate governance. Citizens must elect decision-makers to take action and to control on their behalf, knowing this may often lead to decisions that they, as many different individuals, may not concur with, or even know about. As part of this, we entrust whole bodies of knowledge, work, and advice to particular expertise groups – including those 'intellectual experts' the professionals.

Entrusting bodies of knowledge and decision making to professionals restricts access to, and knowledge of, governance; only a few are able to act in respect of these matters. This professional privileging, a sort of societal contract (Rueschemeyer 1986: 41), remains an unavoidable part of actually being able to govern – specialization and work on citizens' behalf is necessary for complex problems to be dealt with. However, these mechanisms also separate citizens from the way in which they rule themselves (Dahl 1982; Mouffe 2000). Ideally, professional planning ethics would be the institutionalized expression of prevailing public morality (Koehn 1994: 150). Yet, many barriers are located between this professional ideal and the reality – it is the institutional barriers in the Victorian planning system which the remainder of the chapter will examine, before suggesting a revised professional ethic as a solution.

Planning must deal with the difficulty of how professionalism can be conceived within democratic planning governance. Many other models of planning do exist, such as Friedmann's household and community-based planning (Friedmann 1973, 1992, 1998), Davidoff's advocacy planning (Davidoff 1965), or Sandercock's alternative planning (Sandercock 1998). However, these approaches suffer a profound contradiction for the maintenance of professional independence – they seek to act outside government – but if they became 'the way' of planning, would become government.

The body of work known as 'communicative planning' (Fischer and Forester 1993; Healey 1996) has had a strong impact, even while considerable scepticism exists. Forester highlighted the discursive and political nature of practice (Forester 1985, 1989). Healey and Innes used Habermasian ideas to develop insights into plans' communicative implications (Healey 1993; Healey and Hillier 1995; Innes 1996). Healey's institutional critique of the links between state and civil society consider the dilemmas of participation and democracy (Healey 1996, 1997, 1999; Innes and Booher 1999; Hillier 2000; Vigar et al. 2000). The reliance on

Habermasian understandings of democracy is critiqued for its foundationalism (Flyvbjerg 2001), while Yiftachel suggests communicative theory elevates process above substance (Yiftachel 2001: 6). Other criticisms include communicative planning's failure to demonstrate any actual examples of true exemplars (Tewdwr-Jones and Allmendinger 1998). This chapter contends that, rather than focus on communicative planning ideals, it is much more useful to draw upon other aspects of Habermas' work. In particular, Habermas' ideal of democratic public spheres, and the impediments to its proper functioning are worthy of examination. The following section sets these out in brief.

Habermas offers an ideal: a democracy must aim to both to 'know itself' *and* to 'steer itself' (Habermas 1996/1987). A people *knowing itself* (accepting differences) must understand its challenges alongside the options available, while seeking rational, inclusive and empowering understanding. To *steer itself*, a people must have the ability and conviction to take collective action to meet the challenges developed in knowing itself. In this view, planners would not simply seek technical knowledge and impartiality. Rather, planners would seek that urban governance *steering* (planning as collective action) is a realization of citizens' full *knowledge*. While this ideal cannot be fully realized, this directs attention to seeking removal of as many impediments to knowing and steering as possible. To this end, the following section examines the impacts of *mediatization* upon democracy, to then explain the difficulties of professionalism within planning governance.

Habermas considers mediatization of the 'steering media' of governance (Habermas 1984, 1987, 1989/1962) to be *the* central impediment to full democracy. Mediatization leads discourse-based mechanisms for collective knowing and steering to be replaced with instrumental logics resistant to the open processes of deliberation (Habermas 1987: 187; Habermas 1996/1987: 35). Instrumental reasoning leads to choices based upon achieving success in a given sphere (Audi 1998: 674–675). Sub-systems, such as the economy and the apparatuses of the state, mediatize public and private spheres (Habermas 1984: 196, 318–323, and passim; Habermas 1987: chapter 12). For example, money coordinates activity so that individuals act for economy or profit – the medium's success criteria. In turn, medium of money overall 'steers' society via markets – many individuals act according to the same logic – making similar decisions. Accordingly, mediatized steering mechanisms exert their internal logics as overall influence, reducing the ability for democratic control (Habermas 1987: 196). In parallel, a particular medium's own logic is used for understanding, modifying the way people individually and collectively understand (Habermas 1996: 357).

Importantly, the duality of media means they are not simply 'bad' – in fact they remain essential to governing of mass societies. However, they divide collective knowledge and action (March and Low 2004):

1. Shared understandings reached in the production of rules and actions (*rational*).
2. Knowledge oriented only to rules themselves and to the use of these rules to achieve success (*instrumental*) (Habermas 1987: 180).

Accordingly, it is to this division of professional planning knowledge and action in the Victoria case studies that attention is now directed. This allows explanation of the disparate opinions and actions of professional planners in an institutional setting, combined with an understanding of the way this erodes potential for desirable collective action.

Institutionalized professionalism

In almost all cases studied, success for planners was seeking to use the existing planning rules and processes to best advantage, to achieve goals according to their affiliation. This oriented professional knowledge not to substantive issues of cities and regions, but to achieving success on a case-by-case basis. The sum of these individual instances results in a profession limited in its knowledge and concern with effecting meaningful change. Professionalism, in this sense, simply becomes a pastiche of the various ways of 'succeeding' in individual planning cases. The following section examines the institutional media of planning in Victoria – law, politics, bureaucracy, and money (March and Low 2004) – in terms of their effects on professional knowledge and action. It is contended that planning professionalism is strongly influenced by the various logics of media, eroding the ability of planners to fully play their roles.

To begin, law provided an immutable framework for urban spatial planners within which the detail of planning occurs. As a medium, it represents 'facts' around which people must work to achieve their desires (Habermas 1996). In all cases studied, planners acknowledged the unchangeable nature of legal rules and processes, directing their energies to achieving results within the rules rather than challenging wider concerns with planning itself. This both empowered and restricted thought and action to that possible within the existing planning VPP scheme blueprint. In preparing new plans, local planners saw legal and policy success in terms of legal compliance – a good scheme was one that achieved local goals while complying with state policy as set out in the VPP blueprint. State-level planners (holding final decision-making powers regarding approval) understood compliance as ensuring that the local scheme did not contradict state policy (Planning and Environment Act 1987: s7). Local councillors saw legal and policy value as the ability to achieve local desires. Accordingly, councillors encouraged the local planner to instrumentally use the 'palette' of state-given VPP controls, but to achieve local ends. Where councillors sought to influence amendment processes, planners interpreted legal conformance in such a way that best reflected council views. However, the potential for VPP schemes to be interpreted and modified to local desires is limited by the privileging of state-level planning understandings of conformance, demonstrated in this statement by a local government planner:

> We can only achieve what is allowed by the VPP scheme format and trying to change that is worse than useless, because also, people think you are unrealistic.

Ultimately, local planners only seek to progress amendments they consider to have a good chance of final state-level approval. They measured their likelihood of success against their knowledge of the legal and 'policy matrix' as a test of:

> Whether or not we can get the thing through at DSE [Department of Sustainability] level.

In exhibition and panel processes, where any interested party is allowed to examine proposed plans and to make comments, almost all submissions related to specific sites and impacts upon these, rather than to wider policy or general provisions. Accordingly, planning consultants narrowly argued about modification of their clients' rights, usually seeking a different zone from the VPP palette that was more favourable to their clients' desires. A government agency (unsuccessfully) argued for modification to the VPP controls, on the basis that it would compromise their ability to undertake their statutory functions. Alternatively, a separate government agency successfully brought errors to the panel's attention, resulting in modifications to the scheme.

Panellists, the appointed planners charged with facilitating the process of hearing submissions regarding proposed plans, strictly maintained conformance with the pre-established VPP policy matrix through panel processes (Planning and Environment Act 1987: part 3). Experienced planners from various government agencies and planning consultants made few efforts to engage with local political or administrative processes prior to panel hearings. They only made brief written submissions during the advertising stage, maintaining their rights to make arguments to the subsequent planning panel hearing. When they did make submissions to the panel, these were based directly upon legal and policy success criteria, on the understanding that this type of knowledge trumped others in the panel process. Explaining his non-engagement with local political processes, a consultant stated:

> It's just not worth getting involved in local political processes like this. [It's] better to just concentrate on the [subsequent] panel.

When it came to local implementation processes (development control), planners were even more focussed on compliance with, and interpretation of, legal rules and processes. Council planners in development control saw success as legal and policy compliance alone – only allowing outcomes in accordance with law and policy and actively discounting other types of information as 'non-planning reasons'. To do this, they would disseminate information, and explain or debate interpretation of policy with various actors. In contrast to Forester's 'making sense together' (Forester 1989) the information disseminated was almost wholly to do with explaining 'rules', policy, and processes. Some council planners were required to prepare tribunal submissions for contested cases, going against their own planning department's recommendation. Planners in this position, known as advocates, instrumentally used the discretionary scope (or 'grey area') of the existing legal and policy framework to develop arguments supporting their councils' positions as their criterion for success, often against personal beliefs:

> Policy success in this case would have been a permit issuing . . . but my role stemming from the Local Government Act means that I had to argue for refusal in VCAT – I built a case accordingly.

Consultants understood professional success as achieving the goals of their affiliation – essentially the entity paying for their expertise, while operating within the 'fact' of legal and policy compliance. They developed arguments accordingly, resulting in several starkly contradictory professional opinions surrounding one application. Agreements were rarely sought with other parties, suggesting that consensus ideals have little meaning in these contested settings. Rather, actors attempt to demonstrate winning arguments to decision-makers, using the tests provided by the Act and planning scheme. No attempts were made to question the legal and policy framework. Tribunal members, both in the hearings and research interviews, held compliance with policy as their primary measure of success. Interpretation of 'grey areas' of policy was undertaken on the basis of balancing the relative weight that ought to be given to various policy considerations.

> Our concern is to see that the objectives of the Act, as manifest in the planning scheme, are achieved – we expressly do not have a role in questioning the content of schemes, aside perhaps from occasional comments.

In the spaces provided by law, the establishment of bureaucratic agencies and processes are inevitable components of any planning system. In the case studies all planners saw success in bureaucratic processes as reliant upon maintaining established and accepted processes, maintaining appropriate forms of input to process at predetermined points. In plan-making, the process encourages involvement only via the standard bureaucratic forms of submissions, and allows only the use of VPP form planning controls and arguments (Planning and Environment Act, 1987: div 2, passim). Government planners manage a process that allows public involvement only during exhibition and panel hearings: involvement beyond standard processes is rare. These forums require the submission of arguments in particular forms, on the basis of compliance with VPP policy. In development control processes, the 'managed' nature of the bureaucratic process by planners, coupled with the clear objective of applicants, encouraged actors to negotiate with planners to a greater degree. However, local planners themselves see success as maintaining due process and using the pre-established tests and forums derived by the Act to be managed by local planners.

> Ultimately – any statutory planner will agree – all the time it is a matter of not making a mistake – the correct advertising, not going over time, using the right tests out of the scheme, and so on . . .

On this basis, local planners had little interest in finding agreements or questioning processes themselves, but rather followed procedure and used the decision criteria derived from the existing planning scheme. Local planners in plan-making were quick to try and quash amendments that went outside what was allowed by the policy framework of the VPPs, knowing that the process of final approval would be based on the test of conformance with the VPP policies.

> It's simply not worth investing scarce energy and money on amendments that won't get through the DSE [state planners].

Knowing this, planning consultants seeking amendments for clients instrumentally used their knowledge of bureaucratic process and of VPP policy to argue that their clients' amendments converged with the local planners' intent to maintain due process and policy integrity. None of the planners questioned bureaucratic process, simply accepting it, even suggesting that 'doubting' accepted procedure eroded their chances of success:

> [Q]uestioning the process too much makes you look like a loose cannon to the decision maker and distracts them from the case you are trying to make.

The immutability of bureaucracy even encouraged local planners to take on an educative role with others, to teach them how to be successful in 'the system'.

> I tried to give all parties as much information as possible, and get everyone talking.

This planner then went on to put the considerable conflicts and disagreements involved in the specific case aside, indicating emphatically that bureaucratic success in this case would simply have been:

A permit issuing, in accordance with its compliance with planning policy.

All planning consultants understood bureaucratic success as using their knowledge of process and policy criteria to achieve their respective clients' desired outcomes, effectively seeking to establish convergence between the success criteria of their clients and of decision-makers. Accordingly, success for one planning consultant was:

. . . understanding the process better than them and being forceful, to do [my] job well. Holding to certain interpretations of the rules can produce poor outcomes . . . We did find agreement [with the planning officer] on some major aspects of the project

As contentious applications progressed to tribunal hearings, consultants understood bureaucracy simply as a vehicle to enable them to make a case in the more legal/policy oriented realm of the tribunal.

I suppose bureaucratic success is 'not screwing up' in following procedures . . . so we have a right to make a case.

The influence of money upon planning professionalism is paradoxical in Victoria. Key decision-makers such as local planners, panellists, tribunal members, and state planners put aside considerations of finance and markets in decision-making, even while many other actors are directly motivated by money. The paid nature of consulting work requires that the consultant adopt the values of the hirer, further testing professional ideals of autonomy and public good. Professionals themselves must derive some benefit from their actions (Koehn 1994: 143), but payment for professional representation represents a contractual obligation. The exercise of decision-making discretion, a cornerstone of professionalism, is compromised by this contractual obligation (Koehn 1994: 40, 52). Private proponents of amendments saw success directly in terms of profit, facilitated by amendments to the planning scheme. On the basis of monetary success criteria, they instrumentally sought outcomes providing financial returns as the ability to develop land. One rezoning proponent stated:

I have invested in this property – so it owes me. I need it to complement my other car dealership down the road to unload less prestigious trade-ins. [Success is] to be able to rezone to Business 1.

Proponents instrumentally sought to use the amendment process for desirable financial outcomes. Proponents engaged planning consultants to assist them. In turn, all of these planning consultants acted on the basis of receiving financial payment for their services. In the case cited above, the planning consultant stated:

> I can't deny it – getting paid is financial success for me – but in the long term it requires me to be successful in getting my clients what they want, whatever that might be, most of the time.

Accordingly, the instrumental nature of the money medium appears to actually be reinforced by its removal from consideration as a planning matter, effectively separating the motivations and understandings of many actors from planning processes, which they seek instrumentally to use for financial ends. The paid nature of consulting work and its attendant requirement that the consultant adopt the values of the hirer, tests ideals of professional autonomy and the public good. To achieve success, consultants instrumentally attempted to achieve their clients' particular goals via the mechanisms of urban planning implementation processes. Consultants working for government saw success as:

> presenting evidence that could be believed in, concisely and understood by all, having an impact on the tribunal's decision . . . to justify [my expert witness] statement to justify council's decision . . .

allied financially with:

> being paid . . . evidence like this is a nice little earner.

Consultants engaged to support applicants' cases, however, also sought to align their financial goal of being paid with their clients' financial success criterion of making a profit, seeing success as being paid for their services, because:

> We're not a charity! . . . The client wasn't driven by a cash flow problem in this case, but we wanted to maximize his profit, balanced with a reasonable development, but we paid for being reasonable [with the objectors subsequently taking the matter to appeal despite consultation].

The instrumental nature of the knowledge consultants rely upon is typically directed to very specific outcomes, based upon being paid to do so:

> I gave advice based on the VPPs and policy matrix, and presented on this basis to the panel, orally and in writing. [Success] would have been to have made

an argument accepted by the panel and Minister's office [for Residential 2 zoning on the site].

However, the real measure of professional success for a planning consultant was in prediction of a client's project's likely success in the ultimate test provided by the tribunal, including the 'rules' being obeyed, such as objectors being allowed to make their cases. Success for him was:

> VCAT [the Tribunal] agreeing with my pre-application advice to the applicant [and developer] . . . and that the process was successful [resulting in a permit for my client], meaning that people got their chance to be heard. [It's] good to have professional judgement okayed by VCAT at the end.

An experienced planner explained his professional reluctance to question processes or rules openly in the tribunal thus:

> No, not in VCAT. My role in VCAT is my opinion.

In sharp contrast to planning consultants, proponents and often objectors, in each case studied planning panellists put aside financial matters as did all state and local planners, and tribunal members.

Political influences on professionalism varied considerably, although no planners were overtly political in their core professional activities. In plan-making, where decisions of council are required for the commencement of scheme amendments, local planners were sometimes 'led' by councillors in their decision-making.

> I was influenced by the Councillors, particularly the ward Councillor . . . I followed his lead in deciding that [the] amendment would have facilitated a development not in keeping with community interests.

Local planners often engaged in plan-making exercises seeking to deal with problems raised by councillors.

> My political goals are simply to try and reflect Council's desires in policy that works.

In some cases consultants lobbied councillors where they considered that it might improve the chances of an amendment being granted council support. However, in cases that were contentious, and which were highly likely to be contested and therefore proceed to panel processes, political lobbying was usually abandoned in favour of spending energy and money on making a case to panellists using 'planning reasons'.

> It's not worth getting involved in local political processes like this; better to concentrate on the Panel [which has the most influence upon ultimate success].

When it came to appraisal of proposed scheme amendments, expert panellists and Department of Sustainability officers paid no heed to political concerns. Development control processes can be highly controversial in Victoria, and local political processes are the focus of considerable contention. Accordingly, many consultants lobbied local politicians and decision-makers when they considered it to be advantageous. Alternatively, if they believed the application would ultimately go to the tribunal anyway, they often made no contact with councillors. In contrast, local government planners actively sought:

> To put aside and resist political considerations, to ensure that decisions are planning reasons [meaning based only on tests provided by the planning scheme].

However, where councils made decisions to refuse (often contrary to planners' advice) which were subsequently appealed by applicants, council planners then had to take on the role of advocate at tribunal hearings to maintain council's political decision, often going against their original position.

> Irrespective of our [the planning department's] previous recommendation to approve this, my role was to try and maintain Council's position, in this case that a refusal should issue.

Notwithstanding all the politics at local level, in tribunal hearings, appointed members paid no heed to political reasons, in favour of legal-policy decision criteria.

> Our [the Tribunal's] role is to maintain the integrity of the policy framework, primarily by using the planning scheme applicable to each case as our primary decision making tool, within the wider processes of the [Planning & Environment] Act

It is contended that the analysis above demonstrates significant levels of alignment between professional planners' values with the logics of various media: money, law, bureaucracy, and politics. The implications of planning being so directly aligned with media in all these cases, is that the sum of planning's knowledge base is stifled from the level of the individual case, through to the whole profession. The final section takes up this theme more fully.

Inward facing planning experts

The analysis above is based upon empirical examination of real cases of planners operating in the institutional setting of Victorian planning – the values of what planners did, rather than their abstract beliefs. The findings suggest that the value of professional planning knowledge and practice is questionable, particularly when one looks to the wider terrain of the planning system.

Victorian professional planning knowledge is oriented to satisfying the mechanisms of planning in the standard processes of plan-production and implementation. As an overall focus for knowledge, the expertise of planning largely represents a series of skills in achieving short-term success, depending upon roles as 'experts'. Further, the institutional setting causes planners' success criteria to derive almost wholly from the media of law, bureaucracy, money, and politics, rather than substantive questions of collective concern oriented to urban and regional planning. These media are an unavoidable part of any planning system, but when the institutional processes cause thinking to be oriented only to achieving success according to these media, deeper independent and intellectual thought suffers. Planning 'thought' and knowledge becomes a series of routine exercises, eroding the professional body's ability to have any meaning or substantive knowledge at its core. Accordingly, Victorian planning professionalism is 'hollow' in terms of acting as a guide and repository of planning knowledge about substantive urban issues, beneath the 'shell' of routine processes and short-term goals.

Of course, being oriented to success in terms of planning processes (as instrumental action) in an individual instance is not in itself irrational – it is essential for the maintenance of coherent policy. The cases demonstrate that the mediatization of professionalism provides benefits in the Victorian case:

- procedural certainty;
- policy certainty;
- efficiency and equity (within the bounds of the existing system); and
- action is possible without consensus.

However, if this instrumental orientation is extended across the entire terrain of professional planning action and knowledge, it is irrational in that there is limited capacity for building a common substantive professional knowledge, or to modify existing urban conditions or directions of movement. For instance, there is little joy to procedural or policy certainty if it is flawed in its ability to achieve desirable collective outcomes. Since the forums in which professionalism is exercised mediatize planners' professional knowledge and action and encourage *only* instrumental action, this militates against re-appraisal of 'the rules' themselves – planners are reduced to experts in a highly constrained realm. Further, the tying of professionalism to roles rather than ideals, ethics, theory, belief, or the development of shared knowledge, distances the individual planner from taking responsibility for the wider 'public good' implications of their actions.

As a result of the case-by-case orientation of professional planners to success according to affiliation, professional development for individuals is peculiarly circular (Figure 13.1).

When the exercise of professional planning knowledge is viewed in the context of the wider planning system, the mediatization of Victorian professionalism has serious disadvantages:

- overall planning is not informed by consideration of wider understandings, but by a succession of individuals' success objectives;
- case-by-case agreements between actors occur only on the basis of shared instrumental goals, not on wider policy directions or deeper understandings;
- governance steering mechanisms cannot be influenced by new mutual understandings between actors, if actors diverge from the established controls;
- no forum exists in standard processes for reappraisal of the overall rules and processes of planning – justification for individual instances only ever relies upon the existing rules as they stand; and
- cumulatively, professional planning knowledge is not oriented to development of new collective knowledge in the profession or wider community, but to ways of achieving success in individual instances.

Conclusions: outward-facing professionals within governance

The chapter uses an institutional approach to operationalize communicative ideals (Healey 1999) providing an alternative model of urban spatial planning professionalism which can acknowledge the context of actual planning practice. It demonstrated that the practice of urban spatial planning is difficult to reconcile with

To achieve success, a planner knows and uses the system 'better', achieving goals on a case-specific basis.

Using the system 'better' discourages development of new knowledge, and strengthens existing steering mechanisms.

13.1 The 'better' system.

traditional views of professionalism, which portray professionals as separate to the state, while complementing its aims and activities as a means to achieving the public good. In this understanding, professionals are allowed particular benefits for complying with rules of conduct endorsed by the profession itself and the state, on the basis that the public good is aligned with the resultant actions professionals undertake. However, urban spatial planners' abilities to address the many challenges of cities and regions are not just based in complying with codes of conduct, or in having skills of the sort measured by accreditation programmes, even while these are important. Rather, the location of planning across and within governance means that urban spatial planning professionalism must incorporate perspectives that acknowledge the varied institutional mechanisms through which planning is undertaken by the individuals comprising the profession. Acknowledging the influence of institutional settings upon professionals provides a starting point for redirecting the ends to which planners' skills are directed, to better achieve the public good.

To meaningfully incorporate institutional understandings with planning professionalism, the chapter posited that urban spatial planning can be understood as a process of developing shared knowledge and translating that knowledge into collective action. The corollary of this understanding of planning as knowing and steering is four interrelated points impacting upon planning professionalism. First, the actual knowledge employed by professional planners in Victoria was revealed to be highly disparate and influenced by the specific roles individuals played. Second, these multiple roles aligned professional values with the various media used in planning institutions, such as money, law, bureaucracy, and politics. The resultant orientation of Victorian planners to the success criteria of these media modified the development and deployment of professional knowledge towards ends only loosely associated with the public good. The analysis demonstrated that in Victoria, planning processes are far from neutral in the way they cause professional expertise (and wider public knowledge) to be developed and applied.

Third, there is an apparent absence of forums in the institutions of Victorian planning where knowledge and action is unaffected by the self-perpetuating logics of media. This means that the profession of planning, as it exists currently in Victoria, is limited in its ability to be a force for changing and improving existing goals, and processes used to achieve these goals. Changes to the institutional processes of planning in Victoria would be required before professionals could meaningfully facilitate change.

Finally, these findings strongly suggest that the concept of planning pro-fessionalism itself requires modification to accommodate a rationale more suited to the institutional location of planning within and across governance. This requires understanding the role of planners as a profession-within-governance, as opposed to naively (or conveniently) assuming neutrality or alternatively diversity, losing any chance to act as a force for change. Planners, in being 'professional', need to consciously acknowledge and take on roles as democratic facilitators of knowledge development via planning processes. This enlarged role would begin with an ongoing appraisal of the existing institutional impediments to this process of knowledge development in a given setting. In the Victorian case, planning knowledge is 'hollow' within a shell of routinized processes and roles. In contrast planning professionalism as 'knowing and steering' would seek to find a nuanced and practically oriented position between the unrealistic ideal of a single shared substantive professional knowledge base, and many separate viewpoints inimical to the development of collective action. Victorian planning processes would be modified so that the output of processes is not just substantive development and change, but would also involve, and indeed require, shared knowledge development. Knowledge developed by planners (and others) in their day-to-day work should be oriented, not just to short term 'role' success, but to promoting the development of collective understandings and action regarding the many substantive issues confronting planning.

References

ABS Australia Bureau of Statistics (2006) 3235.2.55.001-Population by Age and Sex, Victoria. Released 11:30 am (Canberra Time) 30 June 2006, available at www.abs.gov.au/ausstats/abs@.nsf/mf/ 3235.2.55.001.

American Planning Association (2007) Ethics for the certified planner. Available at www.planning.org/ ethics/index.htm.

Audi, R. (1998) *The Cambridge Dictionary of Philosophy*, Cambridge: Cambridge University Press.

Birch, E. L. (2001) "Practitioners and the art of planning," *Journal of Planning Education and Research*, 20: 407–422.

Broadbent, J., Dietrich, M. and Roberts, J. (1997) "The end of the professions?" in J. Broadbent, M. Dietrich and J. Roberts (Eds) *The End of the Professions: The Restructuring of Professional Work*, pp. 1–13, London: Routledge.

Campbell, H. and Marshall, R. (2002) "Utilitarianism's bad breath: a re-evaluation of the public interest justification for planning," *Planning Theory*, 1: 163–187.

Campbell, H. and Marshall, R. (2005) "Professionalism and planning in Britain," *The Town Planning Review*, 76: 191–214.

Dahl, R. A. (1982) *Dilemmas of Pluralist Democracy: Autonomy vs. Control*, New Haven: Yale University Press.

Davidoff, P. (1965) "Advocacy and pluralism in planning," *Journal of the American Institute of Planners*, 31: 331–338.

Davidoff, P. (1973) "Advocacy and pluralism in planning," in A. Faludi (Ed.) *A Reader in Planning Theory*, Oxford: Pergamon.

Davidoff, P. and Reiner, T. (1962) "A choice theory of planning," *Journal of the American Institute of Planners*, 28: 11–39.

Durkheim, E. (1957) *Professional Ethics and Civic Morals*, London: Routledge and Kegan Paul.

Eraut, M. (1994) *Developing Professional Knowledge and Competence*, London: Routledge Falmer.

Evans, B. (1993) "Why we no longer need a town planning profession," *Planning Practice and Research*, 8: 9–15.

Evans, B. and Rydin, Y. (1997) "Planning, professionalism and sustainability," in A. Blowers & B. Evans (Eds) *Town Planning into the 21st Century*, London: Routledge.

Eversley, D. (1973) *The Planner in Society: The Changing Role of a Profession*, London: Faber and Faber.

Fainstein, S. (2000) "New directions in planning theory," *Urban Affairs Review*, 35: 451–478.

Faludi, A. (1973) *Planning Theory*, Oxford: Pergamon.

Fischer, F. and Forester, J. (Eds) (1993) *The Argumentative Turn in Policy Analysis and Planning*, Durham, NC: Duke University Press.

Flyvbjerg, B. (1998) *Rationality and Power: Democracy in Practice*, London: University of Chicago Press (English edition).

Flyvbjerg, B. (2001) *Making Social Science Matter: Why Social Enquiry Fails and How It Can Succeed Again*, Cambridge: Cambridge University Press.

Forester, J. (1985) "Critical theory and planning practice," in J. Forester (Ed.) *Critical Theory and Public Life*, pp. 202–230, Cambridge, MA: Institute of Technology.

Forester, J. (1989) *Planning in the Face of Power*, Berkeley: University of California Press.

Forester, J. (1999) *The Deliberative Practitioner*, Cambridge, MA: MIT Press.

Friedmann, J. (1973) *Retracking America*, New York: Doubleday Anchor.

Friedmann, J. (1992) *Empowerment: the Politics of Alternative Development*, Oxford: Blackwell.

Friedmann, J. (1998) "The new political economy of planning: the rise of civil society," in M. Douglass and J. Friedmann (Eds) *Cities for Citizens*, Chichester: John Wiley and Sons.

Friedson, E. (2001) *Professionalism: The Third Logic*, Chicago: University of Chicago Press.

Geddes, P. (1998/1915) *Cities in Evolution: An Introduction to the Town Planning Movement and to the Study of Civics*, London: Routledge.

Giddens, A. (1984) *The Constitution of Society: Outline of the Theory of Structuration*, Berkeley: University of California Press.

Habermas, J. (1984) *The Theory of Communicative Action V1*, Vol. 1 London: Beacon Press.

Habermas, J. (1987) *The Theory of Communicative Action V2*, Vol. 2 London: Beacon Press.

Habermas, J. (1989/1962) *The Structural Transformation of the Public Sphere*, Cambridge: Polity Press.

Habermas, J. (1996) *Between Facts and Norms: Contributions to a Discourse Theory of Law and Democracy*, Cambridge: Polity Press.

Habermas, J. (1996/1987) "Normative content of modernity," in W. Outhwaite (Ed.) *The Habermas Reader*, pp. 341–365, Cambridge: Polity Press.

Hague, C. (1984) *The Development of Planning Thought*, London: Hutchinson.
Hague, C. (2007) "Re-inventing planning: challenges and skills," *Australian Planner*, 44: 24–25.
Hague, C., Wakely, P., Crespin, J. and Jasko, C. (2006) *Making Planning Work: A Guide to Approaches and Skills*, Rugby: ITDG Publishing.
Healey, P. (1993) "The communicative works of development plans," *Environment and Planning B*, 20: 83–104.
Healey, P. (1996) "The communicative turn in planning theory and its implications for spatial strategy formation," *Environment and Planning B: Planning and Design*, 23: 217–234.
Healey, P. (1997) *Collaborative Planning: Shaping Places in Fragmented Societies*, London: Macmillan.
Healey, P. (1999) "Institutional analysis, communicative planning and shaping places," *Journal of Planning Education and Research*, 19: 111–121.
Healey, P. and Hillier, J. (1995) *Community Mobilsation in the Swan Valley: Claims, Discourses and Rituals in Local Planning*, Newcastle on Tyne: University of Newcastle upon Tyne.
Hillier, J. (2000) "Going round the back? Complex networks and informal action in local planning processes," *Environment and Planning A*, 32: 33–54.
Hoch, C. (1994) *What Planners Do: Power, Politics and Persuasion*, Chicago: American Planning Association.
Howe, E. (1994) *Acting on Ethics in City Planning*, New Brunswick: Center for Urban Policy Research.
Innes, J. (1996) "Planning through consensus building: a new view of the comprehensive planning ideal," *Journal of the American Planning Association*, 62: 460–472.
Innes, J. E. and Booher, D. E. (1999) "Consensus building and complex adaptive systems," *Journal of American Planning Association*, 65: 412–423.
Koehn, D. (1994) *The Ground of Professional Ethics*, London: Routledge.
Krumholz, N. and Forester, J. (Eds) (1990) *Making Equity Planning Work*, Philadelphia: Temple University Press.
Lindblom, C. and Cohen, D. (1979) *Usable Knowledge: Social Science and Problem Solving*, New Haven: Yale University Press.
Long, N. (1959) "Planning and politics in urban development," *Journal of the American Institute of Planners*, 25: 167–169.
Macdonald, K. K. (1995) *The Sociology of the Professions*, London: Sage.
March, A. and Low, N. (2004) "Knowing and steering: mediatisation, planning and democracy in Victoria, Australia," *Planning Theory*, 3: 41–69.
March, J. and Olsen, J. (1995) *Democratic Governance*, New York: The Free Press.
Marquand, D. (1997) "Professionalism and politics: towards a new mentality?" in J. Broadbent, M. Dietrich and J. Roberts (Eds) *The End of the Professions: The Restructuring of Professional Work*, pp. 140–147, London: Routledge.
Mayo, H. B. (1960) *An Introduction to Democratic Theory*, New York: Oxford University Press.
McLoughlin, B. (1973) "The future of the planning profession," in P. Cowan (Ed.) *The Future of Planning*, London: Heinemann.

Mouffe, C. (2000) *The Democratic Paradox*, London: Verso.

Parsons, T. (1968) *The Structure of Social Action: A Study in Social Theory with Special Reference to a Group of Recent European Writers*, New York: Free Press.

Perkin, H. (1989) *The Rise of Professional Society: England Since 1880*, London: Routledge.

Planning and Environment Act (1987) Government of Victoria, Melbourne: State Government Printer)

Planning Institute of Australia (2006) Code of Membership, Canberra: PIA National Secretariat.

Reade, E. (1987) *British Town and Country Planning*, Milton Keynes: Open University Press.

Rittel, H. and Webber, M. (1973) "Dilemmas in a general theory of planning," *Policy Sciences*, 4: 155–169.

Roberts, T. (2002) "The seven lamps of planning," *Town Planning Review*, 73: 1–15.

Royal Town Planning Institute (2007) Code of Professional Conduct.

Rueschemeyer, D. (1986) *Power and the Division of Labour*, Cambridge: Polity Press.

Sandercock, L. (1998) *Towards Cosmopolis*, London: John Wiley and Sons.

Schon, D. (1983) *The Reflective Practitioner*, New York: Basic Books.

Taylor, P. (2007) "The United Nations perspective on reinventing planning," *Australian Planning*, 44: 22–23.

Tewdwr-Jones, M. and Allmendinger, P. (1998) "Deconstructing communicative rationality: a critique of Habermasian collaborative planning," *Environment and Planning A*, 30: 1975–1989.

Tribe, L. H. (1972) "Policy science: analysis or ideology?" *Philosophy and Public Affairs*," 2, 1: 66–110.

Tugwell, R. G. (1974/1948) "The study of planning as scientific endeavour," in S. M. Padilla (Ed.) *Tugwell's Thoughts on Planning*, San Juan: University of Puerto Rico Press.

Vigar, G., Healey, P. Hull, A. and Davoudi, S. (2000) *Planning, Governance and Spatial Strategy in Britain: An Institutional Analysis*, London: Macmillan.

Weber, M. (1978/1914) *Economy & Society*, Berkeley: University of California Press.

Yiftachel, O. (1999) "Planning theory at a crossroad: the third Oxford conference," *Journal of Planning Education and Research*, 18: 267–270.

Yiftachel, O. (2001) "Introduction: the power of planning," in O. Yiftachel, J. Little, D. Hedgcock and I. Alexander (Eds) *The Power of Planning: Spaces of Control and Transformation*, pp. 1–20, Dordrecht: Kluwer.

Index

1000 Friends of Florida 191–2

Abrams, R. 121
acceptability 265
access 183–4, 188
accountability 24; large-infrastructure projects 241–4, 245
action, without consensus 291
actor networks 254–5
Adam, B. 259
adjustment packages 215–16
advisory role of planning 5
affordable housing 187; maintaining affordability 190–3
African Americans 194
African National Congress (ANC) 58–9, 69, 79
agency, human 30
aging population 4
Alagados de Sepetiba 149
Albrow, M. 44, 46
Alda, E. 138
Alfasi, N. 59
Alibhai-Brown, Y. 46–7, 48
Alliance for Progress (AfP) 212
ambivalent political will 263–5
amenities 194–5
American Association of Nurserymen 117
American Dream 115
American Institute of Certified Planners (AICP) Code of Ethics and Professional Conduct 241–2
American Planning Association (APA) 236–7, 271
Amin, A. 42–6, 54
Amsterdam 123–4
Andean Pact (1969) 218

Angel, S. 143
Anti-Privatization Forum (APF) 60, 69–72, 76, 77
appraisals: French transport policy 256–8; innovative methods 258; large-infrastructure projects 233–5; overhaul of methods 257–8
Aravena, P. 143
Argentina 206, 207, 209, 216, 217
Australia 50; cities in an international context 86–8; homeownership and oil vulnerability in cities 8, 83–111; planning professionalism 9–10, 270–97
Austria 258
autonomy 77

back lanes between shophouses 162–4, 165
Badcock, B. 93
Bangkok Skytrain 227
Barbour, H. 171, 196
Barksdale, J. 189
Barthélemy, J.-R. 257
Bay St. Louis 173, 176
Baye, E. 257–8
Bayer, P. 194
Bekkersdal mass meeting 70–1
Beliz, G. 138
belonging, sense of 49, 52–4; see also community
benefit shortfalls 224–35; examples 226–7; explanations 229–35; size 224–6; why they are a problem 228–9
Bentham, J. 152
Berlin 124
Berube, A. 182
Better Buildings Programme (BBP) 73–4
BHP Billiton picket 71

Bilbao Guggenheim Museum 227
Biloxi 176, 185, 186, 190, 191, 195, 199
Biloxi Bridge 175, 185, 196
Biscaia, A.C. 139
Blake, P. 112–13, 119
Blunkett, D. 46, 47
Bogotá 208
Boiteux group report 256–7
Boston Central Artery/Tunnel Project
226, 245
Botswana 16–17
Brazil 206, 207, 210, 215, 217; public
security and urban design 8, 136–54
Breda 129
bridges 224–5
Brisbane 98–100
Brown, L. ('Capability') 114, 116
Brundtland Commission report 3
Buenos Aires 208, 210
buffer planting 116, 117–18, 119
bureaucracy 285–7, 291
bureaucratic-authoritarian regimes 213
Burnley, I. 87, 93
Burrows, R. 87
Burt, R. 64

Cachoeirinha 149
Campbell, H. 274
Canada 50, 55–6; see also Vancouver
car dependence 3, 83–4, 86–7, 105;
spatial differences in 90–2; VAMPIRE
index 96–103, 104, 105
Caracas 208
Cardoso, F.H. 214
cars per household 96–103, 104, 105
case-by-case orientation to success
283–90, 292
casinos 185–6
Castro, F. 212
Cattell, V. 167
Central American Common Market
(MCC) 218
central government, and local governance
in Zimbabwe 29–30
Centre for Applied Legal Studies (CALS)
60, 72–6, 77
Centre on Housing Rights and Evictions
(COHRE) 74, 75, 76
challenges for planning 2–6

Channel Tunnel 226, 260
Chaplain, C. 260
charter schools 194
Chicago School 143
Chile 207, 209, 213, 215–16
China 4
Cidade de Deus 150–1
City of Johannesburg 59, 69, 72–6
civic associations: categories of 60;
Johannesburg 7, 58–82; rebuilding
after Hurricane Katrina 193–5
civil disturbances 42–4, 47
civil society see civic associations
Claremont estate 114, 115
Clark, I. 15
climate change 3
Cohen, D. 278
Cohen, V.F. 158–9
collaborative planning 18–19, 32, 61, 62,
63, 281–2; Planact 65–9
collective surveillance 143, 144, 145,
146, 152
Collingwood Neighbourhood House
(CNH) 38, 45, 54–5, 56
Colombia 140, 207, 210, 217
commercial multiculturalism 51
Commission for Architecture and the
Built Environment (CABE) 129
common good 53
communal activities 157–8
communicative planning see collaborative
planning
community 143, 145; cosmopolitan
urbanism 44, 46, 49, 52–4;
involvement and satisfaction with Safe
Urban Spaces project 148–9, 150–1;
rebuilding after Hurricane Katrina
180–2, 185–6; virtual 146
community development 65–9
community development corporations
(CDCs) 197
Community-Driven Housing Initiatives
(CDHIs) 68
competition 242, 243–4
compliance, legal and policy 283–5
condos 185–6
conflict 40–1, 55
Congress for the New Urbanism (CNU)
8–9, 172, 173, 192–3

300 Index

Congress of South African Trade Unions (COSATU) 69, 79
Connolly, W. 48
conservation: historic living cities and 8, 155–70; policy implications of public space use 167–8; urban and the socio-cultural function of public space 158–9
conservative multiculturalism 51
Constitutional Court (South Africa) 75–6
constitutional reforms 209
consultants 233–4, 235, 279
consultation 255–6; *see also* public participation
content analysis 251, 252–61
contingent valuation 258, 267
contradictory objectives 262–3
contrasts 5
control: market control 242, 243–4; public control 242–3; society of 141–2
coordination between transport and urban planning 254
corporate multiculturalism 51
cosmopolitan urbanism 7, 38–57
cost-benefit analysis 9, 256–7; limitations 266
cost overruns 224–35, 246; examples 226–7; explanations 229–35; size 224–6; why they are a problem 228–9
cost recovery policy 69
COWI 224, 233–4, 237, 238
Crime Prevention Through Environmental Design (CPTED) 142, 144, 145
criminal penalties 243
crisis of effectiveness 263
crisis of legitimacy 254–5, 263
critical democratic professionalism 281–3
critical ('revolutionary') multiculturalism 51
critical questioning 245
Cuba 209; Revolution of 1959 212
Cuel, S. 256
cultural diversity 52, 53, 159–60
culture 51; of planners/planning 29, 33; public space in historic living cities 167, 168–9; urban conservation and 158–9
curriculum planning 237–8

Damart, S. 257
Davidoff, P. 278
Davis, K. 211, 219
Davis, M. 198
Davis, R. 192
Debizet, G. 257
debt, foreign 215–16
deception 9, 223, 230, 232–5, 244–5
decision-making: crisis in French transport policy 266; procedures and processes 253–5; scale 6; tensions between contradictory principles of public action 261–3
defensible spaces 144
deindustrialisation 121
Del Pozo, J. 215, 219
delays 228
deliberative democracy 18
democracy: civic associations and 62, 64, 77; critical democratic professionalism 281–3; deliberative 18; guided 18–19, 20
demographic challenges 4
Denmark 50
density 180; high-density housing 106, 180
Denver International Airport 226
Department of Sustainability and Environment (DSE) (Victoria) 274, 279
Dependence theory 214–15
design, connected to policy 190–2
destabilization of policy, planning and implementation 228–9
developing countries 207
development 205–22
development control 275, 276
Development Works 74, 75, 76
D'Iberville 176, 184
dictatorships 213
difference: ethical indifference 41–2; living with 40–6; thinking through 46–9; togetherness in 40–1
disciplinary society 141, 152–3
disparate knowledges 277–80
distributive equality 178–9
diversity: cultural 52, 53, 159–60; socio-economic 182–3, 186–8
Dodson, J. 89, 91, 96

Donald, J. 41–2, 54
drive to maturity 206–7
Du Plessis, J. 73–4
Dublin 124, 125
Dupuy, G. 260

East Kilbride 118
ecological landscape movement 121
economic challenges 5
Economic Commission for Latin America (ECLA) 210–11, 214
economic nationalism 210–11
economic policy 105
effect, research on 259–60
effectiveness, crisis of 263
efficiency 291
Ekurhuleni Metropolitan Municipality 59, 66, 67
Emirbayer, M. 64–5
enclosure of green space 112–13, 121–30; and functionality 125–9
energy dependence 84–5; see also car dependence
energy security 8, 83–111
environmental appraisal 258
environmental challenges 3–4
equity 291; social equity and rebuilding after Hurricane Katrina 8–9, 171–204
Erskine, R. 121
ethics: ethical indifference 41–2; planners' 241, 278
ethnic minorities 43–4, 49, 176, 177
Europe 86
European Archive of Contemporary Urban Space 122–3
European Union research projects 252
evaluation, in transport policy research 250, 265; ex post 260–1
evicted tenants 72–6
ex post evaluations 260–1
expertise, technical 28, 32, 33
experts, professionals as 280, 291–2

Faburel, G. 258
failure of modernization 212–15
Fainstein, S. 63
Faletto, E. 214
favelas 8, 147–52

Fazenda Viegas 149–50
Flyvbjerg, B. 18, 19, 224, 230, 233–4, 237, 238, 242
forecasting: accountability in 241–4, 245; explanations of inaccuracy 229–35; French transport policy 256–7; inaccuracy 225–6; reference-class 236–41
foreign debt 215–16
Forester, J. 279
fossil fuel dependence 3; see also oil vulnerability
Foucault, M. 32, 141, 152
frames of reference 253, 265
France 50, 115, 123, 137
French transport policy 9, 249–69; ambivalent political will 263–5; content analysis of research 251, 252–61; quantitative analysis of body of research 251–2; tensions between contradictory principles of public action 261–3
Frère, S. 254–5
Friedson, E. 272–3
fuel prices see oil prices
functionalist sociology 205, 214
functionality, enclosure and 125–9
Furtado, C. 214

garbage can model 250
gated communities 217
Gautier 176, 190
Germany 115, 124
Gilovich, T. 239
Gilroy, P. 49
global oil supplies, sustainability of 89
Global Planning Education Association Network (GPEAN) 2
global urban transition 4
globalization 43, 44
goals/objectives: contradictory objectives 262–3; meeting project goals in the Safe Urban Spaces project 149–51; misunderstandings about 148–9
Goldfield, D. 198–9
'good practices' 259
governance 6; good and large-infrastructure projects 244–5; local participatory 6, 12–37; planners as a profession-within-governance 293–4;

302 Index

planners' views on local governance 26–8; and planning practice 15–17; planning professionalism 9–10, 270–97
Governance Outreach Programme 12, 20, 22, 23, 24, 29, 30, 31, 32
Governance Outreach Task Force (GOTF) 12–13, 14, 21–2, 23–6, 29, 30–1
governance-sensitive planning approach 18–19
government programmes 192–3
grants 242
Greater Johannesburg 59
green space 8, 112–35; re-enclosure movement 112–13, 121–30; social and environmental critiques 119–21
Greenwich Village, New York 40
Grenoble 123; north bypass project 256
guided democracy 18–19, 20
Gulf Regional Planning Commission growth projections 178
Gulfport 176, 178

Habermas, J. 62, 282–3
Habitat for Humanity 187
Hadenius, A. 64, 66, 70, 77
Haida people 38–9
Hall, S. 50–1
Handsworth Community Fire Station 45
Harris, N. 18
Havana 208
Healey, P. 62
heritage conservation 8, 155–70
Hernandez, F. 254
high-density housing 106, 180
Hillier, B. 146
historic living cities 8, 155–70
Holloway, A.J. 191
Holloway, D. 95
homeownership: and oil vulnerability 8, 83–111; spatial allocation 92–4
Hough, M. 121, 122
households' financial behaviour 106–7
housing: green spaces in social housing projects 117; high-density 106, 180; 'Katrina Cottages' 187; People's Housing Process (PHP) in Johannesburg 66–8, 76, 77, 78–9; tenure 84, 85; see also evicted tenants, homeownership

Howe, E. 278
human agency 30
human rights 60, 72–6
Hurricane Katrina see Katrina Hurricane
hyper-ghetto 141–2
hyper-urbanization 211–12, 214

ideals, professional 272–4
identity 46–9
Idid, S.Z. 159
immature take-offs 9, 205–22
immigration 43–4; and national identity 46–7
impact, measurement of 259–60
import substitution industrialization (ISI) 210–11, 212, 213
incentives, improved 236, 241–4
income: diversity on Mississippi Gulf Coast 176, 177; inequality 5, 216; and mortgage debt 94–5; VAMPIRE index 96–103, 104, 105
independent review 242–3, 273
indifference, ethical 41–2
industrialization 9, 205–22; imbalance between urbanization and 210–12
inflation 89, 95–6; VAMPIRE index 96–103, 104, 105
informal economy 5, 216, 217
Inner City Regeneration Charter 75–6
Inner City Regeneration Strategy 73, 75
inner city tenants, evicted 72–6
inside view 238, 239–40
Institute for Community Economics 190
Institute La Vivienda (INVI) 137, 142, 153
institutional challenges 5–6, 10; planning professionalism 9–10, 270–97
institutional settings 278–9
institutionalized professionalism 283–90
instrumental rationality 16–17, 31–2, 282–3
insurgent planning 63, 69–72
integrated development planning (IDP) 59
integrated transport system 262
integration 146
Inter-American Development Bank (IDB) 138–9
intercultural interaction 40–1, 45, 52, 54–5

intercultural political culture 46, 55–6
interculturalism 7, 51–6
interdisciplinary approach 266
International Monetary Fund (IMF)
  215–16
INTRAMUROS project 258
inward-facing planning experts 291–2
Italy 4

Jacobs, J. 126, 143, 157
Jalan Hang Kasturi 162, 163
Japan 4
Jardim do Itá 148
Jeffrey, R. 145
Johannesburg 7, 58–82
Johannesburg High Court 75
Jones, C.O. 250
Jonker Street 162
Jouve, B. 256
Juntas Comunales 137
'Just City' theory 63, 72–6

Kahneman, D. 230, 236, 237–8, 240
Kampung Pantai Street 166, 167
Katrina Hurricane 173, 174, 175;
  rebuilding after 8–9, 171–204
'Katrina Cottages' 187
Katz, B. 182
Kenworthy, J. 87
Klein, O. 259–60
knowledge: disparate knowledges
  277–80; power and 32; professionalism
  and 272–3, 277–80, 291–2, 293–4;
  technical expertise 28, 32, 33
Krizek, K. 87
Kupke, V. 95

La Cava, G. 95
labour market polarization 5
large-infrastructure projects 9, 223–48;
  characteristics 224; forecasting
  inaccuracy 225–35; problems 224–9;
  reform of planning and implementation
  236–44; towards better practice 244–6
Latin America 9, 205–22
Latin American Association of Free Trade
  (ALALC, later ALADI) 218
Latin American model of development
  215, 218, 219

Latz, P. 121–2, 123
LAURE (law on air and rational energy
  use) 254
Laurie, I. 121
law 283–5, 291
Le Corbusier 120
LEDA 259
legitimacy, crisis of 254–5, 263
liaison group for young self-employed
  workers 21, 25–6, 29
liberal multiculturalism 51
lifestyle 158
Lima 208
Lindblom, C. 278
linear spaces 162–5, 166, 168
living heritage cities 8, 155–70
Lloyd-Evans, S. 213
local government, changing role of 5–6
local liveability, politics of 42–6
local participatory governance 7, 12–37
local planners 279
local social programmes 139–40
locational disadvantage 93–4
Lolive, J. 258
Long Beach 176
lost decade 215–17
Louisiana 172
Lovallo, D. 230, 237, 240
lying 9, 223, 230, 232–5, 244–5
Lyons, T.J. 87

Macala, G. 67, 68
Maher, C. 93
maintenance teams 120
Major Projects Association 242
Malavé Mata, H. 214–15
Malaysia 50; Melaka 8, 155–70
Manchester 112, 113, 124, 125, 126
Mangcu, X. 58
Marano, W. 95
market control 242, 243–4
Marshall, R. 274
Martens, K. 59
Marxism 212–15
massification 205, 208–10
Mathews, R. 186
matrix reversal 122
maturity, drive to 206–7
McHarg, I. 120, 121

304 Index

mediatization 271–2, 282–3, 291–2, 293;
 institutionalized professionalism
 283–90
Melaka, Malaysia 8, 155–70
Melbourne 90, 93, 102–3
methodology 260
Mexico 206, 207, 210, 216, 217;
 Revolution of 1910 209
Mexico City 208, 210
micro-publics 45–6
military juntas 213
mill towns 42–4, 45
'million-cities' 216–17
Milton Keynes 117, 124–5, 127, 128
minorities, ethnic 43–4, 49, 176, 177
misrepresentation, strategic 9, 223, 230,
 232–5, 244–5
Mississippi Department of Transportation
 (MDOT) 196
Mississippi Gulf Coast towns 8–9,
 171–204
Mississippi Renewal Forum 187–8;
 planning charrette 173, 174, 191, 195–6
Mobiscopie research project 253
Modern Architectural Research Group
 (MARS) 114
modernism 8, 112, 113–19
modernization 205–22; failure of 212–15
money 287–9, 291
Morris, J. 91
Morro do Borel 149
mortgages 8, 83–111; socio-spatial
 patterns in mortgage debt 94–6;
 VAMPIRE index 96–103, 104, 105
Moss Point 176, 180
multicultural cities 7, 38–57; approaches
 to living together 40–6; Melaka
 159–60
multiculturalism 7, 56; reconsidering
 49–54
multi-layered organizations 64, 66, 70,
 74, 77
Mumford, L. 114
Municipal Development Partnership
 (MDP) 20
Muzema 151

Naples 129
NASA space shuttle 226–7

National Building Regulations and
 Building Standards Act (1977) (South
 Africa) 73, 75
national economies, mega-projects and
 229, 244–5
National Housing Code (South Africa) 75
national identity 46–9
National Public Security and Citizenship
 Program (PRONASCI) 136–7,
 138–41
natural surveillance 143, 144, 145, 146,
 152
neighbourhood planning 186, 197–8
neo-liberalism 215–17
Netherlands, The 50, 121, 123–4, 245
networks 254–5
new approaches to urban and regional
 planning 6
new social movements 60, 69–72
New Urbanism 8–9, 171–204
Newman, O. 8, 144, 145
Newman, P. 87, 106
Ngwane, T. 70
Nicholson-Lord, D. 120
nodal spaces 166–7, 168
Non-Governmental Organisations Bill
 (NGO Bill) (Zimbabwe) 20
non-routine projects 240–1
North America 86; *see also* Canada,
 United States of America (US)
Nova Sepetiba 148
Ntuli, L. 70

Ocean Springs 176, 187
Offner, J.-M. 254, 260–1
oil crisis of the 1970s 212–13
oil prices 83–4, 88–90, 95–6, 105
oil supplies, global 89
oil vulnerability 8, 83–6, 96; VAMPIRE
 index 96–103, 104, 105
open-plan landscapes 8, 114–19, 120
optimism bias 230, 231–2, 245
Organization of American States (OAS)
 210
organizational structure 64, 66, 70, 74,
 77
other countries' experiences 264–5
outside view 237–41
outward-facing professionals 292–4

Pan-American Games 140, 147
Panerai, P. 123
panopticon 141, 152
Parekh, B. 41, 53
parking spaces 166
parks 119, 120, 126–7
parkway movement 113, 115
Parque Itambé 149
participatory local governance 7, 12–37; in action 19–26
Pascagoula 173, 176
Pass Christian 173, 174, 176, 181, 190
payment, professionalism and 287–9, 291
pedestrian sheds 183, 184
People's Housing Partnership Trust (PHPT) 68
People's Housing Process (PHP) 66–8, 76, 77, 78–9
Perkin, H. 272
Perón, J. 217
Perth 103, 104
Peru 210
Phillips, S. 195
picturesque landscape theory 114–21
plan-implementation processes 275, 276
plan-preparation processes 275
Planact 60, 65–9, 77
planners' views on local governance 26–8
Planning and Environment Act (1987) (Australia) 275
planning fallacy 230
Planning Institute of Australia (PIA) 270–1
planning practice, governance and 15–17
planning professionalism 9–10, 270–97
Planning Sustainable Cities (UN-Habitat) 2–3
planning theories 61–3
'plans de déplacements urbains' (PDUs) 253–5, 262–3
Plassard, F. 260
pluralism 64, 77
pluralist multiculturalism 51
policy: compliance with 283–5; connected to design 190–2; defining 250; French transport policy 9, 249–69; sequential model of public policy analysis 250
policy certainty 291

policy implementation 250
PolicyLink 190
political will, ambivalent 263–5
politicians, role of 27
politics: of local liveability 42–6; political-economic explanations of forecasting inaccuracy 230, 232–5; political instability 217; political theory 64–5; and professionalism 289–90, 291
population aging 4
post-colonialism 43
postmodern urbanism 8, 112–35
Potter, R.B. 213
poverty: Latin America 217; Mississippi Gulf Coast 175, 176, 182–3
power 64; pervasiveness of 18; relations between youths and planners in Zimbabwe 30–2
Prevention of Illegal Eviction and Unlawful Occupation of Land Act (PIE Act) (South Africa) 74, 75
price inflation *see* inflation
prison system 139–40
private sector 189; accountability 241–4
procedural certainty 291
process, planning as 197–8
professional ideals 272–4
professional organizations 280
professional penalties 243
professionalism 9–10, 270–97
programme development 250
programme identification 250
project appraisal *see* appraisals
project organization 244
Promotion of Administrative Justice Act (South Africa) 74, 75
PRONASCI 136–7, 138–41
psychological explanations of forecasting inaccuracy 230, 231–2
public control 242–3
public facilities 193–5
public participation 6; civil society and 60, 65, 66, 77–8; conservation of historic living cities 167–8, 169; French transport policy 255–6, 257, 262; participatory governance in Zimbabwe 7, 12–37; rebuilding after Hurricane Katrina 195–8; tensions between decision making and 262

Public Order and Security Act (POSA)
  (Zimbabwe) 20
public policy *see* policy
public-sector accountability 241–4
public-sector investment 105–6, 185,
  189
public security *see* security
'Public Security Project for Brazil' 139
public spaces: enhancing and regenerating
  167; importance in historic living cities
  157–8; living city conservation 8,
  155–70; post-Katrina rebuilding
  180–2; significance in the revitalization
  of cities 156–7; socio-cultural function
  of 158–9; usage patterns in Melaka
  162–7
public transport: Australia 91–2, 100,
  103, 105–6; *see also* rail transport
PUC-Chile 142, 153
Putnam, R. 178

Quitungo 149

'race riots' 42–4, 47
radical planning 63
rail transport 252; large-infrastructure
  projects 224–6, 238–9, 240; regular
  hourly scheduling 264–5
Randolph, B. 87, 95
rational process model of planning
  16–17, 23
Rau, M. 143, 144, 145, 146, 153
rebuilding after Hurricane Katrina 8–9,
  171–204
re-enclosure of green space 112–13,
  121–30
reference-class forecasting 236–41
regulation, new instruments of 264
Reichow, H.B. 115
Reid, B. 38
Reiner, T. 278
Reissman, L. 211, 219
relationships 15–16, 29
Relph, E. 159
Reserve Bank of Australia 95–6
resistance, strategic 25–6, 31
revitalization of cities 156–7
'Reviving the Renaissance' initiative 191
Richardson, T. 18, 19

rights: civic associations and human rights
  60, 72–6; right to the city 46, 52; right
  to difference 46, 52
Rio de Janeiro 208, 210; Safe Urban
  Spaces project 8, 147–51
river walkway 162, 164
roads 224–6; modernism and green
  space 117, 118; widening in
  Mississippi Gulf Coast 185; *see also*
  car dependence
Rocinha 149
Rogers, R. 40
roles, and professionalism 278–9,
  283–90, 291–2, 293
Romero, J.L. 209
Rostow, W.W. 206–7, 214, 218
Rousselot, M. 257
Route 90 widening 185
Roy, B. 257
Royal Commission on Environmental
  Pollution (RCEP) 129
Royal Town Planning Institute (RTPI)
  241, 271
rural crisis 211
Ruys, M. 120

safe urban spaces 141–4; theories of
  142–4
Safe Urban Spaces Project (SUSP) 136–7,
  146–52
safety *see* security
San Francisco–Oakland Bay Bridge 226
Sandercock, L. 63
Santiago 208
São Paulo 208, 210
Sassen, S. 43
scale 6
Scandinavia 129–30
scheme amendment process 275
Schmitt, G. 185–6
School of Dependence (SoD) 214–15
schools 194
scientism 17
Scotland 129
Seaside Conference on policy and design
  192
sectorisation 262
security 120, 195; Brazil and safe urban
  spaces 8, 136–54

'Segurança Cidadã' project 138–9, 140, 141, 147, 151
self-regulation 272
semi-nodal spaces 166, 168
Sennett, R. 40–1, 42
sequential model of public policy analysis 250
services 194–5
Shanghai Statement 2
Sheller, M. 64–5
Silva Michelena, H. 214–15
Simmel, G. 157
Simon, J. 95
Singapore 50
Sipe, N. 89, 91, 96
situational theory 144, 145
'smart growth' 87
SmartCodes 187–8
Soares, L.E. 139
social equity 8–9, 171–204
social housing projects 117
social inclusion 55
social interaction 156–8, 167
social mixing 182–3, 186–8
social mobilization 60, 69–72
social transformation 62–3, 69–72
society of control 141–2
socio-economic diversity 182–3, 186–8
socio-economic effects of transport 258–60
socio-spatial challenges 5
socio-spatial differentiation/patterns 84; car dependence 90–2; home purchase 92–4; mortgage debt 94–6
Sonntag, H.R. 214–15
Sorkin, M. 191
South Africa 17; civic associations in Johannesburg 7, 58–82
South African Communist Party (SACP) 69, 70–1, 79
Soviet Union 115
space left over after planning (SLOAP) 117–18, 125
space shuttle 226–7
space syntax 144, 146, 153
spatial equity 183
spatial investment strategies 105–6
speed 259
Spirit of Haida Gwaii, The 38–9

stadtlandschaft concept 115
stages of development theory 205–7, 214
state: and civil society 61; crisis of effectiveness 263; crisis of legitimacy 254–5, 263
state-level planners 279
steering mechanisms 282–3, 292
strategic misrepresentation 9, 223, 230, 232–5, 244–5
strategic resistance 25–6, 31
streets 162, 163, 168
structural holes 64, 68, 71
structure, organizational 64, 66, 70, 74, 77
subgroups 64, 66, 74
summative evaluations 261
supervisory evaluations 261
Supreme Court (South Africa) 75
surface-water management 127, 128
surveillance 141–2; collective/natural 143, 144, 145, 146, 152
Switzerland 258
Sydney 90, 91, 92, 93, 100, 101

take-offs, immature 9, 205–22
Tan Kim Seng Bridge 166
technical expertise 28, 32, 33
technical explanations of forecasting inaccuracy 230, 231
tenants, evicted 72–6
Third World syndrome 213
togetherness in difference 40–1
Tooting, South London 44
tourism 8, 155–6, 167
Townsend, C.L. 197
traditional societies 206
traffic-demand forecasts 225–6
training, and local governance 27–8
tram–train 264
TRANSLAND 259
transparency 242–3
transport: French transport policy 9, 249–69; home ownership and oil vulnerability in Australia 8, 83–111; large-infrastructure projects 9, 223–48; socio-economic effects of 258–60; *see also* rail transport, roads
trees 126, 127
Tricot, A. 258

308 Index

Tripartite Alliance 69, 70, 79
'trouble-causing' youths 6, 13, 19, 20–6, 29, 30–1
Tully, J. 39
Tunnard, C. 114, 116, 118
tunnels 224–5
Tversky, A. 230

Uggla, F. 64, 66, 70, 77
UN-Habitat, *Planning Sustainable Cities* 2–3
unemployment 216, 217
unions 71–2
United Kingdom (UK) 115, 116, 129, 245; living in multicultural cities 42–5; 'Planning Out Crime' 137; 'race riots' 42–4, 47
United Nations Development Programme (UNDP) 15
United States of America (US) 50, 87; modernism and road construction 117; parkway movement 113, 115; post-Katrina rebuilding 8–9, 171–204
urban conservation 158–9
urban consolidation policy 106
urban renaissance 41
urban South 198–9
Urban Task Force 125, 126
urban transition, global 4
urbanization 9, 205–22; early 208–10; imbalance between industrializaition and 210–12
Uruguay 209

Vallingby 129, 130
Vancouver: Collingwood Neighbourhood House 7, 38, 45, 54–5, 56; International Airport 38–9
Venezuela 206, 207, 209, 210, 213, 216, 217
Victoria, Australia 9–10, 274–94
Victoria Planning Provisions (VPPs) 274, 284–5
Victorian Civil and Administrative Tribunal (VCAT) 274, 275
Vila Comari 149, 151

virtual community 146
Vosloorus PHP project 66–8, 76, 77
vouchers 194
Vulnerability Assessment for Mortgage, Petrol and Inflation Risks and Expenditure (VAMPIRE) index 96–103, 104, 105, 107

Wachs, M. 230, 234–5, 241
Wacquant, L. 141–2
wages 89
walkable access 183–4, 188
WalMart 181
Warrington 121, 122
waste 228
Waveland 173, 176, 187
Weber, M. 280
Williamson, C. 214
Wilson, S. 73–4
Winter, W. 175
work journeys by car 90, 91, 96–103, 104, 105
workers' committee 21
workers' unions 71–2
World Bank 15–16, 215–16
World Planning Schools Congress (WPSC) 1–2
World Summit on Sustainable Development, demonstrations against 70
Worpole, K. 130

Xhasa Accounting and Technical Centre 68

Yacobi, H. 59–60
Yates, J. 106
Yerpez, J. 254
youths, 'troublesome' 6, 13, 19, 20–6, 29, 30–1
Yrigoyen, H. 217
Yu, X. 87

Zambia 16–17
Zimbabwe 7, 12–37
zoning 187–8
Zurich 126, 127